Handbook of Experimental Pharmacology

Volume 263

The *Handbook of Experimental Pharmacology* is one of the most authoritative and influential book series in pharmacology. It provides critical and comprehensive discussions of the most significant areas of pharmacological research, written by leading international authorities. Each volume in the series represents the most informative and contemporary account of its subject available, making it an unrivalled reference source.

HEP is indexed in PubMed and Scopus.

More information about this series at http://www.springer.com/series/164

Scott M. Whitcup • Mark Hallett
Editors

Botulinum Toxin Therapy

Springer

Editors
Scott M. Whitcup
Akrivista and Whitecap Biosciences
Irvine, CA, USA

UCLA Stein Eye Institute
David Geffen School of Medicine
at UCLA
Los Angeles, CA, USA

Mark Hallett
Bethesda, MD, USA

ISSN 0171-2004 ISSN 1865-0325 (electronic)
Handbook of Experimental Pharmacology
ISBN 978-3-030-66305-6 ISBN 978-3-030-66306-3 (eBook)
https://doi.org/10.1007/978-3-030-66306-3

This Springer imprint is published by the registered company Springer Nature Switzerland AG.
The registered company address is: Gewerbestrasse 11, 6330 Cham, Switzerland

Preface

In 1989, the US FDA first approved botulinum toxin for the treatment of strabismus and blepharospasm. Today, there are five commercially available botulinum toxins in the USA and for therapeutic indications from head (chronic migraine) to toe (lower limb spasticity). Importantly, botulinum toxins have been approved for multiple indications globally. This tremendous progress in the development of botulinum toxins for medical uses has been fostered by both increased understanding of the molecular mechanisms of how these substances interact and affect the tissues in the body and by astute clinicians and researchers observing the effects in patients. Despite being one of the most potent and potentially deadly substances in nature, our ability to reliably manufacture botulinum toxins and to deliver extremely small quantities locally to the site of action has allowed safe treatment of thousands of patients.

Botulinum Toxin Therapy is divided into two parts: a section on the basic science and a section on clinical practice. The basic science section starts with a chapter on the history of botulinum toxins in medicine (Dr. Whitcup), from the recognition of food-borne illnesses over a thousand years ago, to the regulatory approval of botulinum toxins for medical therapy over the last 3 decades. Drs. Dong and Stenmark then review the structure and classification of botulinum toxins, and Dr. Rossetto and colleagues highlight the progress we have made on understanding the molecular biology of the mechanism of action. Of course, the therapeutic use of botulinum toxins requires the manufacture of material that meets good manufacturing practices (GMP) criteria to ensure a safe and reliable source of the drug for patients, and the science of toxin production is discussed by Dr. Hasan. There are currently two botulinum toxin serotypes approved for human use (type A and type B); however, other serotypes and novel botulinum neurotoxins are in development. These novel, native, and engineered botulinum toxins are discussed by Dr. Steward and collaborators.

Part II of the book focuses on the use of botulinum toxins in clinical practice. This section starts with a chapter by Dr. Dressler on the general pharmacologic principles for clinical use including dosing and pharmacokinetics. The rest of the section consists of reviews of the major clinical uses of botulinum toxins by experts in the field. Although over a hundred of clinical uses of botulinum toxins have been attempted or discussed, these chapters predominantly cover approved indications and major indications currently in randomized clinical trials. Drs. Berardelli and Conte discuss the use of botulinum toxins in dystonia, and Drs. Hunter and Wan review uses in ophthalmology; two important clinical areas where botulinum toxins were first studied in humans. Dr. Sheng describes botulinum toxin use in spasticity. Dr. Wang and colleagues review uses in dermatology including aesthetic indications and hyperhidrosis. Dr. Chancellor and Dr. Smith review uses in the genitourinary system, and Dr. Cariati et al. review gastrointestinal uses. The clinical part of the book ends with three chapters on the use of botulinum toxins for neurological and psychiatric diseases. Drs. Yuan and Silberstein review headache disorders and Dr. Lackovic discusses pain. Finally, Dr. Wollmer and colleagues cover the use of botulinum toxin for the treatment of depression.

Our primary goal in putting this book together is to provide an updated review of the science of botulinum toxin therapy to help basic scientists, clinical researchers, and practitioners in the study and use of currently available and future neurotoxins. Our hope is that these chapters provide both a detailed scientific description of the field and a practical guide to applying science. But there are a couple of subplots to *Botulinum Toxin Therapy* that make the story both fascinating and applicable to other medical therapies. One is that an incredibly potent and lethal substance can be studied and applied for beneficial purposes. The potential medical uses of botulinum toxin were recognized well over a century before their potential use as biological weapons. Fortunately for the patients who have benefited from their medical use, biological weapons programs have been largely abandoned or curtailed allowing the manufacture and use of botulinum toxins in research labs and medical practices for societal good. The second interesting part of the story is the importance of clinician scientists in the progress of medical science. Justinus Kerner was a physician and poet who studied cases of food poisoning in the early 1800s in Germany. He not only published the first case history of botulism but also hypothesized on the potential medical uses of the toxin including movement disorders and hypersecretion of body fluids, both approved medical indications for botulinum toxins today. Advancement in the field of medical toxin therapy will require both skilled basic scientists and dedicated and astute clinicians, and hopefully, as potential patients, we may all benefit.

Irvine, CA, USA Scott M. Whitcup
Bethesda, MD, USA Mark Hallett

Contents

Part I

Basic Sciences

The History of Botulinum Toxins in Medicine: A Thousand Year Journey

Scott M. Whitcup

Contents

Abstract

Botulinum toxin is one of the most potent and deadliest substances on earth. Because of its unique mechanism of action at the synaptic junction and the ability to precisely deliver the toxin locally to where it is needed, botulinum toxin has been used as an effective treatment for a plethora of diseases from head to foot, from chronic migraine to ankle spasticity. Unlike systemic drugs, botulinum toxin is delivered by injection to the site of disease. As we will see from the history of botulinum toxin, the ability to deliver the drug locally to minimize the amount of botulinum toxin needed and thereby minimizing systemic exposure has been key to its medical utility. Botulinum toxin was first approved by the US Food and Drug Administration in 1989 for the treatment of blepharospasm and strabismus, but the history starts long before this, with outbreaks of food poisoning in the tenth century. Importantly, the development of botulinum toxins for medical use continues today with the engineering of novel toxins to treat disease.

Keywords

Alan Scott · Botulism · History · Justinus Kerner · Strabismus

S. M. Whitcup (✉)
Clinical Faculty, UCLA Stein Eye Institute, David Geffen School of Medicine at UCLA, Los Angeles, CA, USA

Akrivista and Whitecap Biosciences, Irvine, CA, USA

S. M. Whitcup, M. Hallett (eds.), *Botulinum Toxin Therapy*,
Handbook of Experimental Pharmacology 263, https://doi.org/10.1007/164_2019_271

1 Introduction

Ancestors of modern bacteria probably appeared on earth approximately 3–4 billion years ago. A recent study presented direct fossil evidence life on land 3,220 million years ago in the form of terrestrial microbial mats in South Africa (Homann et al. 2018). Today it is estimated that there are about 5×10^{30} bacteria on earth (Whitman et al. 1998). Botulinum toxin is produced by *Clostridium botulinum*, a rod-shaped, gram-positive, anaerobic bacterium. There are seven serotypes (A–G), but types A, B, and E are the serotypes commonly involved in human disease and are also the three serotypes approved or being developed to treat human disease. Biological activity of the toxin occurs at approximately 1 ng, a billionth of a gram. As a result, about a tablespoon of toxin could supply the world for all the medical and cosmetic uses for a year.

2 History

The modern, written history of botulinum toxin begins with the recognition of food poisoning and probably starts over a millennium ago in Byzantium. Sausage plays a prominent role in the story of the medical use of botulinum toxin. Sausage has been a delicacy for centuries, and in the Byzantine era, blood sausage was commonly made by taking animal blood, fat, and organs, cooking them for varying amounts of time, and then stuffing them into "cleaned" animal stomach or intestine. Although the bacterial etiology of food poisoning was centuries away, the Byzantine emperor Leo VI embraced the association of blood sausage with food-related illness and later signed an edict that forbid the making and eating of blood sausage prepared in pig stomachs.

Further advancements into understanding the relationship between sausage and illness came during the Napoleonic War which took place from 1795 to 1813. The war leads to poor sanitary conditions in rural food production, and many deaths in Europe were associated with eating smoked blood sausages. These sausage-related deaths were studied, and in the early 1800s, the Department of Internal Affairs of the Kingdom of Wurttemberg attributed this food poisoning to a substance they called prussic acid. However, it was the physician and poet, Dr. Justinus Andreas Christian Kerner (Fig. 1), who made strong scientific inroads into our understanding of food poisoning, the role of botulism, and even the potential medical uses of botulinum toxin (Erbguth and Naumann 1999). Kerner was born in 1786 in Ludwigsburg, Germany. He studied at the University of Tubingen and received his medical degree in 1808. Kerner was a practicing physician who went on to publish the first case study of botulism and on the presumed fatty toxin from sour sausages. He described experiments, including those he performed on himself by eating small amounts of so-called sour sausage, and documented the signs and symptoms of botulism including vomiting and intestinal spasms, mydriasis, ptosis and strabismus, dysphagia, flaccid paralysis, and respiratory failure. He also noted that sausage poison develops under anaerobic conditions, interrupts motor signal transmission in the

Fig. 1 Justinus Kerner 1786–1862. From: Wikimedia Commons; adapted from Lee Byron Jennings: Justinus Kerners Weg nach Weinsberg. Die Entpolitisierung eines Romantikers. Camden House, Columbia, SC 1982, ISBN 0-938100-00-9, frontispiece

peripheral and autonomic nervous system, and is lethal in small doses. Incredibly, Kerner also proposed that the toxin could be used for therapeutic purposes. He hypothesized that this toxin could be used to lower sympathetic nervous system activity associated with movement disorders and decrease the hypersecretion of body fluids.

The bacteria that produces botulinum toxin was finally isolated around 1895 by Emile Pierre-Marie van Ermengem, a bacteriologist at the University of Ghent (van Ermengem 1897). This advance occurred again as a result of a recognized food-poisoning epidemic. However, this time the culprit was not sour sausage but bad ham.

The epidemic occurred on December 14, 1895, in a small town in Belgium where 34 musicians had a meal following a funeral where they played (Devriese 1999). Following the meal, musicians developed signs and symptoms of botulism, and three of the musicians died. Ham served at the meal was suspected as the cause of the illness. The ham was sent to Van Ermengem who performed a detailed scientific analysis where he isolated an anerobic bacteria and injected small pieces of the ham into animals leading to an illness similar to that experienced by the musicians. Van Ermengem named the bacterium *Bacillus botulinum* which stems from the Latin word botulus meaning sausage.

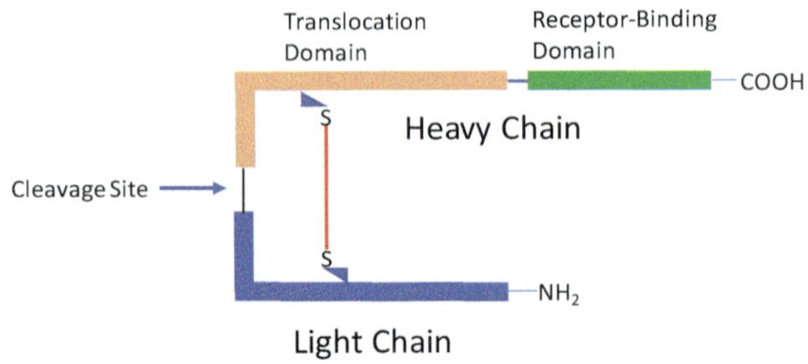

Fig. 2 Schematic drawing showing the structure of botulinum toxin A. The heavy chain and light chain are linked by a single disulfide bond. The heavy chain (approximately 100 kD) contains a receptor-binding domain and a translocation domain. The light chain (approximately 50 kD) acts as an endopeptidase and proteolytically cleaves protein involved in vesicle fusion at the inner cell membrane

Subsequent work leads to the early classification of botulinum serotypes (Burke 1919), and the bacterial exotoxin was first purified and crystalized in the 1920s (Snipe and Sommer 1928). Scientists then began to focus on the mechanism of action of botulinum toxin. Edmunds and colleagues showed that the toxin of botulism caused a complete curare-like action on the endings of the motor nerves to the voluntary muscles producing a paralysis (Edmunds and Keiper 1924). They also noted that the respiratory muscles were affected early and could lead to respiratory depression and death. Guyton and colleagues expanded on these studies describing the peripheral site of action of botulinum toxin (Edmunds and Keiper 1924). Experiments in the late 1940s then showed that the toxin acted by blocking neuromuscular transmission (Edmunds and Keiper 1924). Over the last 50 years, continued progress has been made into understanding the molecular biology of botulinum toxin activity. We now know that botulinum toxin consists of a heavy and light chain held together by a single disulfide bond and that the heavy chain has both a binding domain and a translocation domain Fig. 2. After the toxin binds to the cell, the light chain is internalized into the cell where it binds to a complex of proteins involved in neurotransmitter release (Rizo and Sudhof 1998). As an endopeptidase, the light chain then cleaves proteins involved in transmitter vesicle fusion to the inner cell membrane leading to chemical denervation. More recently scientists have imaged the crystal structure of botulinum toxin (Lacy et al. 1998) and identified the receptor for the toxin (Edmunds and Keiper 1924). The mechanism of action of botulinum toxin will be described in further detail in other chapters in this book.

Of course, the development of botulinum toxin as a medical therapy required highly pure and good manufacturing practices level material. Much of the work on the manufacture of botulinum toxin was spearheaded by Dr. Edward Schantz, a biochemist who worked in the Department of Defense laboratories at Fort Dietrich before continuing his career at the University of Wisconsin in Madison (Schantz et al. 1960). Schantz and colleagues not only worked out a manufacturing process for botulinum toxin but also supplied toxin to researchers. This included both basic scientists and clinicians who first used the toxin to treat disease in humans (Schantz and Johnson 1992).

The first human use of botulinum toxin was for strabismus. Alan Scott, an ophthalmologist, was looking for a surgical alternative to the treatment of strabismus. He started by injecting local anesthetic into the extraocular muscles but later decided to look for something with a longer duration of action. Dr. Scott stated that he initially put this at the bottom of the list because of concerns about toxicity and the thought that it would never be approved by the FDA. Dr. Scott then became aware of the work of Daniel Drachman, who injected small amounts of botulinum toxin into the hind limbs of chick embryos and noted atrophy of skeletal muscle consistent with denervation (Drachman 1964). Dr. Drachman also told Dr. Scott that he received the toxin from Ed Schantz, who was now at the University of Wisconsin. This allowed Dr. Scott to start work on the use of botulinum toxin in experimental models and paved the way to start clinical trials in patients with strabismus and blepharospasm. Dr. Scott was first to describe the beneficial effects of botulinum toxin type A in patients, publishing on its use in strabismus in 1980 (Scott 1980a, b).

These studies paved the way for the first regulatory approval of botulinum toxin for a therapeutic use. Dr. Scott's original product was called Oculinum and was approved by the US FDA on December 29, 1989, for the treatment of strabismus and blepharospasm associated with dystonia including benign essential blepharospasm or (seventh) nerve disorders in patients 12 years of age and above. The product was initially marketed and later sold to Allergan who changed the name to Botox in 1991. The nonproprietary name, onabotulinumtoxinA was given in 2011.

Clinicians from around the world heard about Dr. Scott's research and flew to California to learn how to inject the extraocular and periorbital muscles with botulinum toxin. A group of physicians at the Moorfields Eye Hospital in London received some of the Dr. Schantz botulinum toxin and studied its use in 85 adults with strabismus (Elston et al. 1985). This research fostered a collaboration with the Centre for Applied Microbiology and Research (CAMR) at Porton Down in the United Kingdom who began providing their own botulinum toxin, produced with a different manufacturing process to the researchers, and lead to the creation of Porton International, a biotechnology company that was later purchased by Ipsen. Their commercialized product called Dysport comes from combining dyes from dystonia and port from Porton Down (Monheit and Pickett 2017).

Since the US FDA approval of onabotulinumtoxinA for strabismus and blepharospasm in 1989, there have been many additional FDA approvals for the use of onabotulinumtoxinA for other indications and for other botulinum toxin type A products and a botulinum toxin type B product. In December 2000, the US FDA approved the type B serotype, rimabotulinumtoxinB, under the brand name of Myobloc by Solstice Pharmaceuticals for the treatment of patients with cervical dystonia to reduce the severity of abnormal head position and neck pain associated with cervical dystonia. AbobotulinumtoxinA, under the brand name of Dysport by Ipsen, was initially approved in the United States in 2009 for the treatment of cervical dystonia. In July 2010, incobotulinumtoxinA, under the brand name Xeomin by Merz, was first approved by the FDA for the treatment of cervical

Table 1 US Food and Drug Administration (FDA) approval of botulinum toxins

Botulinum toxin (brand name)	Abbreviated indication[a]	FDA approval (year)
AbobotulinumtoxinA (Dysport)	Cervical dystonia	2009
	Glabellar lines	2009
	Adult upper limb spasticity	2015
	Pediatric lower limb spasticity	2016
	Adult lower limb spasticity	2017
IncobotulinumtoxinA (Xeomin)	Cervical dystonia	2010
	Blepharospasm	2010
	Glabellar lines	2011
	Adult upper limb spasticity	2015
	Sialorrhea	2018
OnabotulinumtoxinA (Botox and Botox cosmetic)	Strabismus	1989
	Blepharospasm	1989
	Cervical dystonia	2000
	Glabellar lines	2002
	Axillary hyperhidrosis	2004
	Adult upper limb spasticity	2010
	Chronic migraine	2010
	Urinary incontinence due to detrusor overactivity	2011
	Overactive bladder	2013
	Lateral canthal lines	2013
	Adult lower limb spasticity	2016
	Forehead lines	2017
	Pediatric upper limb spasticity	2019
PrabotulinumtoxinA-xvfs (Jeuveau)	Glabellar lines	2019
RimabotulinumtoxinB (Myobloc)	Cervical dystonia	2000

[a]Please read FDA labeling for the complete labeled indication

dystonia and blepharospasm. In February 2019 prabotulinumtoxinA, under the brand name Jeuveau by Evolus, was approved by the FDA for temporary improvement in the appearance of moderate to severe glabellar lines associated with corrugator and/or procerus muscle activity in adults.

Many of the botulinum toxins are approved for multiple uses in dozens of countries around the world. Table 1 lists all of the botulinum toxins currently approved in the United States and the indications they are approved to treat. When people hear about the use of botulinum toxin in patients, they usually think of wrinkles. New therapeutic uses of botulinum toxins have been predominantly driven by astute clinicians who understood the science behind both the treatment and other potential disease states or who recognized beneficial effects in a second condition in patients being treated for a separate disease. In 1989, botulinum toxin received FDA approval for strabismus. That same year, Clark and Berris reported on the use of botulinum toxin as a treatment for facial asymmetry caused by a facial nerve paralysis (Clark and Berris 1989). The patient was noted to experience satisfactory relief of the asymmetry caused by one-sided forehead wrinkling and brow elevation. Jean and Alister Carruthers, an ophthalmologist and dermatologist practicing in Canada, noticed that their patients treated for blepharospasm had resolution of their frown lines and published on the treatment of glabellar frown lines with botulinum toxin in 1992 (Carruthers and Carruthers 1992). A number of other clinicians also noted the aesthetic use of botulinum toxin injections and began conducting randomized clinical trials to study the safety and efficacy of this approach for facial wrinkles (Keen et al. 1994). Botulinum toxin was first FDA approved for the treatment of moderate to severe glabellar lines in April 2002.

Fortunately for patients, the history of botulinum toxin for medical use does not stop here. Academic laboratories, pharmaceutical companies, and clinicians continue to conduct research on the use of botulinum toxins in medicine, both on additional indications for currently approved botulinum toxins and for new botulinum toxins in development. We continue to learn more about the molecular structure of botulinum toxin and its mechanism of action. These advancements should allow us to better achieve the goal to provide additional therapeutic options for patients with the hope of improving efficacy and minimizing adverse effects.

References

Burke GS (1919) The occurrence of Bacillus botulinus in nature. J Bacteriol 4:541–553

Carruthers JD, Carruthers JA (1992) Treatment of glabellar frown lines with C. botulinum-A exotoxin. J Dermatol Surg Oncol 18:17–21

Clark RP, Berris CE (1989) Botulinum toxin: a treatment for facial asymmetry caused by facial nerve paralysis. Plast Reconstr Surg 84:253–255

Devriese P (1999) On the discovery of Clostridium botulinum. J Hist Neurosci 8:43–50

Drachman DB (1964) Atrophy of skeletal muscle in chick embryos treated with botulinum toxin. Science 145:719–721

Edmunds CW, Keiper GF (1924) Further studies on the action of botulinus toxin. JAMA 83:495–502

Elston JS, Lee JP, Powell CM et al (1985) Treatment of strabismus in adults with botulinum toxin A. Br J Ophthalmol 69:718–724

Erbguth FJ, Naumann M (1999) Historical aspects of botulinum toxin: Justinus Kerner (1786-1862) and the “sausage poison”. Neurology 53:1850–1853

Homann M, Sansjofre P, Van Zuilen M et al (2018) Microbial life and biogeochemical cycling on land 3,220 million years ago. Nat Geosci 11:665–671

Keen M, Blitzer A, Aviv J et al (1994) Botulinum toxin A for hyperkinetic facial lines: results of a double-blind, placebo-controlled study. Plast Reconstr Surg 94:94–99

Lacy DB, Tepp W, Cohen AC et al (1998) Crystal structure of botulinum neurotoxin type A and implications for toxicity. Nat Struct Biol 10:898–902

Monheit GD, Pickett A (2017) AbobotulinumtoxinA: a 25-year history. Aesthet Surg J 37:S4–S11

Rizo J, Sudhof TC (1998) Mechanics of membrane fusion. Nat Struct Biol 5:839–842

Schantz EJ, Johnson EA (1992) Properties and use of botulinum toxin and other microbial neurotoxins in medicine. Microbiol Rev 56:80–89

Schantz EJ, Stefanye D, Spero L (1960) Observations on the fluorescence and toxicity of botulinum toxin. J Biol Chem 235:3489–3491

Scott AB (1980a) Botulinum toxin injection into extraocular muscles as an alternative to strabismus surgery. Ophthalmology 87:1044–1049

Scott AB (1980b) Botulinum toxin injection into extraocular muscles as an alternative to strabismus surgery. J Pediatr Ophthalmol Strabismus 17:21–25

Snipe P, Sommer H (1928) Studies on botulinus toxin. 3. Acid precipitation of botulinus toxin. J Infect Dis 43:152–160

van Ermengem E (1979) Classics in infectious diseases. A new anaerobic bacillus and its relation to botulism. E. van Ermengem. Originally published as "Ueber einen neuen anaëroben Bacillus und seine Beziehungen zum Botulismus" in Zeitschrift für Hygiene und Infektionskrankheiten 26: 1-56, 1897. Rev Infect Dis 1:701–719

Whitman WB, Coleman DC, Wiebe WJ (1998) Prokaryotes: the unseen majority U S A. Proc Natl Acad Sci 95:6578–6583

The Structure and Classification of Botulinum Toxins

Min Dong and Pål Stenmark

Contents

Abstract

Botulinum neurotoxins (BoNTs) are a family of bacterial protein toxins produced by various *Clostridium* species. They are traditionally classified into seven major serotypes (BoNT/A-G). Recent progress in sequencing microbial genomes has led to an ever-growing number of subtypes, chimeric toxins, BoNT-like toxins, and remotely related BoNT homologs, constituting an expanding BoNT superfamily. Recent structural studies of BoNTs, BoNT progenitor toxin complexes,

M. Dong (✉)
Department of Urology, Boston Children's Hospital, Boston, MA, USA

Department of Microbiology, Harvard Medical School, Boston, MA, USA

Department of Surgery, Harvard Medical School, Boston, MA, USA
e-mail: min.dong@childrens.harvard.edu

P. Stenmark (✉)
Department of Biochemistry and Biophysics, Stockholm University, Stockholm, Sweden

Department of Experimental Medical Science, Lund University, Lund, Sweden
e-mail: pal.stenmark@med.lu.se

S. M. Whitcup, M. Hallett (eds.), *Botulinum Toxin Therapy*,
Handbook of Experimental Pharmacology 263, https://doi.org/10.1007/164_2019_342

tetanus neurotoxin (TeNT), toxin-receptor complexes, and toxin-substrate complexes have provided mechanistic understandings of toxin functions and the molecular basis for their variations. The growing BoNT superfamily of toxins present a natural repertoire that can be explored to develop novel therapeutic toxins, and the structural understanding of their variations provides a knowledge basis for engineering toxins to improve therapeutic efficacy and expand their clinical applications.

Keywords
Bacterial toxins · BoNT · BoNT-like toxins · Botox · Botulinum neurotoxin · Botulinum toxin · Tetanus neurotoxin · X-ray crystal structure

Botulinum neurotoxins (BoNTs) are a family of bacterial protein toxins that cause the human and animal disease botulism (Fig. 1) (Dong et al. 2019; Pirazzini et al. 2017; Montal 2010; Rossetto et al. 2014). Together with the related tetanus neurotoxin (TeNT), they are known as clostridial neurotoxins. These toxins are composed of two chains and three functional domains (Fig. 2a): the light chain (LC, ~50 kDa), which is a zinc-dependent metalloprotease that cleaves the target proteins in neurons, and the heavy chain (HC), which can be further divided into the N-terminal membrane translocation domain (H_N, ~50 kDa) and the C-terminal receptor-binding domain (H_C, ~50 kDa). These toxins are initially produced as a single polypeptide known as the pro-toxin. The linker region between the LC and HC needs to be

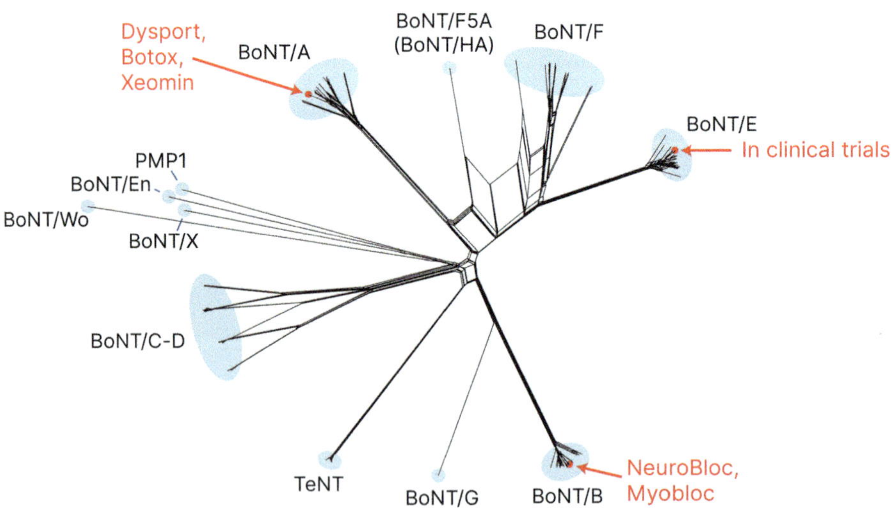

Fig. 1 A phylogenetic split network of BoNT and BoNT-like toxins. The diagram illustrates the potential evolutionary relationships based on comparing protein sequences of all known BoNT subtypes, chimeric toxins, BoNT-like toxins, and BoNT/Wo. BoNT/A1 (with such brand names as Dysport, Botox, and Xeomin from different companies) and BoNT/B1 (with the brand names NeuroBloc or Myobloc) have been approved by the FDA for medical and cosmetic uses, while BoNT/E1 is under clinical trials. These three toxins are marked in red

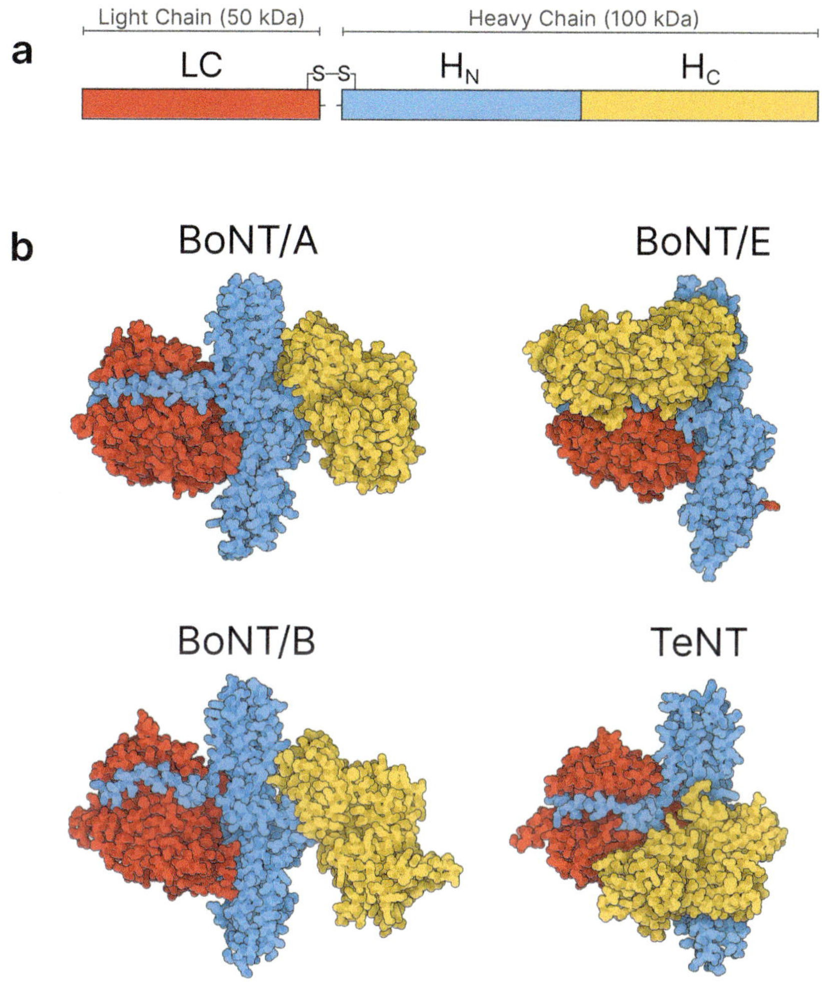

Fig. 2 The three-domain architecture of BoNTs. (**a**) A schematic drawing of the di-chain and three-domain architecture of BoNTs. The light chain (LC) is colored red, the translocation domain (H_N) blue, and the receptor-binding domain (H_C) yellow. The two chains are connected via a disulfide bond. (**b**) The crystal structures of full-length BoNT/A, BoNT/B, BoNT/E, and TeNT. LC is colored red, the H_N blue, and the H_C yellow. The protein structures are shown in a space filling representation. BoNT/A and BoNT/B display a linear arrangement for the three domains, with the LC and the H_C on each side of the H_N. BoNT/E and TeNT have their LC and H_C located on the same side of the H_N. PDB: 3BTA, 3FFZ, 1S0D, and 5N0B

cleaved by bacterial or host proteases, which converts the inactive pro-toxin to a di-chain active form. The LC and HC remain connected via a single disulfide bond. Once the H_C recognizes the receptors on nerve terminals, the toxin enters neurons via receptor-mediated endocytosis. The H_N then mediates translocation of the LC across endosomal membranes into the cytosol. The LC cleaves neuronal substrate proteins, including Syntaxin 1, SNAP-25, and VAMP1, 2, 3, which are required for neurotransmitter release (Jahn and Scheller 2006; Sudhof and Rothman 2009), thus blocking neurotransmission.

1 BoNT Serotypes, Chimeric Toxins, and Subtypes

The classifications of BoNTs are traditionally based on their antigenicity and are known as serotypes, meaning that anti-sera generated against one toxin cannot recognize and neutralize another toxin (Fig. 1). The first BoNT was identified in 1897 (van Ermengem 1897). A serologically distinct BoNT was recognized in 1904, and hence serotypes A and B were designated to differentiate these two toxins (Burke 1919a, b; Leuchs 1910). This was followed by recognition of serotypes C in 1922, D in 1928, E in 1936, and F in 1960. The latest serotype, BoNT/G, was isolated from soils in Argentina and reported in 1970 (Gimenez and Ciccarelli 1970). This traditional serological classification has provided a way to distinguish diverse BoNT members and played a key role in developing vaccines and neutralizing antibodies against BoNTs.

The DNA and protein sequences for the prototypes of the seven BoNTs, as well as related TeNT, were resolved by the early 1990s, revealing ~37–70% variation in protein sequences among different serotypes. A phylogenic tree can be constructed based on protein sequences (Fig. 1). The two pairs BoNT/B versus BoNT/G and BoNT/E versus BoNT/F show the highest sequence identity (57% and 63%, respectively) among BoNTs. BoNT/A and BoNT/B have been approved by the FDA for use in humans (Schantz and Johnson 1992; Johnson 1999), and BoNT/C and BoNT/F have been investigated for potential medical use (Eleopra et al. 1997, 2006). BoNT/E is currently under clinical trials, which has a faster onset and shorter paralysis duration than BoNT/A (Fig. 1) (Eleopra et al. 1998).

The limitations of serological classification were recognized as early as the 1920s, when inconsistent neutralization efficacy was observed while serotyping "type C" toxins from different bacterial strains. Sequence information later revealed that this was due to the existence of naturally occurring chimeric toxins. For instance, there is a chimeric BoNT/CD with its LC-H_N derived from BoNT/C and its H_C from BoNT/D (Moriishi et al. 1996b). Anti-sera raised against BoNT/CD can neutralize BoNT/C, but anti-sera against BoNT/C are not effective in neutralizing BoNT/CD (Pfenninger 1924). There is also a chimeric toxin BoNT/DC, which is composed of a LC-H_N that is 98% identical with the corresponding region in BoNT/D and a H_C that shares 77% identity with BoNT/C-H_C (Moriishi et al. 1996a). Serologically, because this toxin can be recognized and neutralized by anti-sera against BoNT/D, it has been considered a BoNT/D (the bacteria strain is known as the strain D-5995, D-SA, or D-4947) and has been supplied as BoNT/D by a commercial vendor (Metabiologics Inc. Madison, WI, USA); however, its H_C is clearly distinct from BoNT/D-H_C.

Sequencing toxin genes has also revealed a growing number of subtype toxins with significant protein sequence variations from known toxin sequences (Hill et al. 2007; Peck et al. 2017). These variations could significantly reduce the efficacy of the standard anti-sera. For instance, there are at least eight BoNT/A subtypes (A1-A8).

The prototype is referred to as BoNT/A1, which is the only FDA-approved BoNT/A type for use in humans (Peck et al. 2017). Among BoNT/A subtypes, BoNT/A3 contains the greatest sequence variations from BoNT/A1 (15.4%). BoNT/F subtypes contain the most variation among all seven serotypes from their prototype BoNT/F1: as high as 30.2% for BoNT/F5 and 26.3% for BoNT/F7. The difference between BoNT/F5 and BoNT/F7 is 36.2%, the highest variation among all subtypes. These sequence differences help explain the significant variations in neutralization efficacy observed when BoNT/A and BoNT/F were serotyped from different bacterial strains.

The limitations of the traditional serotyping approach are further illustrated by the recent controversial naming of BoNT/H. This BoNT was identified in 2013 from a bacterial strain isolated from an infant botulism case (Dover et al. 2014; Barash and Arnon 2014). This toxin was not neutralized by anti-sera against other BoNTs following the established serotyping protocol. Thus, it was proposed as a new serotype. However, sequencing the toxin gene revealed that its LC shares ~80% identity with the LC of BoNT/F5, while its H_C shares ~84% identity with BoNT/A1-H_C (Maslanka et al. 2016). As the LC of BoNT/F5 (F5-LC) has a relatively high sequence variation from BoNT/F1-LC (only ~47% sequence identity) (Kalb et al. 2012), it is not surprising that anti-sera against BoNT/F1 failed to neutralize this toxin. Later studies showed that this toxin can be neutralized by antibodies against BoNT/A1, albeit a higher antibody titer is required than the standard serotyping protocol (Maslanka et al. 2016). Thus, this toxin is also considered a chimeric toxin BoNT/FA or more precisely BoNT/F5-A. To make the matter even more complicated, the H_N of this toxin does not appear to be close to either BoNT/F5-H_N or BoNT/A1-H_N. Thus, it has also been speculated that the LC-H_N of this toxin might be derived from a yet-to-be-identified BoNT.

Most subtypes likely cleave the same substrate protein at the same site and utilize the same receptors as their prototypes. However, exceptions with altered functional specificity have been reported. For instance, BoNT/F5 cleaves a site on VAMP1/2/3 that is distinct from all other BoNTs, indicating that its sequence variation is large enough to shift its cleavage site (Kalb et al. 2012). Similarly, sequence variations between the H_Cs of BoNT/DC and BoNT/C result in BoNT/DC utilizing a protein receptor that is not the receptor for BoNT/C (Peng et al. 2012).

Interestingly, even relatively low levels of sequence variations among subtypes, which may not alter the cleavage site on their substrates or switch receptors, could have measurable impacts on in vivo efficacy and pharmacological properties. For instance, BoNT/A2, which has 90% sequence identity with BoNT/A1, showed faster onset than BoNT/A1 on cultured neurons and in animal models (Torii et al. 2011; Pier et al. 2011; Whitemarsh et al. 2013; Pellett et al. 2015). It has been suggested that this faster onset time is because BoNT/A2 has an overall faster translocation process across the membrane than BoNT/A1 (Pier et al. 2011; Whitemarsh et al. 2013). As faster onset is clinically beneficial, BoNT/A2 has been explored for clinical uses. Additionally, it was recently reported that BoNT/A6, which has 4.3% sequence variation from BoNT/A1, also showed faster entry into neurons in culture (Moritz et al. 2018). Another example is that sequence variations in the LC of BoNT/A3 from BoNT/A1-LC result in a shorter duration of paralysis induced by BoNT/A3 compared with that induced by BoNT/A1 (Pellett et al. 2018). Thus, sequence variations among the growing number of subtypes provide valuable resources for developing a new generation of therapeutic toxins and for optimizing toxin sequences to improve their pharmacological properties.

While traditional serotyping has value as a framework for categorizing toxins and identifying their distinct antigenic properties, describing toxins with seven serotypes is clearly insufficient to capture the growing diversity among BoNTs. Given the current ease of determining the exact toxin sequences, it will be important to note the specific subtype information when discussing a particular BoNT. To manage the naming of the growing number of subtypes, a guideline was proposed in 2017, which officialized the previous proposed threshold for a toxin sequence to be considered a new subtype as >2.6% variations at protein sequence levels from any known BoNT sequences (Peck et al. 2017). A few previous defined subtypes with <2.6% variations are grandfathered in, such as BoNT/B2, B3, and B6, which encompass only 1.5–1.9% variations, and BoNT/E2, E3, and E7, which differ from BoNT/E1 by only 1.0 to 2.1%.

To avoid duplication in numbering new subtypes, an email address has been set up at the Centers for Disease Control and Prevention (CDC, bontsubtype@cdc.gov) to receive requests for designation of new subtypes (Peck et al. 2017). There are also other efforts to develop a unified reporting system and database. One such a database, BoNTbase (https://bontbase.org) developed by Dr. Jonathan Davies in the laboratory of Prof. Stenmark, contains all reported BoNT subtypes as well as BoNT-like sequences along with associated research publications.

2 TeNT

TeNT, produced by *Clostridium tetani*, shares the same overall domain structures and mode of actions with other BoNTs; in fact, sequence alignment places TeNT in the middle of the family (Fig. 1). However, TeNT is not classified as a BoNT because it causes tetanus, a disease that is clinically distinct from botulism. TeNT and BoNTs both target and enter peripheral motor neurons. Unlike BoNTs, which block neurotransmitter release from motor neurons, thus causing muscle relaxation (flaccid paralysis), TeNT undergoes retrograde transport and transcytosis: it moves along the axons of motor neurons into the cell body in the spinal cord and is then released from motor neurons and enters the connecting inhibitor neurons where TeNT blocks neurotransmitter release (Lalli et al. 2003; Surana et al. 2018). Loss of inhibitory input leads to overactivity of motor neurons, resulting in spastic paralysis. Interestingly, it has been suggested that at least a small fraction of some BoNTs such as BoNT/A1 may also undergo long-range transport and transcytosis along peripheral neuronal axons into connecting neurons (Restani et al. 2011, 2012; Antonucci et al. 2008; Bomba-Warczak et al. 2016). The molecular basis for the different traffic pathways utilized by TeNT versus BoNTs remains unknown.

3 BoNT-Like Toxins

Rapid progress in sequencing microbial genomes in recent years has fundamentally changed how novel toxins are discovered. In 2015, a new toxin gene was recognized through bioinformatic analysis of the genome of a *Clostridium botulinum* strain. It encodes a protein containing the same three functional domains and key motifs found in BoNTs, with ~28–30% of sequence identity compared with the seven BoNTs (Fig. 1) (Zhang et al. 2017). This toxin was named BoNT/X because of varying opinions on what naming convention to utilize. Subsequent functional characterization confirmed that BoNT/X is capable of cleaving VAMP1, 2, 3 at a novel cleavage site distinct from all known cleavage sites for BoNTs. Interestingly, BoNT/X is a unique toxin that can also cleave VAMP family members VAMP4, VAMP5, and Ykt6, although the physiological consequences of these noncanonical cleavage events remain to be determined. Because BoNT/X was not recognized by antisera raised toward any of the seven BoNTs, it could be considered a novel serotype. However, unlike the seven classic BoNTs, BoNT/X showed only a low level of toxicity in mice. These findings suggest that BoNT/X may not naturally target mice and other vertebrates. The host species targeted by BoNT/X remains to be established.

In 2017, sequencing the genome of an *Enterococcus faecium* strain collected from cow feces revealed another BoNT-like toxin, designated BoNT/En (Zhang et al. 2018; Brunt et al. 2018). It too shares the same three-domain arrangement and key motifs found in BoNTs, with 24–27% protein sequence identity with the seven classic BoNTs. It is most closely related to BoNT/X, sharing 37% sequence identity. Functional validation demonstrated that BoNT/En is capable of cleaving VAMP1, 2, 3 and SNAP-25 in cultured neurons. Because BoNT/En is not recognized by any anti-sera against the seven BoNTs and BoNT/X, it too can be considered a new serotype, but BoNT/En showed no toxicity in mice. This is largely due to the lack of appropriate receptors in mice, as a chimeric toxin containing the LC-H_N of BoNT/En fused with the H_C of BoNT/A showed high neuronal toxicity and induced muscle paralysis. Thus, BoNT/En does not appear to target mouse motor neurons, and the host species naturally targeted by BoNT/En remains unknown.

In 2019, another BoNT-like toxin, PMP1 (paraclostridial mosquitocidal protein 1), was reported (Contreras et al. 2019). It was identified by screening and analyzing bacteria that can kill *anopheles* mosquito larvae. The toxin gene is located on a plasmid found in two strains with mosquitocidal activity: *Paraclostridium bifermentans malaysia* isolated from a mangrove swamp in Malaysia and *Paraclostridium bifermentans Paraiba* isolated in Brazil. PMP1 shares 36% sequence identity with BoNT/X and 34% identity with BoNT/En. These three toxins form a distinct branch in the BoNT superfamily (Fig. 1). Functional analysis showed that PMP1 is capable of cleaving mosquito Syntaxin 1 and has no toxicity in mice. PMP1 is the first known neurotoxin that naturally targets *anopheles* mosquito larvae. Its insecticidal toxicity and selectivity have the potential to be harnessed for developing novel mosquito control agents.

The crystal structure of the PMP1-H_C has been solved, revealing features distinct from the classic BoNT-H_Cs (Contreras et al. 2019). For instance, there are a dozen aromatic residues exposed on the surface of PMP1-H_C, forming unique hydrophobic patches. Mutations at these hydrophobic patches reduced toxicity, suggesting that they may contribute to receptor binding. The receptors for PMP1 and other BoNT-like toxins remain unknown, which likely dictate the species targeted by each toxin, although other barriers may also exist. It is possible that BoNT/X and BoNT/En may also target insects or other invertebrates, and we expect that additional members of this group will continue to be discovered, which may form a group of neurotoxins targeting invertebrates. As the H_C can be switched between BoNT and BoNT-like toxins, chimeric toxins utilizing the LC-H_N part of BoNT-like toxins may provide an additional toolbox for designing new therapeutic toxins with unique properties.

4 BoNT Homologs

Bioinformatic analysis also revealed a growing number of sequences bearing various degrees of homology to BoNT, defined as BoNT homologs. The first was discovered in the genome of *Weissella oryzae*, a gram-positive anaerobe isolated from fermented rice in Japan (Mansfield et al. 2015). The protein was later named BoNT/Wo (Zornetta et al. 2016). The protein sequence of BoNT/Wo can be divided into LC, H_N, and H_C based on homology analysis with BoNTs, and it contains a few key conserved moieties found in BoNTs. However, BoNT/Wo is significantly different from BoNTs and BoNT-like toxins. First, the sequence identity of BoNT/Wo to other BoNTs and BoNT-like toxins is only 14–16% (Fig. 1). Second, there is no cysteine located at the linker region between BoNT/Wo-LC and HC. Third, while all BoNTs and BoNT-like toxins are located within a similar gene cluster (discussed in Sect. 6), BoNT/Wo is not in such a cluster. Thus, BoNT/Wo is only a distant homolog of BoNTs. It has been reported that BoNT/Wo-LC is capable of cleaving VAMP2 in vitro, but its physiological function remains to be established (Zornetta et al. 2016).

Three more BoNT homologs were recently reported in the genome of *Chryseobacterium piperi* (Mansfield et al. 2019). They showed low levels of sequence identity to BoNTs. For instance, one of these proteins, designated Cp1, shares ~17% identity with BoNT/A1. Cp1 can be divided into LC, H_N, and H_C based on homology analysis with BoNTs, and there are two cysteine residues located at the linker region between its LC and HC, suggesting an inter-chain disulfide bond. The function of these BoNT homologs remains to be fully characterized.

5 Three-Domain Architecture

The full-length crystal structures of BoNT/A, BoNT/B, BoNT/E, and TeNT have been determined, clearly demonstrating a three-domain architecture, composed of the LC, H_N, and H_C (Fig. 2b) (Lacy et al. 1998; Swaminathan and Eswaramoorthy 2000; Kumaran et al. 2009; Masuyer et al. 2017). The overall fold of each domain is largely conserved across these toxins, despite their rather low levels of amino acid sequence identity. BoNT/A and BoNT/B both showed a linear domain arrangement, with the LC and H_C located on each side of the H_N, while the LC and H_C in BoNT/E are located on the same side of H_N and interact with each other. Thus, BoNT/E has an overall more-compact globular shape than BoNT/A and BoNT/B. The structure of TeNT has been investigated using multiple approaches: small-angle X-ray scattering analysis showed that TeNT is in a linear domain arrangement (open state) at neutral pH and changes into a compact globular form (closed state) under acidic pH (Masuyer et al. 2017). An intermediate semi-open state was also observed by low-resolution cryogenic electron microscopy (Cryo-EM). The high-resolution X-ray crystal structure of TeNT showed a closed state, with all three domains interacting with each other. Within TeNT, the LC-H_N forms a relatively stable core, while the H_C alters its position under different experimental conditions. The physiological relevance of the domain rearrangement in TeNT and whether similar flexibility exists in BoNTs remain to be determined.

5.1 Translocation Domain

The crystal structures of BoNT/A, B, E, and TeNT all reveal that the LC forms extensive contacts with the H_N. Particularly, the N-terminal region composed of ~50 residues of the H_N, designated the "belt" region, wraps around the LC (Fig. 2b). Because the belt region partially covers the active site of the LC, the LC reaches its full activity only once it is dissociated from the H_N after the disulfide bond connecting the LC and HC is broken (reduced).

The H_N is responsible for translocating the LC across the endosomal membrane. It is well established that the low pH within endosomes triggers conformational changes in BoNTs, leading to translocation of the LC, but the molecular mechanism for this translocation process remains to be elucidated. The H_N domain prominently features two long α-helices of ~105 Å. It remains unclear how these helices may alter their conformations upon encountering the low pH within endosomes.

A potential transmembrane region has been proposed based on analyzing hydrophobicity (e.g., residues 659–681 in BoNT/A), and a similar region (residues 593–686) in BoNT/A has been suggested to contribute to forming a channel in membranes (Montal et al. 1992; Lebeda and Olson 1995; Fischer et al. 2012). Bioinformatic analysis comparing the H_N of BoNTs and BoNT-like toxins revealed similarities between this region and the proposed transmembrane helix of diphtheria toxins, and also identified a conserved K/R. . .PxxG motif (Mansfield et al. 2019).

A recent study reported that the isolated H_N fragment of BoNT/A lacking the belt region can be produced as a soluble protein, and its crystal structure under acidic pH conditions has been resolved (Lam et al. 2018a). The structure highlights major conformational changes in the region from residues 620 to 667. This region is termed the BoNT-switch and contains disordered loops and short helices under neutral pH but switches to β-hairpins containing five β-strands under acidic pH. Interestingly, the sequence of this region, particularly the β2/β3 loop, is highly conserved across all BoNTs and bears an "aromatic-hydrophobic-glycine" tripeptide motif flanked by proline residues, which is similar to the lipid-binding peptide found in viral fusion proteins such as the internal fusion loop of Ebola virus glycoprotein 2. Thus, it was

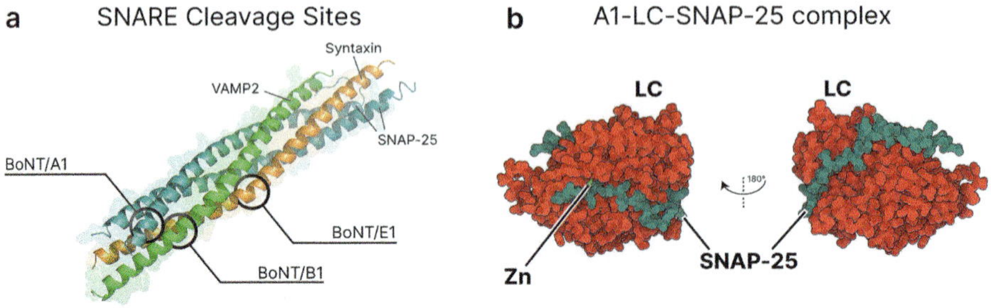

Fig. 3 BoNT-LCs cleave SNARE proteins. (**a**) The three SNARE proteins, Syntaxin 1, SNAP-25, and VAMP1/2/3 form a complex of four alpha helix bundles, which is essential for fusion of synaptic vesicle membranes to the plasma membrane of neurons. Cleavage of any one of these three SNARE proteins is sufficient to block vesicle exocytosis and neurotransmitter release. The cleavage sites for BoNT/A1, B1, and E1 are marked. PDB: 1N7S. (**b**) The crystal structure of a SNAP-25 fragment (colored dark green) in complex with BoNT/A1-LC (red), showing extensive interactions of SNAP-25 with the BoNT/A1-LC (right panel: rotated 180°). PDB: 1XTG

suggested that the BoNT-switch region is responsible for sensing the pH change and initiating membrane penetration via a mechanism similar to that used by viral fusion peptides. These results represent a major advance in our understanding of pH-induced conformational change in H_N. How the changes in the BoNT-switch region leads to further conformational changes in the rest of H_N and the eventual translocation process remains to be determined.

5.2 The Structure of the LC

By aligning the protein sequence of five BoNTs and TeNT, Giampietro Schiavo and Cesare Montecucco recognized a conserved HEXXH motif that is the key feature of metalloproteases, suggesting that BoNTs and TeNT act as proteases (Schiavo et al. 1992b). In their following seminal work published in 1992, they identified BoNT/B and TeNT as zinc-dependent proteases that cleave the synaptic vesicle protein VAMP2 (Schiavo et al. 1992a). Within a few years, it was fully established that BoNT/B, D, F, and G cleave homologous VAMP1, 2, and 3, while BoNT/A, C, and E cleave the peripheral membrane protein SNAP-25 (Fig. 3a). In addition, BoNT/C can also cleave the plasma membrane protein Syntaxin 1. BoNT/B and TeNT both share the same cleavage site on VAMP1, 2, and 3, while all other toxins have their own unique cleavage sites. These three toxin substrates are members of SNARE family proteins. They form the core complex that mediates fusion of synaptic vesicle membranes to plasma membranes, which is essential for releasing neurotransmitters (Jahn and Scheller 2006; Sudhof and Rothman 2009).

The crystal structures of all seven BoNT-LCs have been resolved, revealing an overall conserved globular fold (Jin et al. 2007; Arndt et al. 2005, 2006). The catalytic site with the signature motif HEXXH is conserved in both composition and geometry across all BoNTs. BoNT-LCs are zinc-dependent proteases with remarkable substrate specificity. As their catalytic sites are similar, the specific recognition and cleavage of different substrates must involve regions outside the catalytic site. Indeed, co-crystal structure of an inactive form of BoNT/A-LC (A-LC, containing two-point mutations that abolish its protease activity) in complex with a fragment of its substrate SNAP-25 (residues 141–204) reveals that SNAP-25 wraps around A-LC, forming extensive interactions particularly via an α-exosite bound by the N-terminal region of the SNAP-25 fragment as well as a β-exosite bound by the C-terminal region of SNAP-25 (Fig. 3b) (Breidenbach and Brunger 2004). This requirement of "long stretch" of SNAP-25 to be properly docked into A-LC ensures specificity.

Each BoNT-LC likely possesses its own distinct exosites, whose location and composition determine the selection of the substrate SNARE proteins and the specific cleavage site. The co-crystal structure of BoNT/F-LC (F-LC) in complex with a VAMP2 fragment containing a point mutation that renders it resistant to BoNT/F is the only other toxin-substrate complex that has been resolved (Agarwal et al. 2009). This structure also demonstrated extensive interactions between F-LC and the VAMP2 fragment, with VAMP2 docked onto F-LC through at least three exosites distinct from the exosites in A-LC. The precise locations of exosites in other BoNT-LCs remain to be established. The crystal structure of BoNT/X-LC (X-LC) has been solved (Masuyer et al. 2018). Despite only ~30% sequence identity with other BoNTs, X-LC display a typical BoNT-LC fold with many conserved secondary structural features. The structure further demonstrates that X-LC is a bona fide member of the BoNT-LC family. The crystal structure of the BoNT/Wo-LC has recently been solved as well, showing that it shares a common core fold found in other BoNT-LCs but also revealing several distinct features including an unusually wide and open catalytic site (Kosenina et al. 2019).

Notably, A-LC has been shown to maintain its activity in cultured neurons for several months, which is the major reason for BoNT/A's ability to induce persistent paralysis that lasts several months in humans (Keller et al. 1999; Whitemarsh et al. 2014; Tsai et al. 2017; Pellett et al. 2015; Foran et al. 2003). This is a key pharmacological property that contributes to the success of BoNT/A as a therapeutic agent. Among the seven BoNTs, BoNT/E showed the shortest half-life, with only a few weeks in humans (Foran et al. 2003), a key feature differentiating it from BoNT/A. The molecular basis for the extremely long half-life of BoNT/A remains to be fully established. A-LC has been shown to bind the cytoskeleton protein septin complex, which may shield A-LC from degradation (Vagin et al. 2014). It has also been suggested that A-LC recruits deubiquitinase to reduce its ubiquitination (Tsai et al. 2017). The structural basis for those interactions remains to be solved.

5.3 Structural Basis for Receptor Recognition

The structure of the 50 kDa H_C showed two distinct sub-domains roughly equal in size. BoNTs have extreme specificity toward nerve terminals, which is achieved by recognizing at least two receptor components in a "double-receptor" model (Montecucco 1986). One is a family of glycolipids on cell membranes known as gangliosides, which are comprised of a lipid tail and a glycan headgroup containing various numbers of negatively charged sialic acids (Simpson and Rapport 1971; Hamark et al. 2017). Gangliosides are abundant at nerve terminals and serve as low-affinity receptors to enrich the toxin onto the cell surface. A ganglioside-binding site (GBS) has been identified and is conserved in BoNT/A, B, E, F, and G (Rummel et al. 2003, 2004; Fotinou et al. 2001). This GBS is at the C-terminal region of the H_C and contains the signature residues SXWY. The crystal structures of the H_C in complex with the headgroup of gangliosides have been solved for BoNT/A, B, and F, showing that GBS interacts with the GalNAc-Gal motif as well as sialic acids within gangliosides (Stenmark et al. 2008; Berntsson et al. 2013; Benson et al. 2011).

Besides gangliosides, many BoNTs also require specific neuronal protein receptors. Two sets of synaptic vesicle membrane proteins, synaptic vesicle glycoprotein 2 (SV2) and Synatotagmin I and II (Syt I/II), serve as receptors for multiple BoNTs. For instance, BoNT/A utilizes SV2 (including all three isoforms SV2A, SV2B, and SV2C) as its receptors, while BoNT/B, BoNT/DC, and BoNT/G all utilize homologous Syt I and Syt II as receptors (Dong et al. 2003, 2006; Nishiki et al. 1994; Mahrhold et al. 2006). These synaptic vesicle membrane proteins travel to cell surfaces only transiently, and this entry pathway is thus activity facilitated, as neuronal activity promotes synaptic vesicle exocytosis and endocytosis, leading to enhanced binding and entry of BoNTs.

Syt I and II are single-pass transmembrane proteins, with a short luminal domain (the region inside vesicles). BoNT/B, DC, and G all recognize the same short section of the luminal domain of Syt I/II, located next to the transmembrane domain. The co-crystal structure of the H_C of BoNT/B (B-H_C) in complex with the Syt II fragment containing the toxin-binding site revealed that the toxin-binding segment is induced to form an amphipathic α-helix and dock into a hydrophobic groove within B-H_C (Fig. 4a) (Jin et al. 2006; Chai et al. 2006). The complex is stabilized by highly specific side-chain-to-side-chain interactions. It has been reported that Syt II is expressed in most motor neurons in diaphragm neuromuscular junctions, while Syt I is detectable in only ~40% of motor neurons in mice (Pang et al. 2006). These findings suggest that Syt II is the dominant receptor at diaphragm motor nerve terminals. On the other hand, bladder tissues express mainly Syt I, but not Syt II (Elliott et al. 2019). Since bladder tissues are smooth muscles controlled by autonomic nerves, it is possible that Syt I is the dominant receptor in autonomic nerves.

It has long been clinically observed that higher doses of BoNT/B are required to achieve the same level of paralysis produced by BoNT/A in humans. This is because human Syt II happens to contain a residue change from phenylalanine, which is commonly found in most mammalian species, to leucine at a key position within the toxin-binding region. This single residue change drastically reduces the binding

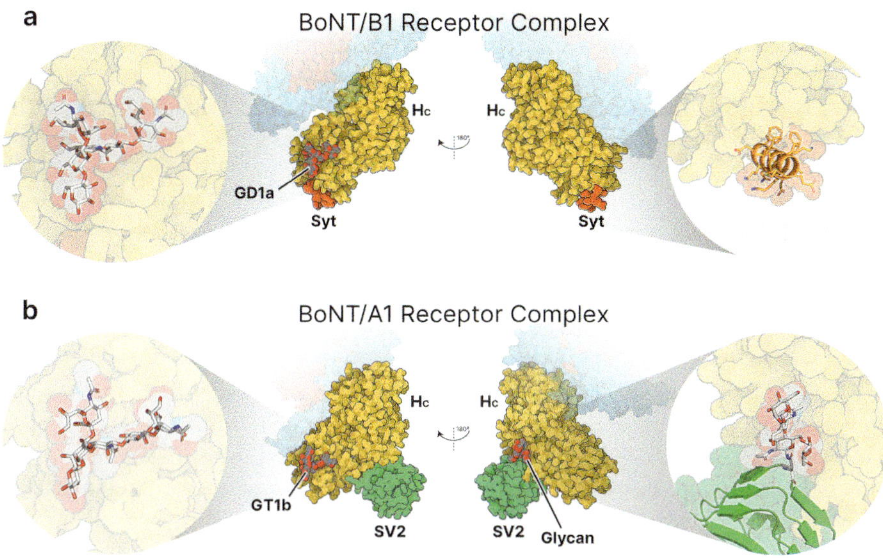

Fig. 4 BoNT/A1 and BoNT/B1 in complex with their receptors. (**a**) The crystal structure of BoNT/B1-H_C (yellow) in complex with its ganglioside coreceptor (colored according to chemical element) and the toxin-binding region of its protein receptor Syt II (orange). PDB: 4KBB. (**b**) BoNT/A1-H_C (yellow) in complex with its ganglioside coreceptor (colored according to chemical element) and the L4 of its protein receptor SV2C (green). PDB: 2VU9 and 5JLV

affinity of BoNT/B, DC, and G to human Syt II compared with Syt II from mice (Peng et al. 2012; Strotmeier et al. 2012).

This "defect" in human Syt II creates the need to engineer BoNT/B to improve its efficacy in humans. This has been achieved by structure-assisted mutagenesis approaches, which eventually identified that mutating residue 1191 (glutamic acid) to methionine, cysteine, glutamine, or valine, in combination with mutating serine 1199 to tryptophan or tyrosine, creates B-H_C mutants that can bind robustly to both mouse and human Syt II (Tao et al. 2017). Two of these toxin variants, E1191M/S1199Y and E1191Q/S1199W, have been produced recombinantly as full-length active toxins in *E. coli*. They were tested on a "humanized" transgenic mouse model, in which the mouse Syt II luminal domain has been replaced with the human version (Elliott et al. 2019). While natural BoNT/B showed a drastically reduced potency in this model, modified BoNT/B mutants showed the same level of potency in humanized mice and control mice. Therefore, these modified BoNT/B are expected to have better therapeutic efficacy in humans compared with natural BoNT/B.

While BoNT/B binds specifically to Syt I/II through side-chain-mediated interactions, BoNT/A recognition of SV2 involves not only the protein part of SV2 but also carbohydrate moieties at a glycosylation site of SV2 (Fig. 4b) (Yao et al. 2016). The three members of SV2 (SV2A, SV2B, and SV2C) are 12-transmembrane-domain proteins, with both the N- and C-termini located in the cytosol.

The fourth luminal domain (SV2-L4) is the longest among all luminal domains, and BoNT/A recognizes its middle portion. The crystal structure of H_C of BoNT/A1 (A1-H_C) in complex with human SV2C-L4 expressed and purified in *E. coli* has been solved, revealing that SV2C-L4 folds into a right-handed, quadrilateral β-helix pattern, similar to pentapeptide-repeat proteins (Benoit et al. 2014). The overall architecture of SV2C-L4 is similar to the structure of amyloid fibrils, forming a stack of β-strands. BoNT/A1 recognizes the top of this structure by stacking two of its own β-strands in the middle of A1-H_C (Fig. 4b). This is an unusual type of toxin-receptor recognition, as most interactions are through backbone-backbone hydrogen bonds. The crystal structure of BoNT/A2 in complex with human SV2C L4 has also been solved (Benoit et al. 2017; Gustafsson et al. 2018). The complexes are highly similar to BoNT/A1 – SV2C L4.

SV2 is heavily glycosylated, and there are three conserved N-linked glycosylation sites within the L4, with one located right in the middle of the BoNT/A1-SV2 interface. These three sites are not glycosylated in SV2C-L4 purified from *E. coli*. Mutagenesis studies showed that abolishing the glycosylation site at the BoNT/A1-SV2 interface reduced the potency of BoNT/A1 on neurons, suggesting that glycosylation at this site contributes to toxin-SV2 interaction (Yao et al. 2016; Dong et al. 2008).

In 2016, the crystal structure of A1-H_C in complex with a glycosylated SV2C-L4 (expressed in eukaryotic cells) was elucidated (Yao et al. 2016). The overall structure of glycosylated SV2C-L4 and the protein-protein interaction interface between SV2C-L4 and A1-H_C are the same as shown in the structure containing un-glycosylated SV2C-L4. The new discovery is that the base of the N-linked glycan, including two GlcNAc, a mannose, and a fucose, is directly recognized by A1-H_C. These interactions with glycans significantly reduce the dissociation constant and enhance the overall binding affinity between A1-H_C and SV2C-L4 (Mahrhold et al. 2016; Yao et al. 2016). The location of this N-linked glycosylation site is highly conserved across all three SV2s as well as in different vertebrate species.

BoNT/E also utilizes SV2A and SV2B as its receptors, and the recognition is even more dependent on the presence of glycosylation at the same site of L4 than it is with BoNT/A (Dong et al. 2008; Mahrhold et al. 2013). Though structural basis for BoNT/E-SV2 interactions remains to be established, it has been shown that BoNT/E cannot recognize SV2C in neurons, which is a major difference from BoNT/A. SV2B and SV2C are both widely expressed at motor nerve terminals, while SV2A is expressed only in a subset of motor neurons controlling slow muscle fibers (Chakkalakal et al. 2010). It has also been reported that trigeminal sensory neurons are insensitive to BoNT/E, as they express only SV2C. This insensitivity can be overcome by an engineered chimeric BoNT/E-A toxin in which the H_C of BoNT/E is replaced with A1-H_C (Meng et al. 2009).

6 BoNT Gene Cluster and Progenitor Toxin Complex

BoNTs exist naturally in protein complexes when produced from *Clostridium* bacteria. BoNT/A1 was first purified in 1946 in crystalline form (Lamanna et al. 1946). It was later discovered that this purified form can be further separated into a toxic and a nontoxic component, with the latter capable of inducing aggregation of erythrocytes (hemagglutinin activity). Further studies revealed that BoNT/A1 can exist in three different complexes, known as 12S, 16S, and 19S based on ultracentrifugation analysis of their molecular weight. The 12S form contains a BoNT/A and a nontoxic molecule of similar size known as NTNHA (nontoxic non-hemagglutinin protein, also known as NTNH). The 16S contains the 12S plus a nontoxic protein complex with hemagglutinin activity (known as HA complex). The 19S appears to be a dimer of 16S, and this is the crystalline form first put into clinical use. Current generation of therapeutic BoNT/A and BoNT/B are both in complex form, containing NTNHA and HA proteins. The exception is that the product from Merz is the BoNT/A molecule alone, isolated through additional purification from the original complex form.

The genes encoding BoNTs are located within two kinds of gene clusters, both containing a gene encoding NTNHA next to the toxin gene (Fig. 5a). One cluster, designated as the HA cluster, encodes three proteins (HA17, HA33, and HA70) that form a complex with hemagglutinin activity. BoNT/A1, B, C, D, and G all contain the HA cluster and produce both 12S and 16S complexes. The other cluster is known as the OrfX cluster, encoding four proteins (OrfX1, OrfX2, OrfX3, and P47) with unknown functions. BoNT/A2, E, and F contain the OrfX cluster, and only 12S complexes were purified for these toxins.

The reason why BoNTs are produced in complexes can be understood from their mode of action: they are oral toxins. They naturally enter human and animal bodies via ingestion of contaminated food, and the toxin must be able to pass through the harsh environment of the gastrointestinal tract and be absorbed into the circulatory system as an intact molecule. Because the isolated toxin can easily be degraded by proteases, its oral toxicity is low, whereas the BoNT-NTNHA complex is extremely resistant to proteases. Thus, BoNTs are always combined with their corresponding NTNHA into a minimally functional progenitor toxin complex (M-PTC, Fig. 5b), whose formation is pH dependent. For instance, BoNT/A1 and its NTNHA form a complex under acidic pH conditions, and the toxin dissociates from NTNHA at pH 7.3. This provides a mechanism to release the toxin once the complex enters the circulatory system. In clinical applications, BoNT/A1 is expected to dissociate from the complex once it is injected into the tissue; the toxin itself is responsible for the therapeutic effect.

The structural basis for BoNT-NTNHA complexes has been established by recent work revealing the crystal structures of BoNT/A1 with its NTNHA (NTNHA-A) and BoNT/E with its NTNHA (NTNHA-E) (Gu et al. 2012; Eswaramoorthy et al. 2015). The two complex structures share a high degree of similarity and confirm that the toxins and the NTNHAs form interlocked "handshake" complexes, protecting a large portion of the solvent-accessible areas (Fig. 5b). The NTNHAs also contain three domains, termed nLC, nH_N, and nH_C, which are homologous to the LC, H_N,

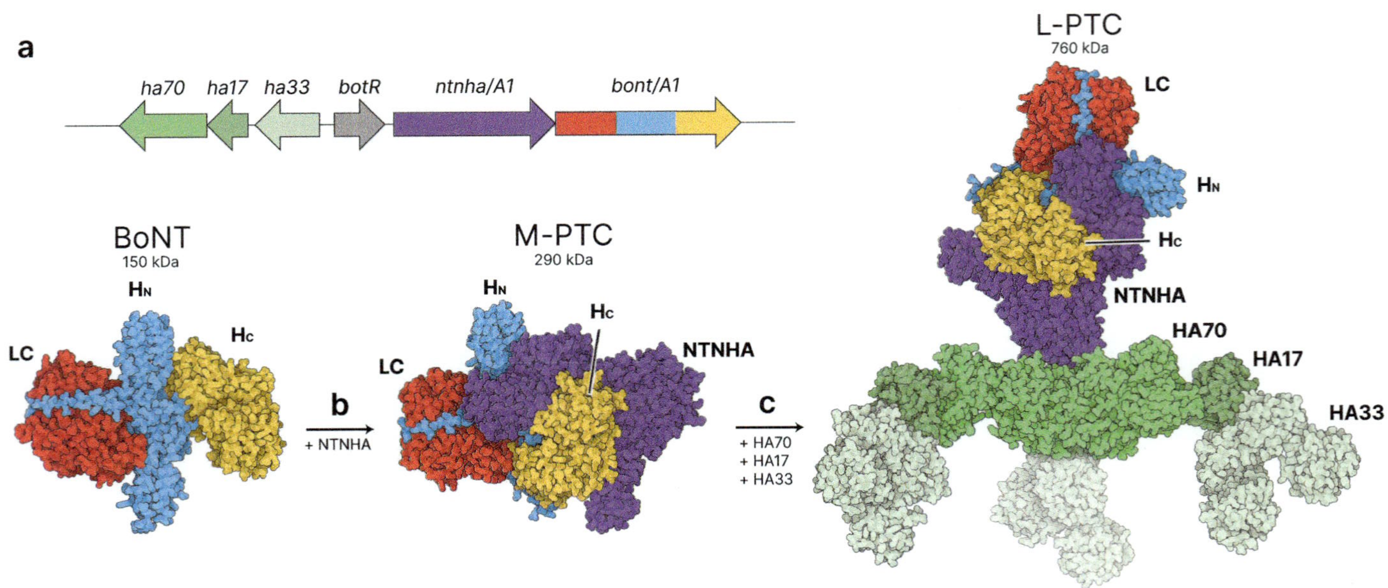

Fig. 5 The BoNT gene cluster and progenitor toxin complex. (**a**) Schematic drawing illustrating the HA gene cluster of BoNT/A1. (**b**) Structure of the M-PTC composed of a BoNT in complex with its NTNHA. The BoNT-LC is colored red, the H_N blue, H_C yellow, and NTNHA purple. PDB: 3BTA and 3V0A. (**c**) Structural model of the large progenitor toxin complex (L-PTC) composed of a BoNT, a NTNHA, three HA70, six HA33, and three HA17. HA proteins are colored in different shades of green. PDB: 4LO7 and 3V0A

and H_C in BoNTs, but NTNHAs do not contain the protease motif HEXXH in their nLCs, and there is no disulfide bond between nLC and nH_N. The H_C is at the center of the complex, forming interactions with all three domains of NTNHA, suggesting that H_C is the key region being protected. On the other hand, the LC and nLC are both pointing outward from the complex, and the LC of BoNT/A1 does not form any interactions with NTNHA-A. It has been proposed that key clusters of acidic residues at the interface of BoNT and NTNHA dictate the pH dependency of the complex formation. These acidic residues are not charged at low pH but become negatively charged (deprotonated) in neutral or basic conditions, creating a repulsive force that disassembles the complex.

The major difference between NTNHA-A and NTNHA-E is that NTNHA-A contains an extra loop within the nLC (termed nLoop), which is conserved in the NTNHAs from BoNT/A1, B, C, D, and G, but missing from all toxins associated with the OrfX cluster. This nLoop binds to the HA proteins and serves as a linker to dock the BoNT-NTNHA onto a complex of HA proteins, which comprises three HA70, three HA17, and six HA33 molecules (Fig. 5c). Together, the entire complex is defined as large progenitor toxin complex (L-PTC). The crystal structure of the HA complex of BoNT/B has been determined, showing a triskelion shape with three HA70 forming a central hub (Amatsu et al. 2013). HA17 serves as a linker, binding to one HA70 on one side and two HA33 on the other side. A similar structure has also been constructed for the HA complex of BoNT/A (Lee et al. 2013). By docking crystal structures into the shape of the entire toxin complex observed from negative-stain electron microscopy, a structural model of BoNT/A1 L-PTC has been constructed (Fig. 5c) (Lee et al. 2013).

The major function of the HA complex appears to be facilitating the absorption of the toxin complex in the intestine. First, the complex contains up to nine carbohydrate binding sites, one on each HA70 and HA33 (Lee et al. 2013). HA-carbohydrate interactions have been shown to contribute to the initial absorption of the toxin complex through microfold cells (M-cells) in the intestine (Matsumura et al. 2015). Second, the HA complexes of both BoNT/A and BoNT/B recognize cell surface adhesion molecule E-cadherin (Sugawara et al. 2010). Structural studies revealed that one HA complex binds to three E-cadherin molecules. The interaction blocks the trans-dimerization of two E-cadherin on neighboring cells, which is essential for forming cell-cell junctions, thus opening the tight junction in the intestinal epithelial barrier for the large toxin complex to pass through (Lee et al. 2014).

The function of the OrfX proteins and whether they form complexes with each other and/or with BoNT-NTNHA remain unknown. The crystal structures of OrfX2 and P47 show that both proteins contain a tubular lipid-binding (TULIP) fold, suggesting lipid-binding activity (Gustafsson et al. 2017; Lam et al. 2018b). All three BoNT-like toxins, BoNT/X, BoNT/En, and PMP1, are located in gene clusters homologous to the OrfX cluster. It has been shown that deletion of OrfX proteins reduces the toxicity of PMP1 to *anopheles* mosquito larvae, demonstrating that these OrfX proteins contribute to the oral toxicity of PMP1 (Contreras et al. 2019).

7 Concluding Remarks

Botulinum neurotoxins have captivated researchers for well over 100 years. The extreme potency of these toxins, and the impressive number of medical conditions they can be used to treat, continues to fascinate both scientists and clinicians. Our growing understanding of these toxins, with rapid advances in solving their structures and discovering new toxin variations and homologs, will lead to development of a new generation of therapeutic toxins with improved efficacy and expanded clinical applications.

Acknowledgments We thank Jonathan Davies for preparing all figures and for valuable discussions during the preparation of this chapter. This work was supported by the Swedish Research Council and the Swedish Cancer Society to P.S. and by grants from NIH (R01NS080833, R01AI132387, R01AI139087, and R21NS106159), Intelligence Advanced Research Projects Activity (IARPA, grant number W911NF-17-2-0089), and the Investigator in the Pathogenesis of Infectious Disease award from the Burroughs Wellcome Fund to M.D.

References

Agarwal R, Schmidt JJ, Stafford RG, Swaminathan S (2009) Mode of VAMP substrate recognition and inhibition of Clostridium botulinum neurotoxin F. Nat Struct Mol Biol 16(7):789–794

Amatsu S, Sugawara Y, Matsumura T, Kitadokoro K, Fujinaga Y (2013) Crystal structure of Clostridium botulinum whole hemagglutinin reveals a huge triskelion-shaped molecular complex. J Biol Chem 288(49):35617–35625

Antonucci F, Rossi C, Gianfranceschi L, Rossetto O, Caleo M (2008) Long-distance retrograde effects of botulinum neurotoxin A. J Neurosci 28(14):3689–3696

Arndt JW, Yu W, Bi F, Stevens RC (2005) Crystal structure of botulinum neurotoxin type G light chain: serotype divergence in substrate recognition. Biochemistry 44(28):9574–9580

Arndt JW, Chai Q, Christian T, Stevens RC (2006) Structure of botulinum neurotoxin type D light chain at 1.65 Å resolution: repercussions for VAMP-2 substrate specificity. Biochemistry 45 (10):3255–3262

Barash JR, Arnon SS (2014) A novel strain of Clostridium botulinum that produces type B and type H botulinum toxins. J Infect Dis 209(2):183–191

Benoit RM, Frey D, Hilbert M, Kevenaar JT, Wieser MM, Stirnimann CU, McMillan D, Ceska T, Lebon F, Jaussi R, Steinmetz MO, Schertler GF, Hoogenraad CC, Capitani G, Kammerer RA (2014) Structural basis for recognition of synaptic vesicle protein 2C by botulinum neurotoxin A. Nature 505(7481):108–111

Benoit RM, Scharer MA, Wieser MM, Li X, Frey D, Kammerer RA (2017) Crystal structure of the BoNT/A2 receptor-binding domain in complex with the luminal domain of its neuronal receptor SV2C. Sci Rep 7:43588

Benson MA, Fu Z, Kim JJ, Baldwin MR (2011) Unique ganglioside recognition strategies for clostridial neurotoxins. J Biol Chem 286(39):34015–34022

Berntsson RP, Peng L, Dong M, Stenmark P (2013) Structure of dual receptor binding to botulinum neurotoxin B. Nat Commun 4:2058

Bomba-Warczak E, Vevea JD, Brittain JM, Figueroa-Bernier A, Tepp WH, Johnson EA, Yeh FL, Chapman ER (2016) Interneuronal transfer and distal action of tetanus toxin and botulinum neurotoxins A and D in central neurons. Cell Rep 16(7):1974–1987

Breidenbach MA, Brunger AT (2004) Substrate recognition strategy for botulinum neurotoxin serotype A. Nature 432(7019):925–929

Brunt J, Carter AT, Stringer SC, Peck MW (2018) Identification of a novel botulinum neurotoxin gene cluster in Enterococcus. FEBS Lett 592(3):310–317
Burke GS (1919a) Notes on Bacillus botulinus. J Bacteriol 4(5):555–570
Burke GS (1919b) The occurrence of Bacillus botulinus in nature. J Bacteriol 4(5):541–553
Chai Q, Arndt JW, Dong M, Tepp WH, Johnson EA, Chapman ER, Stevens RC (2006) Structural basis of cell surface receptor recognition by botulinum neurotoxin B. Nature 444 (7122):1096–1100
Chakkalakal JV, Nishimune H, Ruas JL, Spiegelman BM, Sanes JR (2010) Retrograde influence of muscle fibers on their innervation revealed by a novel marker for slow motoneurons. Development 137(20):3489–3499
Contreras E, Masuyer G, Qureshi N, Chawla S, Dhillon HS, Lee HL, Chen J, Stenmark P, Gill SS (2019) A neurotoxin that specifically targets Anopheles mosquitoes. Nat Commun 10(1):2869
Dong M, Richards DA, Goodnough MC, Tepp WH, Johnson EA, Chapman ER (2003) Synaptotagmins I and II mediate entry of botulinum neurotoxin B into cells. J Cell Biol 162 (7):1293–1303
Dong M, Yeh F, Tepp WH, Dean C, Johnson EA, Janz R, Chapman ER (2006) SV2 is the protein receptor for botulinum neurotoxin A. Science 312(5773):592–596
Dong M, Liu H, Tepp WH, Johnson EA, Janz R, Chapman ER (2008) Glycosylated SV2A and SV2B mediate the entry of botulinum neurotoxin E into neurons. Mol Biol Cell 19 (12):5226–5237
Dong M, Masuyer G, Stenmark P (2019) Botulinum and tetanus neurotoxins. Annu Rev Biochem 88:811–837
Dover N, Barash JR, Hill KK, Xie G, Arnon SS (2014) Molecular characterization of a novel botulinum neurotoxin type H gene. J Infect Dis 209(2):192–202
Eleopra R, Tugnoli V, Rossetto O, Montecucco C, De Grandis D (1997) Botulinum neurotoxin serotype C: a novel effective botulinum toxin therapy in human. Neurosci Lett 224(2):91–94
Eleopra R, Tugnoli V, Rossetto O, De Grandis D, Montecucco C (1998) Different time courses of recovery after poisoning with botulinum neurotoxin serotypes A and E in humans. Neurosci Lett 256(3):135–138
Eleopra R, Tugnoli V, Quatrale R, Rossetto O, Montecucco C, Dressler D (2006) Clinical use of non-A botulinum toxins: botulinum toxin type C and botulinum toxin type F. Neurotox Res 9 (2–3):127–131
Elliott M, Favre-Guilmard C, Liu SM, Maignel J, Masuyer G, Beard M, Boone C, Carre D, Kalinichev M, Lezmi S, Mir I, Nicoleau C, Palan S, Perier C, Raban E, Zhang S, Dong M, Stenmark P, Krupp J (2019) Engineered botulinum neurotoxin B with improved binding to human receptors has enhanced efficacy in preclinical models. Sci Adv 5(1):eaau7196
Eswaramoorthy S, Sun J, Li H, Singh BR, Swaminathan S (2015) Molecular Assembly of Clostridium botulinum progenitor M complex of type E. Sci Rep 5:17795
Fischer A, Sambashivan S, Brunger AT, Montal M (2012) Beltless translocation domain of botulinum neurotoxin A embodies a minimum ion-conductive channel. J Biol Chem 287 (3):1657–1661
Foran PG, Mohammed N, Lisk GO, Nagwaney S, Lawrence GW, Johnson E, Smith L, Aoki KR, Dolly JO (2003) Evaluation of the therapeutic usefulness of botulinum neurotoxin B, C1, E, and F compared with the long lasting type A. Basis for distinct durations of inhibition of exocytosis in central neurons. J Biol Chem 278(2):1363–1371
Fotinou C, Emsley P, Black I, Ando H, Ishida H, Kiso M, Sinha KA, Fairweather NF, Isaacs NW (2001) The crystal structure of tetanus toxin Hc fragment complexed with a synthetic GT1b analogue suggests cross-linking between ganglioside receptors and the toxin. J Biol Chem 276 (34):32274–32281
Gimenez DF, Ciccarelli AS (1970) Another type of Clostridium botulinum. Zentralbl Bakteriol Orig 215(2):221–224

Gu S, Rumpel S, Zhou J, Strotmeier J, Bigalke H, Perry K, Shoemaker CB, Rummel A, Jin R (2012) Botulinum neurotoxin is shielded by NTNHA in an interlocked complex. Science 335 (6071):977–981

Gustafsson R, Berntsson RP, Martinez-Carranza M, El Tekle G, Odegrip R, Johnson EA, Stenmark P (2017) Crystal structures of OrfX2 and P47 from a Botulinum neurotoxin OrfX-type gene cluster. FEBS Lett 591(22):3781–3792

Gustafsson R, Zhang S, Masuyer G, Dong M, Stenmark P (2018) Crystal structure of botulinum neurotoxin A2 in complex with the human protein receptor SV2C reveals plasticity in receptor binding. Toxins 10(4):E153

Hamark C, Berntsson RP, Masuyer G, Henriksson LM, Gustafsson R, Stenmark P, Widmalm G (2017) Glycans confer specificity to the recognition of ganglioside receptors by botulinum neurotoxin A. J Am Chem Soc 139(1):218–230

Hill KK, Smith TJ, Helma CH, Ticknor LO, Foley BT, Svensson RT, Brown JL, Johnson EA, Smith LA, Okinaka RT, Jackson PJ, Marks JD (2007) Genetic diversity among botulinum neurotoxin-producing clostridial strains. J Bacteriol 189(3):818–832

Jahn R, Scheller RH (2006) SNAREs--engines for membrane fusion. Nat Rev Mol Cell Biol 7 (9):631–643

Jin R, Rummel A, Binz T, Brunger AT (2006) Botulinum neurotoxin B recognizes its protein receptor with high affinity and specificity. Nature 444(7122):1092–1095

Jin R, Sikorra S, Stegmann CM, Pich A, Binz T, Brunger AT (2007) Structural and biochemical studies of botulinum neurotoxin serotype C1 light chain protease: implications for dual substrate specificity. Biochemistry 46(37):10685–10693

Johnson EA (1999) Clostridial toxins as therapeutic agents: benefits of nature's most toxic proteins. Annu Rev Microbiol 53:551–575

Kalb SR, Baudys J, Webb RP, Wright P, Smith TJ, Smith LA, Fernandez R, Raphael BH, Maslanka SE, Pirkle JL, Barr JR (2012) Discovery of a novel enzymatic cleavage site for botulinum neurotoxin F5. FEBS Lett 586(2):109–115

Keller JE, Neale EA, Oyler G, Adler M (1999) Persistence of botulinum neurotoxin action in cultured spinal cord cells. FEBS Lett 456(1):137–142

Kosenina S, Masuyer G, Zhang S, Dong M, Stenmark P (2019) Crystal structure of the catalytic domain of the Weissella oryzae botulinum-like toxin. FEBS Lett 593(12):1403–1410

Kumaran D, Eswaramoorthy S, Furey W, Navaza J, Sax M, Swaminathan S (2009) Domain organization in Clostridium botulinum neurotoxin type E is unique: its implication in faster translocation. J Mol Biol 386(1):233–245

Lacy DB, Tepp W, Cohen AC, DasGupta BR, Stevens RC (1998) Crystal structure of botulinum neurotoxin type A and implications for toxicity. Nat Struct Biol 5(10):898–902

Lalli G, Bohnert S, Deinhardt K, Verastegui C, Schiavo G (2003) The journey of tetanus and botulinum neurotoxins in neurons. Trends Microbiol 11(9):431–437

Lam KH, Guo Z, Krez N, Matsui T, Perry K, Weisemann J, Rummel A, Bowen ME, Jin R (2018a) A viral-fusion-peptide-like molecular switch drives membrane insertion of botulinum neurotoxin A1. Nat Commun 9(1):5367

Lam KH, Qi R, Liu S, Kroh A, Yao G, Perry K, Rummel A, Jin R (2018b) The hypothetical protein P47 of Clostridium botulinum E1 strain Beluga has a structural topology similar to bactericidal/permeability-increasing protein. Toxicon 147:19–26

Lamanna C, Mc EO, Eklund HW (1946) The purification and crystallization of Clostridium botulinum type A toxin. Science 103(2681):613

Lebeda FJ, Olson MA (1995) Structural predictions of the channel-forming region of botulinum neurotoxin heavy chain. Toxicon 33(4):559–567

Lee K, Gu S, Jin L, Le TT, Cheng LW, Strotmeier J, Kruel AM, Yao G, Perry K, Rummel A, Jin R (2013) Structure of a bimodular botulinum neurotoxin complex provides insights into its oral toxicity. PLoS Pathog 9(10):e1003690

Lee K, Zhong X, Gu S, Kruel AM, Dorner MB, Perry K, Rummel A, Dong M, Jin R (2014) Molecular basis for disruption of E-cadherin adhesion by botulinum neurotoxin A complex. Science 344(6190):1405–1410
Leuchs J (1910) Beitraege zur kenntnis des toxins und antitoxins des *Bacillus botulinus*. Z Hyg Infekt 76:55–84
Mahrhold S, Rummel A, Bigalke H, Davletov B, Binz T (2006) The synaptic vesicle protein 2C mediates the uptake of botulinum neurotoxin A into phrenic nerves. FEBS Lett 580 (8):2011–2014
Mahrhold S, Strotmeier J, Garcia-Rodriguez C, Lou J, Marks JD, Rummel A, Binz T (2013) Identification of the SV2 protein receptor-binding site of botulinum neurotoxin type E. Biochem J 453(1):37–47
Mahrhold S, Bergstrom T, Stern D, Dorner BG, Astot C, Rummel A (2016) Only the complex N559-glycan in the synaptic vesicle glycoprotein 2C mediates high affinity binding to botulinum neurotoxin serotype A1. Biochem J 473(17):2645–2654
Mansfield MJ, Adams JB, Doxey AC (2015) Botulinum neurotoxin homologs in non-Clostridium species. FEBS Lett 589(3):342–348
Mansfield MJ, Wentz TG, Zhang S, Lee EJ, Dong M, Sharma SK, Doxey AC (2019) Bioinformatic discovery of a toxin family in Chryseobacterium piperi with sequence similarity to botulinum neurotoxins. Sci Rep 9(1):1634
Maslanka SE, Luquez C, Dykes JK, Tepp WH, Pier CL, Pellett S, Raphael BH, Kalb SR, Barr JR, Rao A, Johnson EA (2016) A novel botulinum neurotoxin, previously reported as serotype H, has a hybrid-like structure with regions of similarity to the structures of serotypes A and F and is neutralized with serotype A antitoxin. J Infect Dis 213(3):379–385
Masuyer G, Conrad J, Stenmark P (2017) The structure of the tetanus toxin reveals pH-mediated domain dynamics. EMBO Rep 18(8):1306–1317
Masuyer G, Zhang S, Barkho S, Shen Y, Henriksson L, Kosenina S, Dong M, Stenmark P (2018) Structural characterisation of the catalytic domain of botulinum neurotoxin X – high activity and unique substrate specificity. Sci Rep 8(1):4518
Matsumura T, Sugawara Y, Yutani M, Amatsu S, Yagita H, Kohda T, Fukuoka S, Nakamura Y, Fukuda S, Hase K, Ohno H, Fujinaga Y (2015) Botulinum toxin A complex exploits intestinal M cells to enter the host and exert neurotoxicity. Nat Commun 6:6255
Meng J, Ovsepian SV, Wang J, Pickering M, Sasse A, Aoki KR, Lawrence GW, Dolly JO (2009) Activation of TRPV1 mediates calcitonin gene-related peptide release, which excites trigeminal sensory neurons and is attenuated by a retargeted botulinum toxin with anti-nociceptive potential. J Neurosci 29(15):4981–4992
Montal M (2010) Botulinum neurotoxin: a marvel of protein design. Annu Rev Biochem 79:591–617
Montal MS, Blewitt R, Tomich JM, Montal M (1992) Identification of an ion channel-forming motif in the primary structure of tetanus and botulinum neurotoxins. FEBS Lett 313(1):12–18
Montecucco C (1986) How do tetanus and botulinum toxins bind to neuronal membranes? Trens Biochem Sci 11(8):314–317
Moriishi K, Koura M, Abe N, Fujii N, Fujinaga Y, Inoue K, Ogumad K (1996a) Mosaic structures of neurotoxins produced from Clostridium botulinum types C and D organisms. Biochim Biophys Acta 1307(2):123–126
Moriishi K, Koura M, Fujii N, Fujinaga Y, Inoue K, Syuto B, Oguma K (1996b) Molecular cloning of the gene encoding the mosaic neurotoxin, composed of parts of botulinum neurotoxin types C1 and D, and PCR detection of this gene from Clostridium botulinum type C organisms. Appl Environ Microbiol 62(2):662–667
Moritz MS, Tepp WH, Bradshaw M, Johnson EA, Pellett S (2018) Isolation and characterization of the novel Botulinum neurotoxin A subtype 6. mSphere 3(5):e00466
Nishiki T, Kamata Y, Nemoto Y, Omori A, Ito T, Takahashi M, Kozaki S (1994) Identification of protein receptor for Clostridium botulinum type B neurotoxin in rat brain synaptosomes. J Biol Chem 269(14):10498–10503

Pang ZP, Melicoff E, Padgett D, Liu Y, Teich AF, Dickey BF, Lin W, Adachi R, Sudhof TC (2006) Synaptotagmin-2 is essential for survival and contributes to Ca2+ triggering of neurotransmitter release in central and neuromuscular synapses. J Neurosci 26(52):13493–13504

Peck MW, Smith TJ, Anniballi F, Austin JW, Bano L, Bradshaw M, Cuervo P, Cheng LW, Derman Y, Dorner BG, Fisher A, Hill KK, Kalb SR, Korkeala H, Lindstrom M, Lista F, Luquez C, Mazuet C, Pirazzini M, Popoff MR, Rossetto O, Rummel A, Sesardic D, Singh BR, Stringer SC (2017) Historical perspectives and guidelines for botulinum neurotoxin subtype nomenclature. Toxins 9(1):38

Pellett S, Tepp WH, Whitemarsh RC, Bradshaw M, Johnson EA (2015) In vivo onset and duration of action varies for botulinum neurotoxin A subtypes 1-5. Toxicon 107(Pt A):37–42

Pellett S, Bradshaw M, Tepp WH, Pier CL, Whitemarsh RCM, Chen C, Barbieri JT, Johnson EA (2018) The light chain defines the duration of action of botulinum toxin serotype A subtypes. MBio 9(2). pii: e00089-18

Peng L, Berntsson RP, Tepp WH, Pitkin RM, Johnson EA, Stenmark P, Dong M (2012) Botulinum neurotoxin D-C uses synaptotagmin I and II as receptors, and human synaptotagmin II is not an effective receptor for type B, D-C and G toxins. J Cell Sci 125(Pt 13):3233–3242

Pfenninger W (1924) Toxico, immunologic and serologic relationship of *B. botulinus*, type C and *B. parabotulinus*. J Infect Dis 35:347–352

Pier CL, Chen C, Tepp WH, Lin G, Janda KD, Barbieri JT, Pellett S, Johnson EA (2011) Botulinum neurotoxin subtype A2 enters neuronal cells faster than subtype A1. FEBS Lett 585(1):199–206

Pirazzini M, Rossetto O, Eleopra R, Montecucco C (2017) Botulinum neurotoxins: biology, pharmacology, and toxicology. Pharmacol Rev 69(2):200–235

Restani L, Antonucci F, Gianfranceschi L, Rossi C, Rossetto O, Caleo M (2011) Evidence for anterograde transport and transcytosis of botulinum neurotoxin A (BoNT/A). J Neurosci 31 (44):15650–15659

Restani L, Giribaldi F, Manich M, Bercsenyi K, Menendez G, Rossetto O, Caleo M, Schiavo G (2012) Botulinum neurotoxins A and E undergo retrograde axonal transport in primary motor neurons. PLoS Pathog 8(12):e1003087

Rossetto O, Pirazzini M, Montecucco C (2014) Botulinum neurotoxins: genetic, structural and mechanistic insights. Nat Rev Microbiol 12(8):535–549

Rummel A, Bade S, Alves J, Bigalke H, Binz T (2003) Two carbohydrate binding sites in the H (CC)-domain of tetanus neurotoxin are required for toxicity. J Mol Biol 326(3):835–847

Rummel A, Mahrhold S, Bigalke H, Binz T (2004) The HCC-domain of botulinum neurotoxins A and B exhibits a singular ganglioside binding site displaying serotype specific carbohydrate interaction. Mol Microbiol 51(3):631–643

Schantz EJ, Johnson EA (1992) Properties and use of botulinum toxin and other microbial neurotoxins in medicine. Microbiol Rev 56(1):80–99

Schiavo G, Benfenati F, Poulain B, Rossetto O, Polverino de Laureto P, DasGupta BR, Montecucco C (1992a) Tetanus and botulinum-B neurotoxins block neurotransmitter release by proteolytic cleavage of synaptobrevin. Nature 359(6398):832–835

Schiavo G, Poulain B, Rossetto O, Benfenati F, Tauc L, Montecucco C (1992b) Tetanus toxin is a zinc protein and its inhibition of neurotransmitter release and protease activity depend on zinc. EMBO J 11(10):3577–3583

Simpson LL, Rapport MM (1971) Ganglioside inactivation of botulinum toxin. J Neurochem 18 (7):1341–1343

Stenmark P, Dupuy J, Imamura A, Kiso M, Stevens RC (2008) Crystal structure of botulinum neurotoxin type A in complex with the cell surface co-receptor GT1b-insight into the toxin-neuron interaction. PLoS Pathog 4(8):e1000129

Strotmeier J, Willjes G, Binz T, Rummel A (2012) Human synaptotagmin-II is not a high affinity receptor for botulinum neurotoxin B and G: increased therapeutic dosage and immunogenicity. FEBS Lett 586(4):310–313

Sudhof TC, Rothman JE (2009) Membrane fusion: grappling with SNARE and SM proteins. Science 323(5913):474–477

Sugawara Y, Matsumura T, Takegahara Y, Jin Y, Tsukasaki Y, Takeichi M, Fujinaga Y (2010) Botulinum hemagglutinin disrupts the intercellular epithelial barrier by directly binding E-cadherin. J Cell Biol 189(4):691–700

Surana S, Tosolini AP, Meyer IFG, Fellows AD, Novoselov SS, Schiavo G (2018) The travel diaries of tetanus and botulinum neurotoxins. Toxicon 147:58–67

Swaminathan S, Eswaramoorthy S (2000) Structural analysis of the catalytic and binding sites of Clostridium botulinum neurotoxin B. Nat Struct Biol 7(8):693–699

Tao L, Peng L, Berntsson RP, Liu SM, Park S, Yu F, Boone C, Palan S, Beard M, Chabrier PE, Stenmark P, Krupp J, Dong M (2017) Engineered botulinum neurotoxin B with improved efficacy for targeting human receptors. Nat Commun 8:53

Torii Y, Kiyota N, Sugimoto N, Mori Y, Goto Y, Harakawa T, Nakahira S, Kaji R, Kozaki S, Ginnaga A (2011) Comparison of effects of botulinum toxin subtype A1 and A2 using twitch tension assay and rat grip strength test. Toxicon 57(1):93–99

Tsai YC, Kotiya A, Kiris E, Yang M, Bavari S, Tessarollo L, Oyler GA, Weissman AM (2017) Deubiquitinating enzyme VCIP135 dictates the duration of botulinum neurotoxin type A intoxication. Proc Natl Acad Sci U S A 114(26):E5158–E5166

Vagin O, Tokhtaeva E, Garay PE, Souda P, Bassilian S, Whitelegge JP, Lewis R, Sachs G, Wheeler L, Aoki R, Fernandez-Salas E (2014) Recruitment of septin cytoskeletal proteins by botulinum toxin A protease determines its remarkable stability. J Cell Sci 127(Pt 15):3294–3308

van Ermengem E (1897) Ueber Einem Neuen anaeroben Bacillus und seine Beziehungen zum Botulismus. Z Hyg Infekt 26:1–56

Whitemarsh RC, Tepp WH, Bradshaw M, Lin G, Pier CL, Scherf JM, Johnson EA, Pellett S (2013) Characterization of botulinum neurotoxin A subtypes 1 through 5 by investigation of activities in mice, in neuronal cell cultures, and in vitro. Infect Immun 81(10):3894–3902

Whitemarsh RC, Tepp WH, Johnson EA, Pellett S (2014) Persistence of botulinum neurotoxin a subtypes 1-5 in primary rat spinal cord cells. PLoS One 9(2):e90252

Yao G, Zhang S, Mahrhold S, Lam KH, Stern D, Bagramyan K, Perry K, Kalkum M, Rummel A, Dong M, Jin R (2016) N-linked glycosylation of SV2 is required for binding and uptake of botulinum neurotoxin A. Nat Struct Mol Biol 23(7):656–662

Zhang S, Masuyer G, Zhang J, Shen Y, Lundin D, Henriksson L, Miyashita SI, Martinez-Carranza M, Dong M, Stenmark P (2017) Identification and characterization of a novel botulinum neurotoxin. Nat Commun 8:14130

Zhang S, Lebreton F, Mansfield MJ, Miyashita SI, Zhang J, Schwartzman JA, Tao L, Masuyer G, Martinez-Carranza M, Stenmark P, Gilmore MS, Doxey AC, Dong M (2018) Identification of a botulinum neurotoxin-like toxin in a commensal strain of *Enterococcus faecium*. Cell Host Microbe 23(2):169–176 e166

Zornetta I, Azarnia Tehran D, Arrigoni G, Anniballi F, Bano L, Leka O, Zanotti G, Binz T, Montecucco C (2016) The first non Clostridial botulinum-like toxin cleaves VAMP within the juxtamembrane domain. Sci Rep 6:30257

Botulinum Neurotoxins: Mechanism of Action

O. Rossetto, M. Pirazzini, F. Fabris, and C. Montecucco

Contents

Abstract

Botulinum neurotoxins (BoNTs) are a growing family of bacterial protein toxins that cause botulism, a rare but often fatal animal and human disease. They are the most potent toxins known owing to their molecular architecture, which underlies their mechanism of action. BoNTs target peripheral nerve terminals by a unique mode of binding and enter into their cytosol where they cleave SNARE proteins, thus inhibiting the neurotransmitter release. The specificity and rapidity of binding, which limits the anatomical area of its neuroparalytic action, and its reversible action make BoNT a valuable pharmaceutical to treat neurological and

O. Rossetto · M. Pirazzini · F. Fabris
Department of Biomedical Sciences, University of Padova, Padova, Italy

C. Montecucco (✉)
Department of Biomedical Sciences, University of Padova, Padova, Italy

Institute of Neuroscience, National Research Council, Padova, Italy

S. M. Whitcup, M. Hallett (eds.), *Botulinum Toxin Therapy*,
Handbook of Experimental Pharmacology 263, https://doi.org/10.1007/164_2020_355

non-neurological diseases determined by hyperactivity of cholinergic nerve terminals. This review reports the progress on our understanding of how BoNTs cause nerve paralysis highlighting the different steps of their molecular mechanism of action as key aspects to explain their extreme toxicity but also their unique pharmacological properties.

Keywords

Botulinum neurotoxins · Botulism · Neuromuscular junction · Neuroparalysis

1 Introduction

The two proteins most largely used in human therapy are the immunoglobulins (Ig) and the botulinum neurotoxins (BoNTs). Apart from sharing the molecular weight of 150 kDa, they have different structures and mechanisms of action. The Ig act by binding specifically their target molecules present on the cell surface or dispersed in body fluids, whilst the BoNTs bind to cholinergic nerve terminals and then enter the cytosol where they catalytically inactivate by proteolysis three protein, dubbed SNAREs, involved in the release of neurotransmitter, thus causing neuroparalysis. Accordingly, minute doses of BoNTs are capable of counteracting diseases caused by hyperfunctioning nerve terminals, and their action is based on a unique set of molecular properties that will be described in the present chapter.

BoNTs are produced by bacteria of the genus *Clostridium* though other bacteria of different classes and even phyla may harbour the gene encoding for BoNT and BoNT-like proteins. They consist of two chains (L, 50 kDa and H, 100 kDa) linked by a single SS bridge. They are produced in eight different serotypes (indicated by letters: BoNT/A, BoNT/B, BoNT/C, BoNT/D, BoNT/E, BoNT/F, BoNT/G and BoNT/X) (Rossetto et al. 2014; Dong et al. 2019). Many subtypes of serotypes are known (dubbed with a suffix number: BoNT/A1, BoNT/A2, etc.) plus chimeric neurotoxins (BoNT/CD, BoNT/DC, BoNT/FA) for a total of many dozens of different toxins (Peck et al. 2017). This figure is bound to increase with expanding DNA sequencing of known bacteria and of novel isolates (Doxey et al. 2018, 2019).

The BoNT molecule shown in Fig. 1b is complexed with a homologous protein dubbed NTNHA (in orange in Fig. 1a), devoid of protease activity, forming a heterodimer that is much more stable than BoNT alone to the acidic and proteolytic conditions found in the gastrointestinal tract (Gu et al. 2012). This heterodimer assembles with accessory non-toxic proteins (Fig. 1a) to form progenitor toxin complexes (PTCs-BoNT/A1 $\approx$ 500–900 kDa), more stable at acidic pH. PTCs rapidly dissociate under slightly alkaline physiologic solutions. The accessory non-toxic proteins are believed to mediate the binding of the complex to the intestinal mucus and then to assist the translocation of the BoNT molecule from the intestinal lumen, across the mucus layer and the polarized epithelial monolayer, into the mucosa in the food-borne and infant forms of botulism (Lam and Jin 2015; Sugawara et al. 2010; Rossetto et al. 2014). The free BoNT then diffuses via lymphatic and blood circulations to the entire body and binds preferentially to

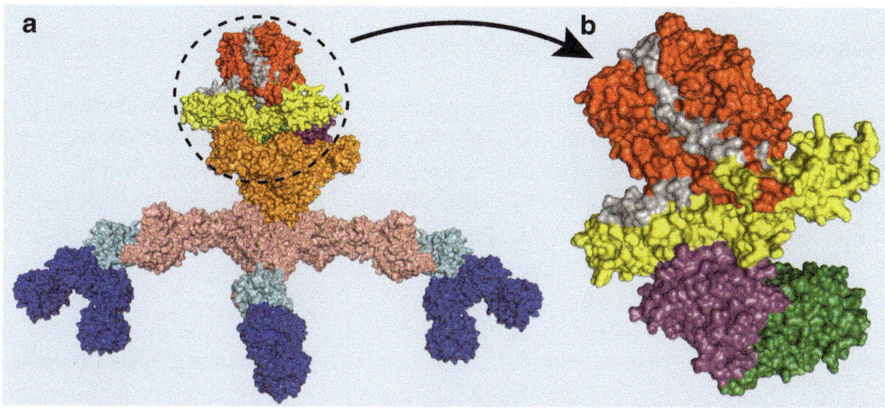

Fig. 1 Molecular structure of BoNT/A1 and its progenitor toxin complex. (**a**) 3D structure of the large progenitor toxin complex (L-PTC) composed of BoNT/A1 on the top, the NTNHA (orange) and the HA complex composed of HA70 (pink), HA17 (cyan) and HA33 (blue). (**b**) Zooming of BoNT/A1 150 kDa molecule showing the organization of the three toxin domains: the neurospecific binding Hc-C sub-domain (green), the lectin-like Hc-N sub-domain (purple), the translocation HN domain (yellow) and the metalloprotease L domain (red). A peptide belt (shown in white) surrounding the L domain and the interchain disulphide bond (orange) linking the L and HN domain are also shown

cholinergic neurons, but do not cross the blood-brain barrier. The intoxication of neurons of the myenteric plexus causes a block of their release of neurotransmitters, thus halting intestinal peristalsis, which is a major symptom in food-borne and in infant botulism (Chatham-Stephens et al. 2018). However, the inhibition of the respiratory function is the major single failure causing death by botulism in humans.

2 Botulism and Toxicity of Botulinum Neurotoxins

Botulism was first described by Kerner in Southern Germany about 200 years ago following episodes of flaccid paralysis and death that afflicted people that had shared contaminated ham and sausages. In 1897 Emile van Ermengem in Belgium demonstrated that the disease was due to a bacterium that produced a powerful poison of peripheral nerves causing the flaccid paralysis characteristic of botulism (Erbguth 2004). Botulism is rare but potentially fatal, and the death rate depends on the capability of identifying the symptoms at hospital admission (Sobel 2005; Fleck-Derderian et al. 2017). The first symptom is the paralysis of cranial nerves with ocular and facial palsy, diplopia and ptosis, dysphagia and dysarthria, followed by a descending flaccid paralysis that includes the neck muscles and the respiratory muscles, which lead the most frequent cause of death: deficient respiration. Additional symptoms are due to paralysis of the autonomic cholinergic nerves with abdominal pain, vomiting, nausea, dry mouth and dizziness. The display of symptoms and the time period intervening between intoxication and development of overt symptoms vary depending on the amount and type of BoNT and on the toxin route of entry (Sobel 2005; Rossetto et al. 2014; Fleck-Derderian et al. 2017; Chatham-Stephens et al. 2018).

The botulism patient is conscious but cannot operate any muscle. If the disease is rapidly diagnosed and respiration is mechanically assisted, the patient survives and recovers almost completely after a variable period of time, depending on dose and type of BoNT: from the 1–2 weeks of BoNT/E to the several months of BoNT/A1 which is the longest acting BoNT so far known. However, damages can result from prolonged external ventilation and variable levels of permanent fatigue may follow the long periods of paralysis.

Botulism is caused solely by the BoNT activity. BoNT is the most poisonous substance known (Rossetto and Montecucco 2019), and this toxicity is due to its neurospecificity and to the neuroparalysis that results from the catalytic action of the metalloprotease L domain in the nerve cytosol. For these reasons, BoNTs are included in the list A of substances with a possible bioterrorist use (Arnon et al. 2001; Bhattacharjee 2011). At the same time, their neurospecific high affinity binding, the reduced spreading of paralysis after injection and the reversibility of the induced neuroparalysis are the basis of the ever-growing therapeutic and aesthetic use of BoNT/A1. In addition, BoNT/B1 has been used in human therapy, and several other BoNTs may follow owing to specific properties useful to treat particular pathologies. Therefore, BoNTs completely fulfil the definition elaborated by Claude Bernard (1866) "Poisons are chemical scalpels to dissect physiological processes" and his prophetic prediction "Powerful poisons will surely become therapeutics, but only after their chemical composition is determined". Indeed BoNT/A1 local injection is the therapy of choice for the treatment of a variety of human pathologies and conditions characterized by hyperfunction of peripheral cholinergic nerve terminals and in plastic surgery (Dressler 2012; Pirazzini et al. 2017; Gart and Gutowski 2016).

Toxicity of botulinum toxins is generally measured as the mouse lethal dose 50% (MLD50), defined as the dose that kills 50% of mice within 4 days, after a single intraperitoneal injection. The MLD50 values vary in the range 0.01–5 ng/kg depending on the BoNT type and in minor proportion on the mice strain. The human lethal dose can be extrapolated from data obtained with primates. For a 70 kg man, the lethal doses are 90–150 ng when injected intravenously, 800–900 ng when inhaled and about 70 μg when introduced orally (Arnon et al. 2001; Rossetto and Montecucco 2019). Recently a large number of in vitro *assays* have been developed to avoid the use of animals in testing BoNT potency and toxicity, and they have been critically discussed by Pellet et al. (2019).

3 The Structural Architecture of Botulinum Neurotoxins

Notwithstanding the large number of serotypes, chimeras and subtypes, BoNTs have a very similar 3D structure, which is strictly linked to their common mechanism of intoxication of nerve terminals. Figure 1b shows the structure of BoNT/A1 (Lacy et al. 1998; Dong et al. 2019), the toxin predominantly used in human therapy. It is organized in four distinct domains endowed with different functions in nerve terminal intoxication and paralysis (Lacy et al. 1998). The crystal structure of other BoNTs has been determined, and it is very similar apart from a displacement of the third and fourth domains in BoNT/E with respect to A1 (Montal 2010; Swaminathan 2011).

BoNTs display a unique mode of binding to the presynaptic membrane to ensure specificity, high affinity and rapidity of binding (Montecucco 1986; Fogolari et al. 2009). This is achieved by the carboxyl terminal domain (Hc-C, 25 kDa, green in Fig. 1b) which contains one conserved binding site for a polysialoganglioside receptor, which is highly enriched in the nerve presynaptic membrane, and to a second receptors that is present on the luminal side of the membrane of synaptic vesicles (Rummel 2013). Hc-C is linked to a lectin like domain (Hc-N, 25 kDa, purple in Fig. 1b) whose role in binding has not been yet clarified though there is evidence that it binds to microdomains of the membrane (Muraro et al. 2009; Zhang and Varnum 2012).

The Hc-N domain is linked, with little protein-protein interaction, to the HN domain (50 kDa, yellow in Fig. 1b) which in turn is linked to the metalloprotease domain termed light chain (L, 50 kDa, red in Fig. 1b) via a long belt that encircles the L domain and by a unique disulphide bond (grey in Fig. 1b). As Hc-C plus Hc-N plus HN make a single polypeptide chain termed heavy chain (H, 100 kDa), this disulphide is termed interchain, and it is important because it is involved in membrane translocation and it prevents the metalloprotolytic activity of L until it is released in the cytosol upon its reduction (see below in Sect. 4.4).

4 The Conserved Mechanism of Nerve Terminal Paralysis by the Botulinum Neurotoxins

4.1 Neurospecific Binding (Step 1)

Biological evolution of these neurotoxins has led to a structural organization designed to deliver the metalloprotease domain into the cytosol of nerve terminals. This remarkable achievement has been attained by exploiting several physiological functions of nerve terminals. On the basis of the presently available experimental notions, the BoNT mechanism of nerve terminal paralysis consists of the five major steps, depicted in Fig. 2: (1) binding to cholinergic nerve terminals, (2) entry inside recycling synaptic vesicles (SV), (3) crossing of the vesicle membrane by the L domain by exploiting the pH gradient (acid inside) across the membrane, (4) release of L in the cytosol by reduction of the interchain disulphide bond and (5) cleavage of one or more of the three proteins that form the SNARE heterotrimeric complex that is essential for the fusion of synaptic vesicle with the presynaptic membrane, thus releasing their neurotransmitter content (Pantano and Montecucco 2014).

After entering the lymphatic and blood circulations, following intestinal absorption or inspiration or injection, the BoNTs rapidly gain access to the perineuronal

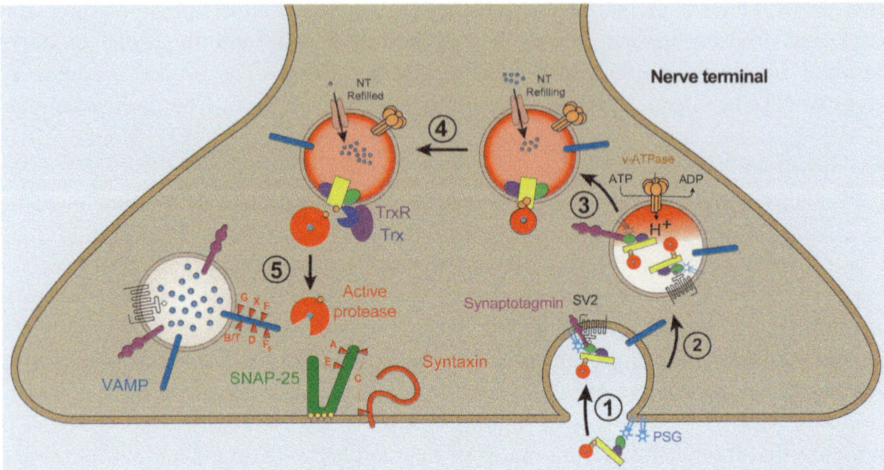

Fig. 2 Mechanism of botulinum neurotoxins entry and paralysis of nerve terminals. The paralysis of nerve terminals by botulinum neurotoxins is a multistep process. The first step (1) is the binding of the HC-C domain (green) to a polysialoganglioside (PSG, light blue) receptor of the presynaptic membrane, followed by binding to a protein receptor of the lumen of synaptic vesicles. The currently known protein receptors are (i) synaptotagmin (magenta) for BoNT/B1, BoNT/DC and BoNT/G; (ii) glycosylated SV2 (black with its attached N-glycan depicted as a hexagon) for BoNT/A1 and BoNT/E1. The BoNT is then internalized inside SVs (step 2). Step 3 begins with the acidification (red) of the vesicle lumen caused by the v-ATPase (orange), which generates a pH gradient that drives the accumulation of neurotransmitter (blue dots) via the vesicular neurotransmitter transporter (pink). The protonation of BoNT leads to the membrane translocation of the L chain into the cytosol which ends step 3. This process is assisted in an unknown way by the HN domain (yellow). Step 4 is the release of the L chain (red with a central light blue dot that represents the active site Zn^{2+} atom) from the HN domain by the action of the thioredoxin reductase – thioredoxin system (TrxR-Trx, violet and dark blue, respectively) which reduces the interchain disulphide bond (orange dots representing the two sulphur atoms). In the cytosol, the L chain displays its metalloprotease activity (step 5): BoNT/B, BoNT/D, BoNT/F, BoNT/G and BoNT/X cleave VAMP (blue); BoNT/A and BoNT/E cleave SNAP-25 (green); and BoNT/C cleaves both SNAP-25 and syntaxin (red). Each of these proteolytic events is sufficient by itself to cause a prolonged inhibition of neurotransmitter release with consequent neuroparalysis. This notion proves that VAMP, SNAP-25 and syntaxin are the core of the nanomachine that drives the release of neurotransmitters within the synaptic cleft

fluid compartment, without crossing the blood-brain barrier (Simpson 2013). The local intramuscular injection of very small doses (few MLD50s) of BoNT/A1 leads to a local paralysis, a property of high therapeutic value (Eleopra et al. 2004; Carli et al. 2009). The BoNTs bind very rapidly and with high affinity the presynaptic plasma membrane of skeletal and autonomic cholinergic nerve terminals. The high affinity is due to a double receptor binding: (a) to the oligosaccharide portion of a polysialoganglioside and (b) to the intra-vesicular domain of synaptic vesicle proteins (SV2 for BoNT/A, BoNT/E and BoNT/F or synaptotagmin for BoNT/B, BoNT/DC and BoNT/G). The neurospecific binding of BoNT/C and BoNT/D is not well characterized, but there is evidence that oligosaccharides of glycolipids or protein-linked N-glycans are involved together with a hydrophobic loop of the toxin that inserts in the lipid bilayer (Nuemket et al. 2011; Rossetto et al. 2014; Zhang and Varnum 2012; Stern et al. 2018; Dong et al. 2019). Once injected, particularly in the case of superficial injections or when small muscle are treated, a competition may be envisaged among entry into the blood or lymphatic circulation via capillaries and binding to the cholinergic terminals followed by the irreversible entry into nerve terminals. If this is the case, the BoNTs shall be capable of rapid membrane binding. In this respect, it is noteworthy that the BoNTs are electrical dipoles with the positive end very close to the PSG binding site and that the PSGs have a strongly negative oligosaccharide head that projects out of the presynaptic membrane like an antenna. In addition, the presynaptic membrane is negatively charged, and this will reorient the BoNT dipole whilst approaching the negatively charged membrane rendering almost any hit with the PSG binding productive (Fogolari et al. 2009). Once this has occurred, the BoNT-PSG complex may move laterally on the membrane to find the second receptor.

4.2 Entry into Nerve Terminals (Step 2)

The internalization of BoNTs is driven by the binding of BoNTs to their second receptor which is localized on the luminal side of the membrane of synaptic vesicles (SV). This binding occurs after the fusion of SV with the presynaptic membrane (Dong et al. 2019). This leads to the exposure of the SV lumen to the plasma membrane surface, whilst BoNT is already bound to PSG; in the case of BoNT/DC and BoNT/B, the hydrophobic loop present between the PSG and SV receptors binding sites additionally contributes to the nerve surface binding (Nuemket et al. 2011; Zhang et al. 2017; Stern et al. 2018).

After intramuscular injection without electrical stimulation, one/two molecules of BoNT/A1 are rapidly taken up and found by electron microscopy inside the lumen of SV at the neuromuscular junction and in neurons in culture (Colasante et al. 2013; Harper et al. 2016). SV exocytosis is strictly coupled to endocytosis, and this explains the fact that BoNT/A1 paralyses is faster in a synaptic terminals stimulated electrically or by exercise, whilst the lowering of synaptic activity prolongs the time of paralysis development (Hughes and Whaler 1962). Recent findings indicate that high activity levels of SV neurotransmitter release leads to SV fusion with incorporation of the SV membrane into the presynaptic membrane (Chanaday et al. 2019). In turn, this would result in an increased extent of exposure of the BoNT luminal SV receptors with a consequent increase of the internalized BoNT (not shown in Fig. 2). This BoNT would end in the lumen of a bulk endosome, rather than in the SV lumen, but SV will form rapidly by clathrin-mediated budding of SV from endosomes. Clearly, the recent novel findings on endocytosis at nerve terminals call for further studies to clarify the different forms of vesicular/endosomal trafficking of the different BoNTs into the nerve terminal. Such studies could lead to improved modes of delivery of BoNT to patients.

4.3 Synaptic Vesicle Membrane Translocation (Step 3)

SV are used as "Trojan horses" by BoNTs to enter inside nerve endings (Fig. 2). However, BoNTs have to exploit another physiological function of the synapse in order to perform the third step of intoxication leading to neuroparalysis. Indeed, BoNT parasitizes the refilling of neurotransmitter inside empty vesicles powered by the action of an ATPase proton pump present on the SV membrane which injects protons inside to create a transmembrane pH gradient that drives the uptake of neurotransmitter from the cytosol into the lumen (Fig. 2). The acidic pH also induces a structural change of BoNT, which falls on the membrane surface and then, in an unknown mode requiring the H chain, leads to the L domain crossing the SV membrane to the cytosolic surface where it remains attached via the SS bond. For more information on step 3, the reader is referred to Pirazzini et al. (2016).

4.4 Reduction of the Disulphide Interchain Bond (Step 4)

The SS interchain bond is exposed to the cytosol after translocation of the L domain, and it is specifically reduced by the NADPH-thioredoxin reductase-thioredoxin redox system (Trx-Tx), bound to the cytosolic surface of SV (Pirazzini et al. 2014). As shown in Fig. 2, this releases the metalloprotease L domain, which then exerts its catalytic activity on the three SNARE proteins that are essential for the SV fusion followed by neurotransmitter release. This notion led to an important translational potential application because inhibitors of the Trx-Tx redox system were found to prevent botulism acting on all BoNTs independently on the serotype and the subtype (Pirazzini et al. 2014; Zanetti et al. 2015; Rossetto et al. 2019). In fact, the different antigenicity of the many known BoNTs speaks against the possibility of using BoNT-specific antibodies to prevent botulism because too many human monoclonal antibodies should be generated. Inhibitors of Trx-Tx do prevent the BoNT-induced nerve terminal paralysis in vivo and are strong candidates for the prevention of botulism in humans and for the treatment of infant botulism and intestinal botulism which imply a continuous production of novel toxin molecules in the intestine (Rossetto et al. 2019).

4.5 SNARE Protein Cleavage (Step 5)

Once released from the cytosolic face of the SV membrane, the L domain is ready to display its Zn^{2+}-dependent proteolytic activity specifically directed to three target proteins: VAMP, SNAP-25 and syntaxin, as shown in Fig. 2. VAMP (vesicle-associated membrane protein, blue) is a protein spanning the SV membrane of and other vesicular cell organelles depending on the isoforms. The two isoforms principally involved in neurotransmitter release are VAMP-1 and VAMP-2. SNAP-25 (synaptosomal nerve-associated protein of 25 kDa, green) is mainly localized on the cytosolic face of the presynaptic membrane via a quartet of Cys residues located in the middle of the protein esterified by four palmitoyl chains acting as hydrophobic anchors. Syntaxin (red in Fig. 2) is present in a number of isoforms spanning the plasma membrane and projecting its mass in the cytosol. These three proteins include a coil domain termed SNARE domain and upon coiling form a heterotrimeric SNARE complex, which is essential for the process of neurotransmitter release (Sutton et al. 1998; Jahn and Scheller 2006; Sudhof and Rothman 2009; Pantano and Montecucco 2014). The formation of the SNARE complex brings the SV close to the active zones of neurotransmitter release, ready to fuse when the [Ca^{2+}] trigger is elicited following the opening of the presynaptic membrane voltage-gated Ca^{2+} channels. BoNT/B, BoNT/D, BoNT/F, BoNT/G and BoNT/X cleave VAMP at single and different sites within the coiling domain. BoNT/C cleaves both syntaxin and SNAP-25, whilst BoNT/A and BoNT/E cleave SNAP-25 at different sites. These cleavages prevent the formation and/or function of the SNARE complex and, consequently, of neurotransmitter release; for detailed information on the specificity of BoNTs for the various isoforms of the three SNARE proteins and the cleaved peptide bonds, please see Pirazzini et al. (2017) and Dong et al. (2019). One amazing aspect of the hydrolytic activity of the BoNT metalloproteases is their absolute specificity for the three SNARE proteins which provided the first and strongest evidence of the essential role of these proteins in neurotransmitter release and vesicular trafficking within cells. This is based on a multiple substrate recognitions chain of the substrate molecule by the L chain which include the peptide bond to be cleaved and several sites located at a distance (Rossetto et al. 1994; Brunger and Rummel 2009; Pantano and Montecucco 2014). So far no other protein substrates of the BoNT L chains have been identified, but it cannot be excluded that ancestral forms of BoNT might cleave other essential cell proteins whose cleavage may provide an evolutionary advantage to the toxin-producing organism.

5 The Different Duration of Action of Botulinum Neurotoxins

The BoNTs do not kill the intoxicated neurons, but they paralyse them, and this paralysis is reversible with time. The duration of paralysis depends on the type of BoNT, on the dose and on the animal species. The proteins present in cells turn over with a half-life time which is characteristic of each protein. Similarly, the BoNT L chains are degraded in the cytosol, and the half-lives of the different BoNT serotypes and subtypes are different. This degradation is considered to be the main determinant of the duration of the BoNT-induced neuroparalysis because the progressive disappearance of the L chain allows for the renewal of its substrate with ensuing neurotransmission recovery. The duration of the BoNT/A1-induced neuroparalysis is the longest among BoNTs (3–4 months for human skeletal terminals, 12–15 months for autonomic cholinergic nerve terminals), whilst the L chain of BoNT/E1 is the shortest living one (paralysis of skeletal terminal lasting about 2–4 weeks). The duration of action of BoNT/A1 in humans is of major importance because it determines the duration of its therapeutic effects. The exceptional length of the paralysis exerted by BoNT/A1 is likely to be supported by effects additional to the L chain degradation. Indeed, there is evidence that the BoNT/A-cleaved SNAP-25 (SNAP-25$^{\#}$), which retains 197 over 206 amino acid residues, is still capable of forming a SNARE heterotrimer with VAMP and syntaxin, which is non-functional in neuroexocytosis but prevents the function of the normal SNARE complex. In other word, SNAP-25$^{\#}$ acts as a dominant negative that causes by itself neuroparalysis as long as it is present inside nerve terminals (Pantano and Montecucco 2014).

6 Long-Distance Effects of Botulinum Neurotoxins

Generalized peripheral neuroparalysis is the most evident symptom of botulism. However, indirect evidence that these neurotoxins could act at a distance from the injection site, i.e. within spinal cord and brain neuronal circuits, was reported long ago. Later on it was experimentally shown that retroaxonal transport of BoNTs does take place, similarly to tetanus neurotoxin (Mazzocchio and Caleo 2015). Compelling evidence of BoNT/A1 retrotransport to the central nervous system (CNS) was provided by tracing the cleavage of SNAP-25 within CNS neurons after peripheral injection of the toxin, using an antibody very specific for the novel epitope generated by the BoNT/A1 cleavage of SNAP-25 (Antonucci et al. 2008; Restani et al. 2012). BoNT/A1 retrograde transport can occur also via sensory neurons, as shown by the injection in the whisker pad which induces the appearance of truncated SNAP-25 in the trigeminal nucleus caudalis (Antonucci et al. 2008; Matak et al. 2011). These long-distance effects are mediated by an active retro-axonal transport of catalytically competent toxins inside motor axons or sensory neurons, and not by passive spread of BoNT/A1 or of SNAP-25$^{\#}$. Moreover, BoNT/A1 can undergo subsequent events of transcytosis and transport, remaining catalytically active (Antonucci et al. 2008; Matak et al. 2011; Restani et al. 2011). The spinal cord contains several cholinergic interneurons whose neurotransmitter release could be inhibited by the L chains of BoNTs (Miles et al. 2007; Zagoraiou et al. 2009; Ramírez-Jarquín and Tapia 2018) with the results of (1) a further peripheral paralytic effect and (2) alteration of the locomotor activity.

7 Future Directions

Although the major aspects of the cellular and molecular mechanism of action of BoNTs have been elucidated, some aspects are not completely understood and are matters of debate. One of the most intriguing topics is the discovery of genes encoding for many novel BoNTs and BoNT-like toxins. Genetic and bioinformatic methods are providing the tools to expand our understanding of the mechanisms underlying this diversity, but the biologic significance of such a large and growing number of BoNTs has not been explained yet. Therefore, ad hoc investigation should be performed in order to answer the fundamental question of the origin and possible role(s) these toxins may have for the producing bacteria within their environments (Montecucco and Rasotto 2015).

Besides the evolutionary significance, the definition of the molecular, cellular, tissue, and pharmacological properties of the many novel botulinum toxins that are being discovered is not fulfilled yet. This is an important goal as, together with engineering of novel BoNTs endowed with specific properties and specificities, it will allow the development of novel therapeutics and protocols that will expand the medical uses of BoNTs. Another issue that deserves attention is that of the long-distance effects of BoNT/A1 consequent to its retro-axonal transport to the central nervous system.

Acknowledgements Work in the author's laboratory is supported by University of Padova, CNR and the Ministry of Defence research project RI.PA.NE.

References

Antonucci F, Rossi C, Gianfranceschi L, Rossetto O, Caleo M (2008) Long-distance retrograde effects of botulinum neurotoxin A. J Neurosci 28(14):3689–3696

Arnon SS, Schechter R, Inglesby TV, Henderson DA, Bartlett JG, Ascher MS, Eitzen E, Fine AD, Hauer J, Layton M, Working Group on Civilian Biodefense et al (2001) Botulinum toxin as a biological weapon: medical and public health management. JAMA 285:1059–1070

Bernard C (1866) Leçons sur les propriétés des tissus vivants. Paris, reproduced by Lighting Source UK, Milton Keynes, UK

Bhattacharjee Y (2011) Biosecurity. Panel selects most dangerous select agents. Science 332:1491–1492

Brunger AT, Rummel A (2009) Receptor and substrate interactions of clostridial neurotoxins. Toxicon 54(5):550–560

Carli L, Montecucco C, Rossetto O (2009) Assay of diffusion of different botulinum neurotoxin type A formulations injected in the mouse leg. Muscle Nerve 40:374–380

Chanaday NL, Cousin MA, Milosevic I, Watanabe S, Morgan JR (2019) The synaptic vesicle cycle revisited: new insights into the modes and mechanisms. J Neurosci 39(42):8209–8216

Chatham-Stephens K, Fleck-Derderian S, Johnson SD, Sobel J, Rao AK, Meaney-Delman D (2018) Clinical features of foodborne and wound Botulism: a systematic review of the literature, 1932–2015. Clin Infect Dis 66(1):S11–S16

Colasante C, Rossetto O, Morbiato L, Pirazzini M, Molgó J, Montecucco C (2013) Botulinum neurotoxin type A is internalized and translocated from small synaptic vesicles at the neuromuscular junction. Mol Neurobiol 48:120–127

Dong M, Masuyer G, Stenmark P (2019) Botulinum and tetanus neurotoxins. Annu Rev Biochem 88:811–837

Doxey AC, Mansfield MJ, Montecucco C (2018) Discovery of novel bacterial toxins by genomics and computational biology. Toxicon 147:2–12

Doxey AC, Mansfield MJ, Lobb B (2019) Exploring the evolution of virulence factors through Bioinformatic data mining. mSystems 4(3):e00162

Dressler D (2012) Clinical applications of botulinum toxin. Curr Opin Microbiol 15:325–336

Eleopra R, Tugnoli V, Quatrale R, Rossetto O, Montecucco C (2004) Different types of botulinum toxin in humans. Mov Disord 19(Suppl 8):S53–S59

Erbguth FJ (2004) Historical notes on botulism, Clostridium botulinum, botulinum toxin, and the idea of the therapeutic use of the toxin. Mov Disord 19(Suppl 8):S2–S6

Fleck-Derderian S, Shankar M, Rao AK, Chatham-Stephens K, Adjei S, Sobel J, Meltzer MI, Meaney-Delman D, Pillai SK (2017) The epidemiology of foodborne Botulism outbreaks: a systematic review. Clin Infect Dis 66(suppl_1):S73–S81

Fogolari F, Tosatto SC, Muraro L, Montecucco C (2009) Electric dipole reorientation in the interaction of botulinum neurotoxins with neuronal membranes. FEBS Lett 583(14):2321–2325

Gart MS, Gutowski KA (2016) Overview of Botulinum toxins for aesthetic uses. Clin Plast Surg 43 (3):459–471

Gu S, Rumpel S, Zhou J, Strotmeier J, Bigalke H, Perry K, Shoemaker CB, Rummel A, Jin R (2012) Botulinum neurotoxin is shielded by NTNHA in an interlocked complex. Science 335 (6071):977–981

Harper CB, Papadopulos A, Martin S, Matthews DR, Morgan GP, Nguyen TH, Wang T, Nair D, Choquet D, Meunier FA (2016) Botulinum neurotoxin type-A enters a non-recycling pool of synaptic vesicles. Sci Rep 6:19654

Hughes R, Whaler BC (1962) Influence of nerve-ending activity and of drugs on the rate of paralysis of rat diaphragm preparations by cl. Botulinum type a toxin. J Physiol 160:221–233

Jahn R, Scheller RH (2006) SNAREs--engines for membrane fusion. Nat Rev 7(9):631–643

Lacy DB, Tepp W, Cohen AC, DasGupta BR, Stevens RC (1998) Crystal structure of botulinum neurotoxin type A and implications for toxicity. Nat Struct Biol 5(10):898–902

Lam KH, Jin R (2015) Architecture of the botulinum neurotoxin complex: a molecular machine for protection and delivery. Curr Opin Struct Biol 31:89–95

Matak I, Bach-Rojecky L, Filipovic B, Lackovic Z (2011) Behavioral and immunohistochemical evidence for central antinociceptive activity of botulinum toxin A. Neuroscience 186:201–207

Mazzocchio R, Caleo M (2015) More than at the neuromuscular synapse: actions of botulinum neurotoxin A in the central nervous system. Neuroscientist 21:44–61

Miles GB, Hartley R, Todd AJ, Brownstone RM (2007) Spinal cholinergic interneurons regulate the excitability of motoneurons during locomotion. Proc Natl Acad Sci U S A 104(7):2448–2453

Montal M (2010) Botulinum neurotoxin: a marvel of protein design. Annu Rev Biochem 79:591–617

Montecucco C (1986) How do tetanus and botulinum toxins bind to neuronal membranes? Trends Biochem Sci 11:314–317

Montecucco C, Rasotto MB (2015) On botulinum neurotoxin variability. MBio 6:e02131–e02114

Muraro L, Tosatto S, Motterlini L, Rossetto O, Montecucco C (2009) The N-terminal half of the receptor domain of botulinum neurotoxin A binds to microdomains of the plasma membrane. Biochem Biophys Res Commun 380(1):76–80

Nuemket N, Tanaka Y, Tsukamoto K, Tsuji T, Nakamura K, Kozaki S, Yao M, Tanaka I (2011) Structural and mutational analyses of the receptor binding domain of botulinum D/C mosaic neurotoxin: insight into the ganglioside binding mechanism. Biochem Biophys Res Commun 411:433–439

Pantano S, Montecucco C (2014) The blockade of the neurotransmitter release apparatus by botulinum neurotoxins. Cell Mol Life Sci 71:793–811

Peck MW, Smith TJ, Anniballi F, Austin JW, Bano L, Bradshaw M, Cuervo P, Cheng LW, Derman Y, Dorner BG, Fisher A, Hill KK, Kalb SR, Korkeala H, Lindstrom M, Lista F, Luquez C, Mazuet C, Pirazzini M, Popoff MR, Rossetto O, Rummel A, Sesardic D, Singh BR, Stringer SC (2017) Historical perspectives and guidelines for botulinum neurotoxin subtype nomenclature. Toxins 9(1):38

Pellet S, Tepp WH, Johnson EA (2019) Critical analysis of neuronal cell and the mouse bioassay for detection of botulinum neurotoxins. Toxins 11(12):E713. https://doi.org/10.3390/toxins11120713

Pirazzini M, Azarnia Tehran D, Zanetti G, Megighian A, Scorzeto M, Fillo S, Shone CC, Binz T, Rossetto O, Lista F et al (2014) Thioredoxin and its reductase are present on synaptic vesicles, and their inhibition prevents the paralysis induced by botulinum neurotoxins. Cell Rep 8:1870–1878

Pirazzini M, Azarnia Tehran D, Leka O, Zanetti G, Rossetto O, Montecucco C (2016) On the translocation of Botulinum and tetanus neurotoxins across the membrane of acidic intracellular compartments. Biochim Biophys Acta 1858:467–474

Pirazzini M, Rossetto O, Eleopra R, Montecucco C (2017) Botulinum neurotoxins: biology, pharmacology, and toxicology. Pharmacol Rev 69(2):200–235

Ramírez-Jarquín UN, Tapia R (2018) Excitatory and inhibitory neuronal circuits in the spinal cord and their role in the control of motor neuron function and degeneration. ACS Chem Neurosci 9 (2):211–216

Restani L, Antonucci F, Gianfranceschi L, Rossi C, Rossetto O, Caleo M (2011) Evidence for anterograde transport and transcytosis of botulinum neurotoxin A (BoNT/A). J Neurosci 31 (44):15650–15659

Restani L, Giribaldi F, Manich M, Bercsenyi K, Menendez G, Rossetto O, Caleo M, Schiavo G (2012) Botulinum neurotoxins A and E undergo retrograde axonal transport in primary motor neurons. PLoS Pathog 8(12):e1003087

Rossetto O, Montecucco C (2019) Tables of toxicity of Botulinum and tetanus neurotoxins. Toxins (Basel) 11(12):E686

Rossetto O, Schiavo G, Montecucco C, Poulain B, Deloye F, Lozzi L, Shone CC (1994) SNARE motif and neurotoxins. Nature 372(6505):415–416

Rossetto O, Pirazzini M, Montecucco C (2014) Botulinum neurotoxins: genetic, structural and mechanistic insights. Nat Rev Microbiol 12(8):535–549

Rossetto O, Pirazzini M, Lista F, Montecucco C (2019) The role of the single interchains disulfide bond in tetanus and botulinum neurotoxins and the development of antitetanus and antibotulism drugs. Cell Microbiol 21(11):e13037

Rummel A (2013) Double receptor anchorage of botulinum neurotoxins accounts for their exquisite neurospecificity. Curr Top Microbiol Immunol 364:61–90

Simpson L (2013) The life history of a botulinum toxin molecule. Toxicon 68:40–59

Sobel J (2005) Botulism. Clin Infect Dis 41(8):1167–1173

Stern D, Weisemann J, Le Blanc A, von Berg L, Mahrhold S, Piesker J, Laue M, Luppa PB, Dorner MB, Dorner BG, Rummel A (2018) A lipid-binding loop of botulinum neurotoxin serotypes B, DC and G is an essential feature to confer their exquisite potency. PLoS Pathog 14(5):e1007048

Sudhof TC, Rothman JE (2009) Membrane fusion: grappling with SNARE and SM proteins. Science 323(5913):474–477

Sugawara Y, Matsumura T, Takegahara Y, Jin Y, Tsukasaki Y, Takeichi M, Fujinaga Y (2010) Botulinum hemagglutinin disrupts the intercellular epithelial barrier by directly binding E-cadherin. J Cell Biol 189:691–700

Sutton RB, Fasshauer D, Jahn R, Brunger AT (1998) Crystal structure of a SNARE complex involved in synaptic exocytosis at 2.4 A resolution. Nature 395(6700):347–353

Swaminathan S (2011) Molecular structures and functional relationships in clostridial neurotoxins. FEBS J 278(23):4467–4485

Zagoraiou L, Akay T, Martin JF, Brownstone RM, Jessell TM, Miles GB (2009) A cluster of cholinergic premotor interneurons modulates mouse locomotor activity. Neuron 64(5):645–662

Zanetti G, Azarnia Tehran D, Pirazzini M, Binz T, Shone CC, Fillo S, Lista F, Rossetto O, Montecucco C (2015) Inhibition of botulinum neurotoxins interchain disulfide bond reduction prevents the peripheral neuroparalysis of botulism. Biochem Pharmacol 98(3):522–530

Zhang Y, Varnum SM (2012) The receptor binding domain of botulinum neurotoxin serotype C binds phosphoinositides. Biochimie 94:920–923

Zhang S, Berntsson RP, Tepp WH, Tao L, Johnson EA, Stenmark P, Dong M (2017) Structural basis for the unique ganglioside and cell membrane recognition mechanism of botulinum neurotoxin DC. Nat Commun 8(1):1637

Manufacturing and Clinical Formulations of Botulinum Neurotoxins

Fauad Hasan

Contents

Abstract

Botulinum Neurotoxins have always existed in nature, but its paralytic effect on humans due to the consumption of poorly preserved food was not recognized until 18th century. There are 8 serotypes of botulinum neurotoxins (A, B, C, D, E, F, G, H). Serotype A have been the most recognized one and was initially developed for large scale production in 1940's. The first batch for clinical use was produced by Edward Schantz, who collaborated with Dr. Alan Scott, an ophthalmologist, evaluating botulinum neurotoxin to treat strabismus. The process Schantz used had variability and led to inconsistent batch production. However, this process is still used by various manufacturers of commercial botulinum neurotoxin products as the foundation. These manufacturers have refined the manufacturing of botulinum neurotoxins by implementing new advanced techniques, including better potency assays. Despite the improvements in the manufacturing process, botulinum neurotoxins are still one of the most potent molecules and therefore, require special handing and additional safety/security measurements during production.

F. Hasan (✉)
Bonti, Irvine, CA, USA

S. M. Whitcup, M. Hallett (eds.), *Botulinum Toxin Therapy*,
Handbook of Experimental Pharmacology 263, https://doi.org/10.1007/164_2019_311

Keywords
Biosafety · Botox · Drug product · Drug substance · Formulation · Manufacturing · Potency · Protein

1 Introduction

The production of botulinum neurotoxin has been perfected in nature, and this phenomenon has taken place for a very long time. Humans have been exposed to them through the consumption of inappropriately preserved food or meat from dead animals. This type of food poisoning is called botulism, caused by the bacteria, *Clostridium botulinum*, an anaerobic, gram-positive, spore-forming rod-shaped, commonly found in soil and water. Botulism can be potentially a fatal condition, characterized by muscle paralysis. However, in ancient times, the connection between food consumption and subsequent death from a paralytic disease in cases of botulism was not generally recognized. Therefore, we have only few historical sources and documents on various causes of food poisoning prior to the eighteenth century (Erbguth 2007). Whenever humans looked for ways to preserve or store food, they created conditions that were conducive to the growth of these bacteria. The awareness around this phenomenon increased, when people in various parts of the world started using cans to store meats and vegetables. Under anaerobic environment, *Clostridium botulinum* produces botulinum neurotoxin as protein complexes of various sizes, with the neurotoxin serotype and protein composition of the complex dependent on the strain of the organism. In ideal conditions, these strains activate the single-chain neurotoxin protein, which forms into a disulfide bond-linked di-chain protein (Brin et al. 2014).

The process used for manufacturing botulinum neurotoxin is very similar to what is used historically to produce wine and other biologic products, such as monoclonal antibodies (mAbs). The typical process for manufacturing biologics is shown in Fig. 1.

As with the production of mAbs, botulinum neurotoxins also are produced in two stages – drug substance and drug product. The drug substance starts when the cells

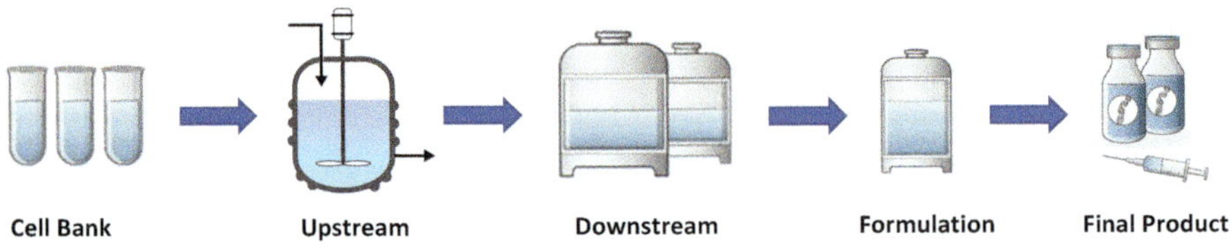

Fig. 1 Typical steps in biologics production. *Cell Bank* – cell line to be used for producing target protein. *Upstream* – protein production by the cells and protein harvest. *Downstream* – purification of the protein and removal of impurities. *Formulation* – making protein into final drug product. *Final Product* – final filled and finished product ready for delivery

are inoculated into the bioreactor, where under ideal conditions, these cells produce the target protein. The protein is then separated from the cells through centrifugation or filtration steps and is prepared for downstream processing. During the next phase, protein goes through purification process in which product or process-related impurities are reduced using various chromatography and filtration steps. At this stage the drug substance, which is also referred to as the active pharmaceutical ingredient (API), is ready. During the drug product stage, the drug substance is formulated, followed by various filling and finishing steps, and packaged into a final drug product that is ready for delivering target protein to the awaiting patients.

Clostridium botulinum produces eight distinct serotypes of botulinum neurotoxin, designated A through H (Keller 2006; Barash and Arnon 2014; Fan et al. 2016) with at least 40 known subtypes (Rossetto et al. 2014). Even though all botulinum neurotoxins have similar modes of action, each serotype has distinct properties. Table 1 summarizes some of the therapeutically important attributes of each neurotoxin subtype.

Of the serotypes listed in Table 1, the majority of the research has been done with botulinum neurotoxin serotype A, which has been successful in treating various conditions. As a result, several botulinum serotype A products have been introduced to the market globally, and they all use the Hall strain of *Clostridium botulinum*. There is one commercial product based on serotype B and a product in clinical development based on serotype E. Table 2 summarized the list of known commercial and clinical products.

In general botulinum neurotoxins have high potency compared to other biological proteins. Therefore, very small dose quantities are needed to achieve the required efficacy, and as a result, the manufacturing scale is much smaller than a typical production batch size of mAb. Most mAb batches yield kilogram quantities of material, whereas only few milligram quantities are needed from a typical botulinum neurotoxin batch.

Table 1 List of botulinum neurotoxin subtypes and their respective attributes

Serotype	Molecular target	Duration	Relative potency	Products
A	SNAP25	~3 months	++++	Botox®, etc.
B	VAMP/synaptobrevin	~2 months	++	MyoBloc
C	Syntaxin/SNAP25	<3 months	++++	None
D	VAMP/synaptobrevin	No human data	–	None
E	SNAP25	~1 months	++++	EB-001 (Phase 2)
F	VAMP/synaptobrevin	2–3 months	+	None
G	VAMP/synaptobrevin	Unknown	Unknown	None
H	VAMP/synaptobrevin?	Unknown	Unknown	None

Table 2 List of current commercial and clinical botulinum neurotoxin products

Product name	Nonproprietary name[a]	Strain	Company
Botox®	OnabotulinumtoxinA	A	Allergan, USA
Dysport®	AbobotulinumtoxinA	A	Ipsen, USA (Therapeutic), Galderma, Switzerland (Aesthetic)
Xeomin®	IncobotulinumtoxinA	A	Merz, Germany
Lantox®/ BTXA	N/A	A	Lanzhou Institute, China
CNBTXA[b]	N/A	A	Nanfeng Medical, China
Jeuveau™/ Nabota®	PrabotulinumtoxinA-xvfs	A	Daewoong Pharm, South Korea, Evolus, USA
Meditoxin®/ Neuronox®	N/A	A	Medytox, South Korea
Innotox®	N/A	A	Medytox, South Korea
Botulax®	N/A	A	Hugel, South Korea
Relatox®	N/A	A	Microgen, Russia
RT-002	DaxibotulinumtoxinA	A	Revance, USA
Myobloc®	RimabotulinumtoxinB	B	US WorldMeds, USA
EB-001	N/A	E	Bonti/Allergan, USA

[a]Established by US FDA (April 2009) and accepted in Canada, EU, and Latin America
[b]Not approved

2 History

One of the early discoverers of the botulism phenomenon was Justinus Kerner, a German medical officer, who in early the 1800s hypothesized that toxin develops under anaerobic conditions and it affects the motor nerves and the autonomic nervous system even in small doses (Erbguth 2007). Later in 1883, a Belgian microbiologist, Emile Pierre van Ermengem, investigated a food poisoning incident and concluded that "it is highly probable that the poison in the food was produced by an anaerobic growth of specific microorganisms during the salting process" (Erbguth 2007). van Ermengem isolated the bacterium in the food and victim's dead bodies. He grew it separately, used it for animal experiments, characterized its culture requirements, and described its toxin in great details. He called it *Bacillus botulinus* (Erbguth 2007), and this pathogen was later renamed as *Clostridium botulinum.*

As the use of canned foods increased in the early 1900s, it led to the prevalence in the cases of food poisoning through botulism. Exposure to botulism was not only coming from the canned meat, but also from canned vegetables. This led to the development of techniques for eradicating bacteria spores that formed due to the anaerobic conditions in the canning process (Erbguth 2007). It was soon discovered that low pH and high osmolarity affect the growth of *Clostridium botulinum* and prevent the production of neurotoxin and that heating leads to toxin inactivation.

The real progress in the development of production of botulinum toxin came during the World War II. In the USA the army began an intensive research program on potential biological weapons, which included botulinum neurotoxins. This research continued after the end of the war. In 1946 Carl Lamanna and Jeff Duff working at the US Army facility at Fort Detrick, Maryland, started developing techniques to concentrate and crystallize the neurotoxin product, so that it can be produced consistently at larger quantities. They were later joined by Edward Schantz, who used Lamanna and Duff's earlier work as foundation, to produce the first batch of botulinum neurotoxin serotype A (Schantz and Johnson 1997). The process Schantz used to produce this first batch became the basis for the manufacturing process for future commercial botulinum neurotoxin products. After the research shut down at Fort Detrick, Schantz continued his research on the use of botulinum toxin at the University of Wisconsin. In the late 1960s, Schantz was approached by an ophthalmologist, Dr. Alan Scott, who wanted to use the botulinum neurotoxin to treat strabismus (Schantz and Johnson 1997).

It was Dr. Scott's research that paved the way for the approval of using botulinum neurotoxin serotype A for the treatment of strabismus by the US Food and Drug Administration (FDA). The material for these studies in humans came from the famous 79-11 batch produced in Schantz lab and included improvement to meet regulatory standards (Truong and Hallet 2013). This product was called Oculinum, and it eventually received a regulatory approval in 1989, as an orphan drug for the treatment of strabismus, blepharospasm, and hemifacial spasm. The product was later acquired by Orange County, California-based company Allergan, which further improved the manufacturing process and changed the history by rebranding it as Botox®.

Across the Atlantic in the UK, botulinum neurotoxin research was spearheaded by the government as well. Research material was produced at Porton Down laboratories, which was part of the military section of the "Centre for Applied Microbiology and Research" (CAMR). The material produced by this laboratory was later used by British clinicians for various therapeutic applications, including the treatment to dystonia. A company formed out of this laboratory and was called Porton International and started producing botulinum neurotoxin as a new drug, called Dysport® (combination of the terms "dystonia" and "Porton") (Erbguth 2007). Later, this company became Ipsen and marketed Dysport®, as a competitor to Botox®.

In the late 1990s and early 2000s, manufacturers in Asia, particularly in China and South Korea, obtained access to Hall strain of *Clostridium botulinum* and started developing botulinum neurotoxin serotype A, similar to Allergan's Botox®. Meditoxin® (also, known as Neuronox®) became the first commercial botulinum neurotoxin serotype A to be manufactured and commercialized by a South Korean company Medytox (Fervert et al. 2018). Several companies also branched off from Medytox and started developing additional botulinum neurotoxin serotype A products. In China, Lanzhou Institute of Biological Products manufactures Lantox® (also known as BTXA), and Nanfeng Medical and Science Technology Company manufactures CNBTXA (Truong and Hallett 2013). Recently Microgen, a company based in Russia, also started producing botulinum neurotoxin serotype A called Relatox® (Fervert et al. 2018).

As with any other successful products, there have been attempts to bring counterfeit and unlicensed botulinum neurotoxin serotype A products in some regions. Some of them are even sold online (Pickett and Mewies 2009). In China Nanfeng Medical started selling CNBTXA without any approval, and the product didn't include any package insert or dosing information. However, the product vials are labeled as containing 55 units, without any supporting documents. When this product was tested in the Allergan biological activity assay, each vial contained an equivalent of 243 Allergan units (Brin et al. 2014). If this product had been mistakenly used at the same doses as an approved product such as Botox® on the assumption of unit interchangeability, it can potentially lead to severe health risks to patients (Walker and Dayan 2014).

3 Schantz Manufacturing Process

According to Eric Johnson, a scientist who worked closely with Schantz at University of Wisconsin, every scientific research project on botulinum neurotoxin that took place around the world between 1950s and 1980s most likely obtained their material from Schantz. This includes material used for clinical proof of concept studies conducted by physicians, such as Dr. Scott. Even today, most commercial and clinical botulinum neurotoxin products are manufactured using Shantz methodology.

Being the sole supplier for a very long period, Schantz took a nontraditional approach to manufacturing botulinum neurotoxin compared to other biologic production. Johnson compared Schantz to a traditional vintner with a unique way of making wine, who added specific technique and methods to his manufacturing process. Although the process Schantz developed became the foundation of botulinum neurotoxins produced globally, it took several years of refinement (Waters 1992).

Compared to the advanced methods used today to produce therapeutic proteins, Schantz's method was very rudimentary. Johnson recalled that Schantz managed to make all the toxins needed for research on a small bench in a combined lab office not much bigger than a coat closet. His typical batch size was 3-gal using a glass carboy, which yielded close to 60 mg of botulinum neurotoxins. The first step of the process is growing colonies of cells using the available *Clostridium botulinum* Hall strain. The colonies are selected, and the cells are inoculated in a fresh medium composed of dextrose, digested milk protein, and yeast extract. This step is called fermentation, and within 18–24 h, cells grow rapidly, turning the medium into a cloudy brown color solution. When the medium runs out of nutrients, the cells lyse, and the neurotoxin which was being produced inside the cell is then released. Usually the fermentation takes about 3 days, resulting in botulinum neurotoxin containing medium. In the following step the botulinum neurotoxin is isolated from the medium. Here Schantz's process uses a very old technique of precipitation. In this step, an acid is used to lower the pH, which makes the neurotoxin insoluble, thereby separating it from the medium. The neurotoxin settles with other substances produced by the bacteria. In subsequent steps, the neurotoxin is further purified from these other substances using the precipitation technique again but this time with the help of alcohol at a very low temperature. This step is repeated until the precipitate consists of highly purified botulinum neurotoxin. Researchers and manufacturers use various biophysical assays to determine the desired purity levels. The final step is to redissolve the solution and add ammonium sulfate (Ton et al. 2015). This causes the botulin to crystallize into microscopic, glasslike needles that are composed of a very pure botulinum neurotoxin. It took Schantz 3 weeks to produce a batch, but overall process time can vary due to limitations such as staff, safety, and regulatory requirements (Waters 1992). Batch 79-11, which became the source material for many clinical studies, and Batch 88-44, which was accepted by European Regulatory agency, were made using this process (Truong and Hallett 2013).

The Schantz process has been successful in producing botulinum neurotoxin, but it has quite a few drawbacks. The use of colonies for inoculation, instead of a cell bank, as is typically used in current production of biologic products, leads to inconsistent cell growth and can affect both the neurotoxin productivity and consistency. The precipitation method is no longer used in the industry because it has a high degree of inconsistency, resulting in variable product yield and quality. Similarly, the crystallization step has quite a bit variability and unnecessary manipulations. Given these limitations, there are always batch to batch variabilities with the Schantz process. For example, in an analysis of potency of different batches from the same manufacturing site, one of the batches, Batch 88-4, had four times more potency than Batch 79-11, even though both were produced with the same process (Truong and Hallett 2013).

4 Current Botulinum Neurotoxin Formulations

Over the years, the bioprocessing industry with the guidance from regulatory authorities has implemented quality by design in drug development and has focused on optimizing the consistency between manufacturing batches. Advancements in bioprocessing technology have been made to improve the manufacturing process efficiency and produce high-quality and safe products for the patients. As the manufacturing field has advanced, this has forced the commercial and clinical manufacturers of botulinum neurotoxin to make necessary improvements to ensure high-quality and consistent products. The manufacturers of botulinum neurotoxin serotype A commercial products made modifications, such as removing the accessory protein or formulation changes, resulting in differentiated products that are not interchangeable (cannot substitute for each other), even though they are based on the same serotype.

When Allergan acquired the Oculinum in 1989, it also inherited the material produced by Shantz in 1979. The company soon needed additional material supply, which had to be made under the regulatory standards at that time. This required further investment in process development and identifying a GMP facility capable of manufacturing a consistent supply of botulinum neurotoxin. After initial struggles, Allergan ended up investing heavily toward building an internal manufacturing infrastructure. Today, Allergan is producing Botox® out of its site in Ireland. Similarly, companies like Ipsen and Merz have also invested heavily in the process development and manufacture of the botulinum neurotoxin products. Ipsen manufactures Dysport®, in the UK, and Merz manufactures Xeomin® in Germany. Revance Therapeutics, a clinical stage botulinum neurotoxin company, has a manufacturing facility in California, whereas Evolus is producing Jeuveau™ in South Korea through their licensing partner Daewoong Pharmaceuticals, Inc. Some of the companies are utilizing contract manufacturing organizations to produce their clinical and commercial products. US WorldMeds, which now markets the only approved botulinum neurotoxin serotype B product, Myobloc®, is manufacturing commercial supply through a third-party manufacturer.

As noted above, the drug substance processes of most marketed botulinum neurotoxin products are based on methodology developed by Schantz. Notable among these products is Botox®, which was first produced by Shantz, when it was known as Oculinum. In the late 1990s, Allergan made improvements to the formulation to reduce the immunogenicity rate in the clinics, which was attributed to product impurities. This prompted others emulating Allergan's success with botulinum neurotoxin serotype A to introduce additional improvements to manufacturing and employ best bioprocess industry practices. Dysport® was the first product manufactured using the chromatography columns for purification replacing the precipitation step, thereby implementing better controls for both quality and quantity (yield). Xeomin® incorporates an additional purification step in its manufacturing process, in which the complexing proteins have been removed from the botulinum toxin complex. Compared to other botulinum neurotoxin serotype A products, Xeomin contains the active pure 150 kD molecule, and given its high-purity formulation, it has been speculated to lead to greater efficacy with reduced risk of sensitization or antibody formation. This is yet to be proven clinically (Walker and Dayan 2014).

Although some of the new manufacturers of botulinum neurotoxin commercial products have emulated the drug product formulation of well-established products, there are some with significant differences. The first step of the drug product manufacturing usually is the dilution of highly concentrated drug substance material. This is done mainly because botulinum neurotoxins have high potency, and as a result, very small quantities of protein are required in the final drug product to achieve the desired effect in the targeted tissue. The dilution step helps bring the final drug product to the appropriate concentration level needed to deliver the required number of units per vial of each product (Klein 1998). However, this does not come without several manufacturing hurdles. To produce small quantities consistently can be challenging, particularly at the time of process validation. The regulatory agencies have also taken strict approach over the years and are forcing the manufacturers to adhere to agreed product specifications, particularly the potency. Given the assay variability of the potency assay, this can be daunting task, and as a result, the manufacturers of botulinum neurotoxin products find themselves producing several batches to ensure they have required material for clinical or commercial use.

The manufacturers also add various excipients to provide stability to the drug product, and these differ in amount and type among the products. The drug product manufacturing process of most botulinum neurotoxins includes a finishing process that involves some method of drying, and this step also differs among the manufacturers. Botox® is vacuum dried, in which the liquid is removed under reduced air pressure without the freezing step. Dysport® is freeze-dried, Xeomin® is lyophilized, and in both of these processes, the liquid is frozen, and the ice evaporated under low pressure. Innotox® produced by Medytox® is the only marketed serotype A product with a liquid formulation. Myobloc® a serotype B also has a liquid formulation but needs to be formulated at a lower pH (5.5–6.5) to maintain stability. However, the drawback is that this acidic formulation is known to cause a stingy discomfort associated with its injections. As a result, Myobloc® is not recommended for facial applications.

Once the finishing step is completed, products are packaged into vials, and potency of the finished drug product is tested prior to release, using the proprietary reference standard. If the potency and associated release specifications are met, product is released for distribution and clinical use. The majority of the botulinum neurotoxin product formulation uses human serum albumin (HSA) as a stabilizer. Relatox® and Lantox® (also known as Prosigne®) formulation contains bovine gelatin protein instead of HSA. The purpose for this is to prevent the neurotoxin from adhering to the wall of the vial or syringe, but because of the bovine source, there is a potential to trigger an immunological response and allergic reactions, including the danger of bovine spongiform encephalopathy (i.e., "mad cow disease") (Walker and Dayan 2014). Revance's RT-002 currently in development does not use HSA as well; instead they use a proprietary peptide.

The underlying difference in manufacturing process and formulation has an impact on the clinical attributes of these products (duration, dose, efficacy, immunogenicity, etc.). Therefore, these botulinum neurotoxin products, specifically the ones that are based on serotype A, cannot be considered interchangeable. Table 3 provides an overview of the drug product formulation of current marketed products worldwide.

All marketed products come with a product insert that provides recommendations for storage, preparation, handling, and additional instructions on delivery. These instructions may vary from product to product due to the intricate formulation and gentle nature of botulinum neurotoxin protein, which can be denatured easily. The products with powdered formulations are reconstituted before delivery and use a recommended preferred diluent. For example, the recommended diluent for Botox® is saline without preservative (Allergan 2013). However, some physicians prefer using sterile saline solution with preservatives, such as benzyl alcohol, because it helps in reducing microbial contamination, making injections less painful, and potentially serving as a local anesthetic (Alam et al. 2002). The final dilution of Botox® is mostly a matter of personal preference. Theoretically, more concentrated solutions reduce reliability in delivering a specific unit dose, and more dilute solutions lead to greater diffusion of the toxin.

Table 3 Formulation comparison of marketed botulinum neurotoxin products

Product name	Protein composition	Formulation	Storage (°C)	Units per vial	Components
Botox®	Complex (900 kDa)	Powder (vacuum dried)	<−5	100, 200	HSA – 0.5 mg, NaCl – 0.9 mg
Dysport®	Complex (500–900 kDa)	Powder (freeze-dried)	2–8	300, 500	HSA – 125 ug, Lactose – 2.5 mg
Xeomin®	Purified toxin (150 kDa)	Powder (lyophilized)	20–25	50, 100	HSA – 1 mg, Saccharose – 4.7 mg
BTXA	Complex (900 kDa)	Powder	2–8	50, 100	Gelatin – 5 mg, Dextran – 25 mg, Sucrose – 25 mg
Jeuveau™/ Nabota®	Complex (900 kDa)	Powder	2–8	100	HSA – 0.5 mg, NaCl – 0.9 mg
Meditoxin®	Complex (940 kDa)	Powder (freeze-dried)	2–8	100	HSA – 0.5 mg, NaCl – 0.9 mg
Innotox®	Complex (900 kDa)	Liquid	2–8	25	Polysorbate (no HSA)
Botulax®	Complex (900 kDa)	Powder	2–8	100	HSA – 0.5 mg, NaCl – 0.9 mg
Relatox®	Complex (900 kDa)	Powder	2–8	100	Gelatin – 6 mg, Maltose – 12 mg
RT-002	Purified toxin (150 kDa)	Powder	2–8	N/A	Peptide (no HSA)
Myobloc®	Complex (700 kDa)	Liquid	2–8	2,500; 5,000; 10,000	HSA – 0.05%, Succinate 0.01 M

There is also a perception that rigorous shaking, bubbling, and foaming reduce the efficacy of botulinum neurotoxins. This has now been refuted by several studies, which showed that the potency and the short- or long-term effects of the product were not affected by foaming during the reconstitution process. In another study the impact of vigorous reconstitution was assessed using vortex touch equipment, and the results came back with similar conclusions (Samizadeh and De Boulle 2018).

For the majority of applications, botulinum neurotoxin products are injected into targeted muscle tissues using a 30-gauge 1-in. needle. However, for the treatment of overactive bladder, since endoscopes are needed, Botox® (only approved botulinum neurotoxin to date) is injected usually with a 25-gauge needle (Shenot and Mark 2014). In general, doses are tailored according to the mode of use and individual patients, and the dose depends on the mass of muscle being injected: The larger the muscle mass, the higher the dose required. However, lower doses may be required in patients with preexisting weakness.

5 Potency Assay and Unit Dosing

Another reason for clinical differences among the botulinum neurotoxins stems from the potency testing methods, which result in distinct unit potencies and dose-response curves for each product. The biological assay used to determine potency of bulk drug substance used to produce each drug product lot has a direct bearing on clinical dosing. The specific potency is explained as the potency per unit weight of toxin protein which means the level of protein administered per injection (Samizadeh and De Boulle 2018). Although international standards for the activity of many biological products are established by the World Health Organization, there are no such standards for botulinum neurotoxin serotype A products. As a result, each manufacturer employs its own proprietary assay methods for testing potency units that include a product-specific reference standard. Typically, biological assays involving animals are sensitive to variations in animal strain, age, sex, diet, temperature, caging, season, and specific experimental procedures such as the liquid used to dilute the product (Brin et al. 2014). A well-established analytical method for assessing potency of the bulk drug substance, recognized by the regulatory agencies, is median lethal dose (LD50) assay. This test usually utilizes the female Swiss-Webster mouse model (Nigam and Nigam 2010).

All major botulinum neurotoxin manufacturers conduct this test for the release of their products but incorporate proprietary steps in their methods, including using specific product reference standards and different dilution techniques. In fact, individual products can be differentially affected by different diluents (Brin et al. 2014). Notably, manufacturers of the major botulinum neurotoxin serotype A products use different diluents for LD50 testing. Allergan uses saline (the diluent also used for clinical reconstitution), and Ipsen uses gelatin phosphate buffer. Merz adds HSA as a stabilizer to its undisclosed diluent. Stabilizers have been shown to enhance the activity of botulinum neurotoxin serotype A products at low concentrations in preclinical tests. A key difference in the activity between the Botox® and Dysport® formulations concerns the potency units established by their manufacturers. Allergan uses Botox units (bU) for Botox®, and Ipsen uses Speywood units (sU) for Dysport® (Scaglione 2016). Both define 1 unit of toxin (1 mouse LD50) as the quantity necessary to kill 50% of a group of mice with an intraperitoneal injection (Samizadeh and De Boulle 2018).

While the mouse LD50 has been the global standard for botulinum neurotoxin serotype A potency testing, efforts have been made in the last decade to reduce the use of animals in product testing. Allergan was the first company to develop a cell-based potency assay (CBPA) for the botulinum neurotoxin, which was approved in for use in the USA in 2011 (Allergan 2012). This method uses a specific cell line and can carry out evaluation of all four phases of botulinum toxin action (binding, internalization, translocation, and SNAP-25 cleavage) (Samizadeh and De Boulle 2018). Allergan cross validated it with its LD50 assay and optimized it specifically for Botox®. This method has no impact whatsoever on the product or product potency. It took several years of development to achieve the required sensitivity to precisely and consistently measure the complex mode of action and a very small amount of botulinum neurotoxin in a strict quality control and high-capacity manufacturing setting (GenomeWeb 2011). This assay is now used for both stability and release testing. Merz now also has as CBPA for Xeomin®, which was approved in 2014. Given the various manipulations steps and the complexities of the potency, the units of biological activity are specific to each botulinum neurotoxin product, and therefore, the unit doses are not interchangeable.

The non-interchangeability of units was demonstrated in a study that examined Botox® (full complex) and Xeomin® (purified toxin) in the Allergan LD50 assay. In this assay, the products were diluted in normal saline and compared against the Allergan 100-unit standard. The results showed that the Xeomin® activity was less than 100 Allergan units (i.e., 69–78 units for 3 different lots). These results were confirmed in several orthogonal assays, including an enzymatic cleavage assay, the Digit Abduction Score assay, as well as replication of the LD50 results. In a separate study that compared these two products in the Merz's LD50 assay, in which the products were diluted with a solution containing added HSA as a stabilizer and were compared against the Merz standard, potency was found to be comparable These results confirm that the potencies of the two botulinum neurotoxin serotype A products were differentially affected by the diluent and stabilizers, indicating that, due to underlying product differences, assay conditions markedly influence potency measurements (Brin et al. 2014).

The differences in biological activity among the botulinum neurotoxin products have been recognized by regulatory agencies. Every approved botulinum neurotoxin product in the USA includes a unit non-interchangeability statement. The regulatory agencies in Europe and most other countries worldwide also require a statement of unit non-interchangeability among botulinum neurotoxin products (Brin et al. 2014). The FDA also mandates that each product would have its own nonproprietary name as listed in Table 2. The FDA-stipulated nonproprietary names help to clearly identify each botulinum neurotoxin product, providing a standardized terminology to minimize the potential for medication errors and enable accurate scientific communication (Brin et al. 2014).

6 Challenges in Manufacturing

Manufacturing is considered a major barrier to entry in botulinum neurotoxin space. One major reason is that these molecules are highly potent, and therefore, they are classified as "select agents," by the Centers for Disease Control (CDC) in the USA. Different regions in the world also have similar designations. As a result, additional complexities in designing and building the facilities for both development and manufacturing of botulinum neurotoxins have to be considered. These facilities not only have to be GMP compliant; they also have to adhere to strict requirements from the CDC as well. Any changes to the facility design or even in the manufacturing process have to be approved by CDC. This includes anywhere from inoculation step to the packaging and shipping.

In addition, given the higher potency of botulinum neurotoxins, the development and manufacturing activities have to be performed in a facility with higher containment and safety levels, such as biosafety level 3 (BSL-3) and above. According CDC guidelines, solutions of sodium hypochlorite (0.1%) or sodium hydroxide (0.1 N) readily inactivate the toxin and are recommended for decontamination of work surfaces and for spills. Employees must go through additional safety training to handle these molecules, including wearing personal protective equipment (PPE). In some countries, employers require their prospective employees to get vaccinated as a safety measure for potential accidental exposure to botulinum neurotoxin. CDC recommends getting pentavalent (A, B, C, D, E) botulinum toxoid vaccine (PBT) injections at 0, 2, 12, and 24 weeks, followed by a booster at 12 months and annual boosters thereafter. Over the years this requirement has eased up in several countries, including the USA (CDC 2009).

To ensure facilities and employees are compliant with the guidelines, CDC performs routine audits, which are very detailed and time-consuming. With the implementation of the US Patriot Act, additional restrictions were put in place, such as conducting background checks on staff members handling the molecules and the associated paperwork. All of these requirements lead to additional burden that most other bioprocessing facilities do not have to deal with, and as a result, the cost of building and maintaining a facility to develop and manufacture botulinum neurotoxin products is very high. An extreme example occurred several years ago when a highly concentrated laboratory preparation of botulinum neurotoxin serotype A was illegally administered to humans at a cosmetic clinic in Florida. Following exposure to this unapproved laboratory preparation, patients developed muscle weakness attributable to systemic distribution of the botulinum neurotoxin serotype A preparation and were hospitalized for up to 14 weeks (Brin et al. 2014). Incident such as these highlights why handling of these molecules requires strict control and oversight.

References

Alam M, Dover JS, Arndt KA (2002) Pain associated with injection of botulinum A exotoxin reconstituted using isotonic sodium chloride with and without preservative. Arch Dermatol 138 (4):510–514

Allergan (2012) Allergan receives positive opinions for first-of-its-kind, fully in vitro, cell-based assay for BOTOX® and VISTABEL®; Allergan Press Release. https://www.businesswire.com/news/home/20120222006363/en/Allergan-Receives-Positive-Opinions-First-of-Its-Kind-Fully-vitro

Allergan (2013) BOTOX® (onabotulinumtoxinA) [prescribing information]. Allergan, Irvine

Barash JR, Arnon SSA (2014) Novel strain of Clostridium botulinum that produces type B and type H. J Infect Dis 209:183–191

Brin M, James C, Matlman J (2014) Botulinum toxin type A products are not interchangeable: a review of the evidence. Biologics 2014(8):227–241

Centers for Disease Control and Prevention (2009) Biosafety in microbiological and biomedical laboratories, 5th edn. US Department of Health and Human Services, Washington

Erbguth FJ (2007) From poison to remedy: the chequered history of botulinum toxin. J Neural Transm 115(4):559–565

Fan Y, Barash JR, Lou J, Conrad F, Marks JD, Arnon SS (2016) Immunological characterization and neutralizing ability of monoclonal antibodies directed against botulinum neurotoxin type H. J Infect Dis 213:1–9

Fervert J, Ahn KY, Park MY, Sunga O (2018) Comparison of botulinum neurotoxin type A formulations in Asia. Clin Cosmet Investig Dermatol 2018(11):327–331

GenomeWeb (2011) FDA approves Allergan's cell-based assay for Botox. https://www.genomeweb.com/archive/fda-approves-allergans-cell-based-assay-botox#.XS_x0OhKhyw

Keller JE (2006) Recovery from botulinum neurotoxin poisoning in vivo. Neuroscience 139:629–637

Klein AW (1998) Dilution and storage of botulinum toxin. Dermatol Surg 24(11):1179–1180

Nigam PK, Nigam A (2010) Botulinum toxin. Indian J Dermatol 55(1):8–14

Pickett A, Mewies M (2009) Serious issues relating to the clinical use of unlicensed botulinum toxin products. J Am Acad Dermatol 61(1):149–150

Rossetto O, Pirazzini M, Montecucco C (2014) Botulinum neurotoxins: genetic, structural and mechanistic insights. Nat Rev Microbiol 12:535–549

Samizadeh S, De Boulle K (2018) Botulinum neurotoxin formulations: overcoming the confusion. Clin Cosmet Investig Dermatol 11:273–287

Scaglione F (2016) Conversion ratio between Botox®, Dysport®, and Xeomin® in clinical practice. Toxins 8(3):65

Schantz EJ, Johnson EA (1997) Botulinum toxin: the story of its development for the treatment of human disease. Perspect Biol Med 40(3):317–327

Shenot PJ, Mark JR (2014) Intradetrusor onabotulinumtoxinA injection: how I do it. Can J Urol 20 (1):6649–6655

Ton J, Patel H, Bates R, Ahmad WM (2015) Process and system for obtaining botulinum neurotoxin; United State Patent US 8,932,827 B2; Patent Date 13 Jan 2015

Truong D, Hallett M (2013) Manual of botulinum toxin therapy, 2nd edn. Cambridge University Press, Cambridge

Walker TJ, Dayan SH (2014) Comparison and overview of currently available neurotoxins. J Clin Aesthet Dermatol 7(2):31–39

Waters T (1992) The fine art of making poison. Discover Magazine, August 1992. discovermagazine.com/1992/aug/thefineartofmaki93

Novel Native and Engineered Botulinum Neurotoxins

Lance Steward, Mitchell F. Brin, and Amy Brideau-Andersen

Contents

Abstract

Botulinum neurotoxins (BoNTs), produced by *Clostridia* and other bacteria, are the most potent toxins known. Their cleavage of the soluble N-ethylmaleimide-sensitive factor activating protein receptor (SNARE) proteins in neurons prevents the release of neurotransmitters, thus resulting in the muscle paralysis that is

L. Steward · A. Brideau-Andersen (✉)
Allergan plc, Irvine, CA, USA
e-mail: Steward_Lance@Allergan.com; Brideau-andersen_Amy@Allergan.com

M. F. Brin
Allergan plc, Irvine, CA, USA

University of California, Irvine, CA, USA
e-mail: Brin_Mitchell@Allergan.com

S. M. Whitcup, M. Hallett (eds.), *Botulinum Toxin Therapy*,
Handbook of Experimental Pharmacology 263, https://doi.org/10.1007/164_2020_351

characteristic of botulism. This mechanism of action has been exploited for a variety of therapeutic and cosmetic applications of BoNTs. This chapter provides an overview of the native BoNTs, including the classical serotypes and their clinical use, mosaic BoNTs, and novel BoNTs that have been recently identified in clostridial and non-clostridial strains. In addition, the modular structure of native BoNTs, which are composed of a light chain and a heavy chain, is amenable to a multitude of novel fusions and mutations using molecular biology techniques. These novel recombinant BoNTs have been used or are being developed to further characterize the biology of toxins, to assist in vaccine production, to serve as delivery vehicles to neurons, and to be utilized as novel therapeutics for both neuronal and non-neuronal cells.

Keywords

AbobotulinumtoxinA · Botulinum toxin · IncobotulinumtoxinA · Neurotoxin · OnabotulinumtoxinA · PrabotulinumtoxinA-xvfs · Recombinant · RimabotulinumtoxinB

1 Introduction

The classical botulinum neurotoxins (BoNTs) are produced by the anaerobic, spore-forming, Gram-positive bacilli *Clostridia.* BoNTs occupy a unique position in that they are the most potent toxins known (Gill 1982) and yet are also used to treat a wide variety of medical conditions (Pirazzini et al. 2017). *Clostridium botulinum* was first discovered in 1895 (Erbguth 2008), and the type A and B toxins (BoNT/A and BoNT/B) were denoted as such in 1919 (Burke 1919). Over the next 50 years, the seven classical serotypes of BoNT were identified, with BoNT/G the last to be identified in 1970 (Giménez and Ciccarelli 1970). Recently, with the advent of genomic sequencing technologies and other biological laboratory techniques, additional naturally occurring BoNT subtypes, mosaics, and non-clostridial BoNT-like-encoding sequences have been identified. In addition, advances in molecular biology and protein engineering technologies have led to the creation of laboratory-engineered chimeras, fusions, and other mutant variants that may ultimately be of clinical use. This chapter will review the novel native and engineered BoNTs following an introduction to the classical BoNTs.

2 BoNT Structure and Mechanism of Action

To understand both the novel native and engineered BoNTs in context, some background information on the classical serotypes is presented. BoNT, initially expressed as a protoxin, is composed of a light chain (LC) and heavy chain (HC) with three functional domains. The LC is in the N-terminus and is a zinc-dependent endoprotease that cleaves soluble N-ethylmaleimide-sensitive factor activating protein receptor (SNARE) proteins, while the HC has two domains:

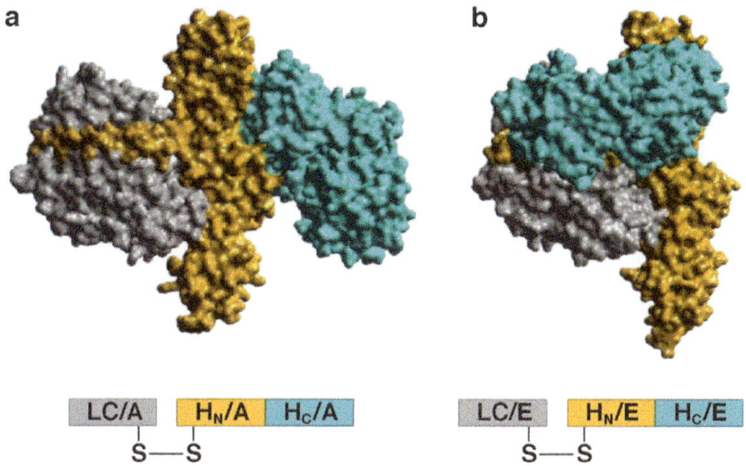

Fig. 1 Structure of botulinum toxins type (**a**) A1 (Protein Data Bank entry 3BTA (Lacy et al. 1998)) and (**b**) E1 (Protein Data Bank entry 3FFZ (Kumaran et al 2009)). H_C, C-terminus of heavy chain; H_N, N-terminus of heavy chain; LC, light chain; S-S, disulfide bond

the N-terminal domain (H_N), responsible for translocation of the LC from the endocytic vesicle into the cytosol, and the C-terminal binding domain (H_C), which recognizes and binds to cell surface receptors (Lacy et al. 1998). The protoxin is converted to the active form by cleavage of a linker between the HC and LC by bacterial (Dekleva and Dasgupta 1990) or host proteases (Duff et al. 1956). Figure 1a is a ribbon diagram of the full-length BoNT/A crystal structure, depicting the three functional domains (protease [LC], translocation [H_N], and binding [H_C] domains) (Lacy et al. 1998). The open modular arrangement of BoNT/A, referred to as the linear butterfly conformation (Pirazzini et al. 2017), is also shared by BoNT/B (Swaminathan and Eswaramoorthy 2000). In contrast, BoNT/E has an alternative, more closed domain structure (Fig. 1b) (Kumaran et al. 2009), which is hypothesized to underlie its faster translocation time and earlier onset of effect compared with BoNT/A and BoNT/B. Despite these differences in the three-dimensional domain arrangement, in all cases the LC and HC remain connected until a disulfide bond between them is reduced and the LC is released into the cytosol, where it cleaves SNARE proteins. This class of proteins includes synaptosomal-associated proteins (SNAP), vesicle-associated membrane proteins (VAMP), and syntaxin (Fig. 2). This cleavage leads to the inability of a functional SNARE complex to form, thus blocking neurotransmitter release from nerve terminals and causing muscle paralysis.

BoNT gene clusters, which also encode several BoNT-associated proteins, are present on the chromosome in some strains and on plasmids or phagemids in others (Smith et al. 2007; Sakaguchi et al. 2015; Woudstra et al. 2018) (Fig. 3). Nontoxic non-hemagglutinin (NTNH) proteins encoded within the gene cluster form a pH-dependent complex associated with BoNT called the progenitor toxin complex, which protects the toxin from the harsh conditions of the host

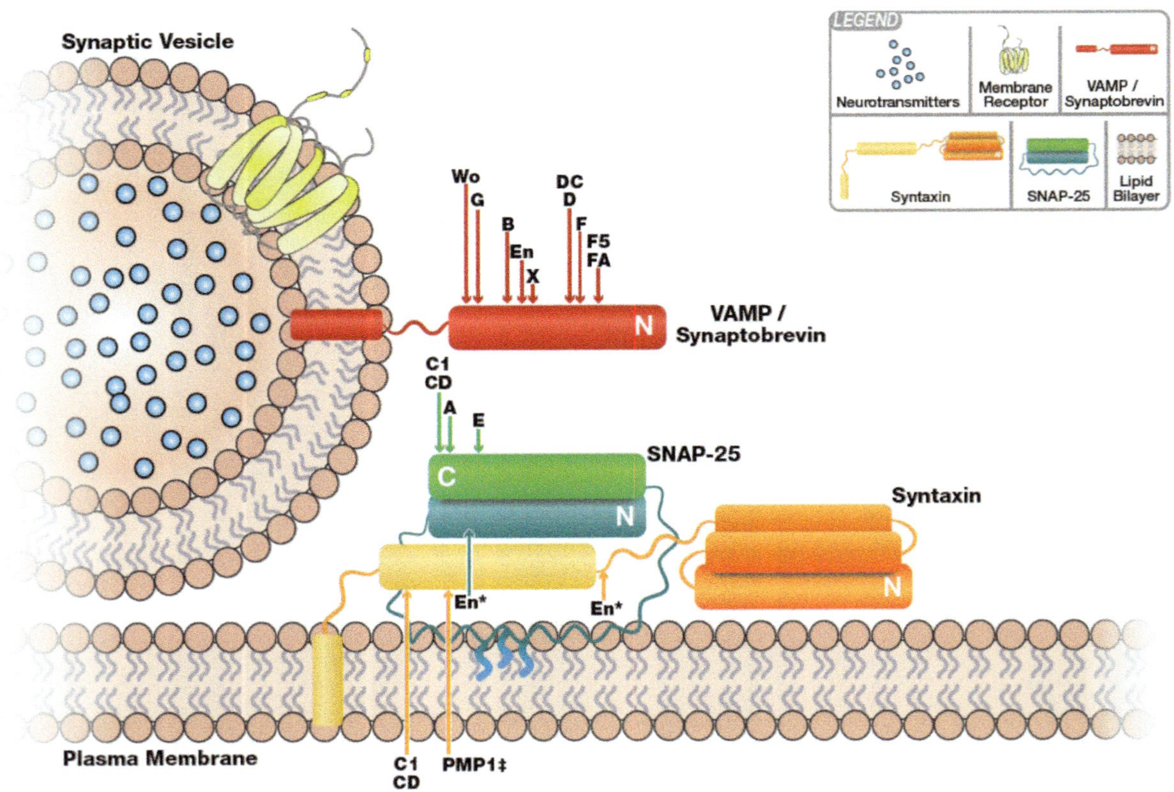

Fig. 2 Botulinum toxin cleavage sites of SNARE proteins. BoNT, botulinum toxin; SNAP-25, synaptosomal-associated protein 25; SNARE, soluble N-ethylmaleimide-sensitive factor activating protein receptor; VAMP, vesicle-associated membrane protein. * Cleavage of syntaxin by BoNT/En only seen at high concentrations in vitro and ‡ cleavage of mosquito syntaxin by PMP1

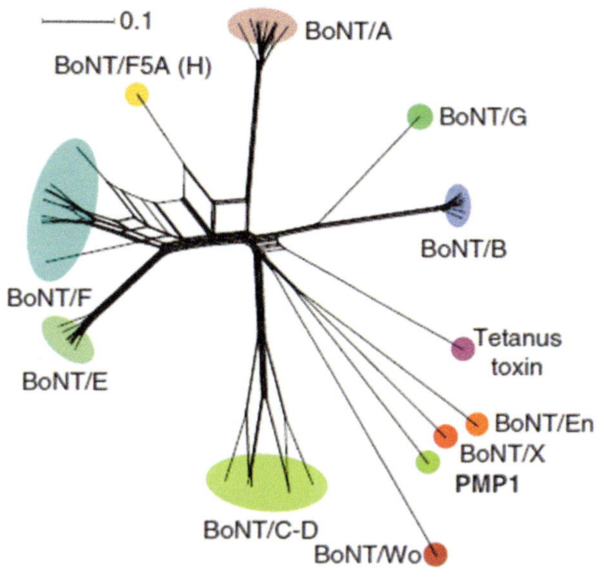

Fig. 3 Phylogenetic network of the botulinum toxins and botulinum toxin-like proteins. The figure is reprinted from Contreras E, Masuyer G, Qureshi N, Chawla S, Dhillon HS, Lee HL, Chen J, Stenmark P, Gill SS. A neurotoxin that specifically targets *Anopheles* mosquitoes. Nat Commun. 2019;10:2869,

gastrointestinal tract (Gu et al. 2012). In addition to *ntnh*, a *bonT* gene cluster always also encodes either an *ha* cluster or an *orfX* cluster. When present, the three hemagglutinin proteins HA17, HA33, and HA70 form a complex with the progenitor toxin complex. Hemagglutinins bind to carbohydrates and cadherins on the cell surface, disrupt cell-cell junctions, enable BoNT to cross the intestinal epithelium, and have been reported to enhance endopeptidase activity (Kukreja and Singh 2007; Sugawara et al. 2010). The *orfX* gene cluster encodes OrfX1, OrfX2, OrfX3, and P47, which are proteins with as-yet unknown function, though there is some evidence that they can bind to lipids (Gustafsson et al. 2017).

3 The Classical Serotypes of BoNT and Their Clinical Use

The initial seven BoNTs were categorized by serological methods, each designated by a letter. There are over 45 subtypes of these seven serotypes, denoted by a number following the serotype letter. Of these, a few serotypes have been commercialized or are in development.

The duration of action of each BoNT varies by subtype and is primarily determined by the LC (Pellett et al. 2018a). BoNT/A1 has the longest known duration of action compared with other serotypes and subtypes, as seen in a variety of head-to-head in vitro experiments and clinical trials (Sloop et al. 1997; Comella et al. 2005; Pellett et al. 2015; Whitemarsh et al. 2014; Eleopra et al. 1997, 1998; Kauffman et al. 1985), and accordingly is the most commonly used commercial subtype. Notably, the clinical duration of action does not directly translate from in vitro experiments and human compound muscular action potential experiments, although the rank order of serotypes generally remains the same. In addition, the duration of clinical benefit of the same BoNT can vary by indication (e.g., BoNT/A1 [onabotulinumtoxinA] has a median duration of 9 months in neurogenic detrusor overactivity (Kennelly et al. 2017) and 4 months in glabellar lines (Glogau et al. 2012)).

There are several commercially available type A1 BoNTs (Table 1), though it is important to note that they are not interchangeable due to different formulations and methods of production (Brin et al. 2014). This is reflected by the creation of unique nonproprietary names for each, as mandated by the Food and Drug Administration (Albanese 2011). OnabotulinumtoxinA, abobotulinumtoxinA, and prabotulinumtoxinA-xvfs are formulated with the core toxin and accessory proteins, though produced and purified under different conditions, while incobotulinumtoxinA lacks the accessory proteins. A few other type A1 BoNT preparations, not all of which have nonproprietary names, are mainly available in Asian and Latin American countries as well as Russia (Table 2). Some of these preparations reportedly do not contain the same amounts of active neurotoxin as others (Frevert et al. 2018), further highlighting the non-interchangeability of toxin preparations.

Other type A BoNT formulations are in development for commercial therapeutic and aesthetic use but are not yet licensed. DaxibotulinumtoxinA (RT002, Revance Therapeutics, Inc.) lacks accessory proteins and is formulated with a peptide (RTP004) that is reported to stabilize it from heat degradation more effectively than human serum albumin (Malmirchegini et al. 2018). A phase 2 study showed that immunogenicity rates are comparable to those of other BoNT/A products (Dressler et al. 2018), and trials in cervical dystonia (Jankovic et al. 2018) and glabellar lines (Carruthers et al. 2017; Kaufman-Janette et al. 2018) have been published. Chintox (Lanzhou Institute of Biological Products Co, Ltd) is another BoNT/A in development, also devoid of accessory proteins, that appeared to have a similar effect on rat gastrocnemius muscle as other type A BoNTs (Wuchao et al. 2018). MT10109L (Medytox and Allergan plc), a liquid formulation of BoNT/A that does not require reconstitution by the injector, is also in development for glabellar lines and lateral canthal lines (Kim et al. 2015).

RimabotulinumtoxinB is the only approved BoNT/B and is indicated for the treatment of cervical dystonia in the United States (Table 1). This type B BoNT is not used as often as the type A BoNTs due to its shorter duration of action and high immunogenicity at therapeutic doses (Comella et al. 2005; Jankovic et al. 2006; Brin et al. 1999).

There is some evidence in the literature showing that a single treatment of BoNT/C1 was effective in patients with dystonia, even in those who had developed neutralizing antibodies against BoNT/A (Eleopra et al. 1997, 2006). However, there are no studies ongoing or available in support of clinical development of this serotype.

A recent study testing BoNT/D in healthy volunteers found that a 110-fold higher protein dose of BoNT/D was required to show a similar clinical effect as incobotulinumtoxinA, as measured by compound muscle action potential of

Table 1 Commercially available BoNTs and their US-approved indications

BoNT type	Product name(s)	Nonproprietary name	Manufacturer	Approved indications (year of approval)
A	Botox®, Botox Cosmetic® (BOTOX® (onabotulinumtoxinA) [prescribing information] 2019; BOTOX Cosmetic® (onabotulinumtoxinA) [prescribing information] 2018)	OnabotulinumtoxinA	Allergan plc	Blepharospasm (1989) Strabismus (1989) Cervical dystonia (2000) Glabellar lines (2002) Axillary hyperhidrosis (2004) Adult spasticity (upper limb (2010); lower limb (2016)) Chronic migraine (2010) Neurogenic detrusor overactivity (2011) Overactive bladder (2013) Lateral canthal lines (2013) Forehead lines (2017) Pediatric upper limb spasticity (2019)
A	Dysport® (DYSPORT® (abobotulinumtoxinA) [prescribing information] 2018)	AbobotulinumtoxinA	Ipsen Ltd	Cervical dystonia (2009) Glabellar lines (2009) Pediatric lower limb spasticity (2016) Adult spasticity (2017)
A	Xeomin® (XEOMIN® (incobotulinumtoxinA) [prescribing information] 2019)	IncobotulinumtoxinA	Merz Pharmaceuticals, LLC	Blepharospasm (2010) Cervical dystonia (2010) Glabellar lines (2011) Adult upper limb spasticity (2015) Chronic sialorrhea (2018)
A	Jeuveau® (JEUVEAU (prabotulinumtoxinA-xvfs) [prescribing information] 2019)	PrabotulinumtoxinA-xvfs	Evolus, Inc.	Glabellar lines (2019)
B	Myobloc® (MYOBLOC® (rimabotulinumtoxinB) [prescribing information] 2009)	RimabotulinumtoxinB	US WorldMeds, LLC	Cervical dystonia (2000)

Table 2 Other botulinum toxins that are not approved in the United States but are commercially available in one or more other countries

BoNT type	Product name(s)	Nonproprietary name	Manufacturer	Indication
A	Botoshot®/Botulax®/Botulim®/Hugel Toxin®/Juvenlife®/Magnion®/Reage®/Regenox®/Zentox®	LetibotulinumtoxinA	Hugel, Inc.	Blepharospasm Glabellar lines Pediatric cerebral palsy with dynamic equinus foot deformity Poststroke upper limb spasticity
A	Botulift®/Cunox®/Meditoxin®/Neuronox®/Siax®	Neu-botulinumtoxinA	Medytox Inc.	Blepharospasm Glabellar lines Lateral canthal lines Pediatric cerebral palsy with dynamic equinus foot deformity Poststroke upper limb spasticity
A	BTXA®/Dituroxal®/Lantox®/Lanzox®/Liftox®/Prosigne®/Redux®	–	Lanzhou Institute of Biological Products (Hengli)	Blepharospasm Glabellar lines Hemifacial spasm Strabismus
A	Coretox®	–	Medytox Inc.	Glabellar lines
A	Hutox	–	Huons Global	Glabellar lines
A	Innotox®	NivobotulinumtoxinA	Medytox Inc.	Glabellar lines
A	Nabota®/Nuceiva®	PrabotulinumtoxinA	Daewoong	Blepharospasm Glabellar lines Lateral canthal lines Upper limb spasticity
A	Relatox®	–	Microgen laboratories	Blepharospasm Facial expression wrinkles Post-apoplectic complications

Note that in some countries, products may be approved for additional indications the extensor digitorum brevis in human volunteers, with a duration of action half that of incobotulinumtoxinA (Kutschenko et al. 2019).

EB-001 is a native 150 kDa BoNT/E that has been tested in a phase 2 study on glabellar frown lines, where it appeared to have a fast onset of action and a duration of 14–30 days (Yoelin et al. 2018). A separate, recombinant BoNT/E exhibited a faster onset of effect, greater peak effect, and good safety profile yet a shorter duration of effect compared with abobotulinumtoxinA in a small double-blind, placebo-controlled trial in which the extensor digitorum brevis muscle was injected in healthy volunteers (Pons et al. 2018).

BoNT/F was assessed in several clinical trials in the 1990s in patients who had developed resistant antibodies to BoNT/A (Ludlow et al. 1992; Greene and Fahn 1993). Although it was effective in treating dystonia, its use was limited by a short duration of effect and was not further developed for commercial use.

4 Mosaic BoNTs

A few naturally occurring mosaic BoNTs have been identified, so named because they appear as hybrids of two serotypes. (The term "mosaic" is generally used for BoNT hybrids occurring in nature, while "chimera" is for those created in a laboratory using molecular techniques.) In particular, mosaics resulting from interserotype recombination between BoNT/C and BoNT/D have been isolated from cases of animal botulism. BoNT/CD has LC and H_N domains similar to those of BoNT/C and an H_C domain similar to BoNT/D (Takeda et al. 2005). In contrast, BoNT/DC has an LC and H_N almost identical to those of BoNT/D and an H_C most similar to that of BoNT/C (Moriishi et al. 1996). Another mosaic toxin (known as BoNT/FA, BoNT/H, or BoNT/HA) has been identified in the last few years and is discussed below.

5 Newly Identified Native BoNTs

With the low cost of sequencing techniques combined with advances in bioinformatic methods, many more native BoNTs and putative BoNTs have been identified across a range of bacterial species. While the seven classical serotypes have all demonstrated lethality in primates (Webb 2018), some of the newly identified BoNTs have no established toxicity, and for some it is unclear whether they are naturally expressed. These newly identified BoNTs and BoNT-like sequences may have useful biomedical applications, e.g., if their LCs have differential specificities for target proteins and/or confer different durations of action or if their HCs recognize novel cellular receptors. Of note, discoveries of some of the same toxins have been made in different laboratories, and there is ongoing discussion on the nomenclature.

5.1 New BoNTs from Clostridial Strains

5.1.1 BoNT/FA (Also Known as BoNT/H and BoNT/HA)

In 2013, an additional serotype designated as BoNT/FA was described from a clostridial strain isolated from a patient with infant botulism that also produced BoNT/B (Barash and Arnon 2014). Like the classical toxins, the toxin was classified as a new serotype due to the inability of antisera against the classical serotypes to neutralize the toxin in the mouse bioassay. Through DNA sequencing, BoNT/FA was revealed as a mosaic toxin, with its LC 86% identical to that of BoNT/F5 and its H_C 85% identical to that of BoNT/A (Dover et al. 2014). Subsequently, it was reported that the toxin could be neutralized with BoNT/A antitoxins (Maslanka et al. 2016). Interestingly, this mosaic toxin does not cleave SNAP-25 like other BoNT/A toxins, but rather cleaves VAMP2 in the same location as BoNT/F5 (Fig. 2) (Kalb et al. 2015). This VAMP2 cleavage activity was very high in cultured human neurons, in contrast with its low potency in mice compared with BoNT/A1 (Pellett et al. 2016). Subsequent work showed that the toxin was also able to cleave VAMP1 and VAMP3, with a duration of action in mice comparable to that of BoNT/B1 (Hackett et al. 2018; Pellett et al. 2018b).

5.1.2 BoNT/X

BoNT/X was the first BoNT serotype identified via bioinformatic searches of published genomic sequences (Zhang et al. 2017). The BoNT/X sequence was found in the genome of *C. botulinum* strain 111 and is more distantly related than the other BoNTs (Fig. 3). This strain, isolated in 1996 from a case of infant botulism in Japan (Kakinuma et al. 1996), was found to also express a plasmid-encoded BoNT/B2 that appeared to be responsible for the virulence of the strain, since strains cured of the plasmid were no longer toxigenic (Hosomi et al. 2014). BoNT/X is not recognized by antisera against known BoNTs and only has 27.5–31.3% sequence identity to the seven classical BoNTs (Zhang et al. 2017). Similar to the classical BoNT serotypes, BoNT/X contains the zinc metalloprotease HExxH motif and the HC SxWY motif involved in ganglioside recognition. The crystal structure of the LC of BoNT/X shows that it contains the same core fold present in all BoNTs (Masuyer et al. 2018). Like BoNT/B/D/F/G, BoNT/X inhibits vesicular release through the cleavage of VAMP1, VAMP2, and VAMP3 (Fig. 2). Furthermore, unlike other known BoNTs, BoNT/X also cleaves VAMP4, VAMP5, and Ykt6 (Zhang et al. 2017). The *bontX* gene is located in a gene cluster encoding an upstream NTNH homolog as well as *p47* and *orfX1*, *orfX2*, and *orfX3* genes, though the latter three are in an opposite orientation than usual. Uniquely, BoNT/X gene cluster also has a second OrfX2 gene (OrfX2b), located downstream of the BoNT/X gene. To our knowledge, it is currently unclear whether BoNT/X is naturally expressed.

5.2 New BoNTs from Non-clostridial Strains

Recently, homologs to BoNTs have been discovered in non-clostridial strains through bioinformatic screens for sequences with similarities to individual BoNT domains in available genomes. Through this process, 161 predicted BoNT sequences have been identified (Mansfield et al. 2019), only a few of which have been characterized and are summarized below.

5.2.1 BoNT/Wo

The first identified non-clostridial BoNT, BoNT/Wo, is encoded by *Weissella oryzae* SG25T, an anaerobe isolated from fermenting rice. It was identified as two adjacent open reading frames (ORFs) via mining of available genomes and metagenomes for multi-domain homology to clostridial neurotoxins (Mansfield et al. 2015). ORF1 (BoNT/Wo) contains the four BoNT domains: the protease domain, which includes the conserved HExxH zinc metalloprotease motif, the translocation domain, and the N- and C-terminal binding subdomains. ORF2 has been characterized as NTNH-like with sequences matching the N-terminal protease domain, residues partially matching to the translocation domain, and the binding domain replaced with two bacterial immunoglobulin-like (Big 3) domains. The sequence identity (~16–19%) of BoNT/Wo is lower than the range for other BoNTs (Mansfield et al. 2015), and the Cys residues that form the interchain disulfide bond in BoNTs are not conserved (Tehran and Pirazzini 2018). Unlike other clostridial gene clusters, there are no genes for *botR*, *ntnh*, *ha*, or *orfx*. ORF1/ORF2 are hypothesized to have originated in *W. oryzae* by lateral gene transfer, which is plausible since *Clostridia* also grow in fermented rice. Recombinant LC and HC of BoNT/Wo were overexpressed and purified from *Escherichia coli*; neither reacted with polyclonal antisera of the seven classical BoNT serotypes (Zornetta et al. 2016). Purified recombinant BoNT/Wo LC containing the zinc metalloprotease domain was found to cleave VAMP2 at a unique site, a Trp-Trp bond in the juxtamembrane segment of VAMP necessary for release of neurotransmitters (Zornetta et al. 2016) (Fig. 2). Thus far, there are no reports in the literature on characterization of the activity of the full-length protein, its expression in *W. oryzae*, or its potential as a foodborne pathogen. The utility of BoNT/Wo to *W. oryzae* is currently unknown, but may be used as a defense mechanism against bacteria-eating amoebae or worms, or it may target SNARE-mediated plant defense systems, all of which contain the juxtamembrane Trp-Trp bond (Mansfield et al. 2015; Zornetta et al. 2016).

5.2.2 Cp1 Toxin

A toxin family from *Chryseobacterium piperi*, a Gram-negative bacterium isolated from a creek in Pennsylvania (Strahan et al. 2011; Wentz et al. 2017), was identified in the aforementioned bioinformatics screen (Mansfield et al. 2019). The putative toxin has low identity to BoNT/A (18%) compared with the identity among BoNT family members (≥28%), but the LC has homology to multiple BoNT motifs, including the HExxH zinc-coordinating active site, the Glu261 zinc ligand, Glu350 in the active site, active site stabilizing motif R363-x-x-Tyr366, and the two Cys that form the disulfide bond between the LC and HC (Mansfield et al. 2019). In the HC, there is significant similarity to the translocation domain in BoNT and diphtheria toxin, and the receptor-binding domain is predicted to have the same fold as the BoNT H_C domain. The putative toxin genes are flanked by putative *ntnh* and *bont* genes, but not *ha*, *p47*, or *orfX*. The LC did not cleave VAMP2, SNAP-25, or syntaxin 1, but expression of the LC in human embryonic kidney HEK 293T cells caused cell death. Expression of LC with point mutations in the HExxH motif did not cause cell death, indicating that the toxicity is dependent on the metalloprotease activity, but the natural targets for this BoNT homolog are currently unknown. As with BoNT/X and BoNT/Wo, it is currently unknown whether these putative toxins from *C. piperi* are naturally expressed.

5.2.3 BoNT/En (eBoNT/J)

The first complete BoNT gene cluster encoding the neurotoxin (named eBoNT/J), accessory proteins, *ntnh* gene, and *orfX* was recently identified from the Gram-positive *Enterococcus* sp. 3G1_DIV0629, a strain isolated from cow feces in South Carolina, through a search of available whole genome sequences (Brunt et al. 2018). Around the same time, another group identified this gene cluster on a plasmid from a commensal strain of *Enterococcus faecium* encoding the same neurotoxin, which they named BoNT/En (Zhang et al. 2018). Recombinantly purified LC and HC of this toxin were not recognized by antisera against BoNT/A-G or BoNT/X, conferring the designation of a novel BoNT serotype (Zhang et al. 2018). This newly identified *boNT* gene cluster contains a lower G+C content than the rest of the *Enterococcus* sp. genome and is bordered by insertion sequence elements, indicating acquisition through horizontal transfer (Brunt et al. 2018). This BoNT has the typical domains for the zinc metalloprotease, translocation, and target cell attachment binding domain and is 39% identical to its closest relative, BoNT/X (Brunt et al. 2018). 3D structure modeling shows that BoNT/En (eBoNT/J) is predicted to closely match the structure of BoNT/A (Brunt et al. 2018). The LC of BoNT/En (eBoNT/J) can cleave VAMP2 and SNAP-25 but at sites unique from those cleaved by known BoNTs (Zhang et al. 2018) (Fig. 2). Full-length BoNT/En, produced through sortase-mediated ligation of the protease/translocation domains with the binding domain, retained ability to cleave VAMP2 and SNAP-25 in rat cortical neurons, yet was unable to induce paralysis in mouse limbs in the digit abduction score (DAS) assay (Zhang et al. 2018). In contrast, when the BoNT/En binding domain was replaced with the corresponding domain from BoNT/A, the sortase-mediated full-length chimeric toxin was more efficient in cleaving its SNARE targets in the rat cortical neurons and successfully induced paralysis in the mouse DAS assay, suggesting a lack of high-affinity BoNT/En receptors on rat/mouse neurons (Zhang et al. 2018). Despite these interesting observations with recombinant versions of BoNT/En, it is not known whether this toxin gene cluster is expressed in *E. faecium* under natural growth conditions, as it is not apparent that cows from which the feces were obtained were exhibiting botulism symptoms. Although the potency of BoNT/En on human neurons is currently not known, the identification of this novel native toxin that cleaves multiple SNARE proteins is cause for concern. Although *E. faecium* serve as commensal bacteria, they are also causative agents of hospital-acquired, multidrug-resistant infections, so the potential ability to produce a toxin is problematic (Zhang et al. 2018).

5.2.4 PMP1

PMP1 is a neurotoxin isolated from a strain of *Paraclostridium bifermentans*, bacteria that colonize *Anopheles* mosquitoes (Contreras et al. 2019). The gene encoding this toxin is located on a plasmid within an operon that also encodes NTNH, OrfX, and P47 proteins, flanked by insertion sequences and transposon elements. PMP1 is 36% and 34% identical to BoNT/X and BoNT/En, respectively. Expression of PMP1 alone in *Bacillus thuringiensis* showed no toxicity to *Anopheles* mosquito larvae, but co-expression of the NTNH and PMP1 proteins led to 33% mortality, while co-expression of NTNH, PMP1, and the OrfX proteins led to 70% mortality. Importantly, this establishes a role of the OrfX proteins in toxicity. Injection of recombinant PMP1 (to bypass the gut barrier) inhibited the ability to fly in adult *Aedes* and *Anopheles* mosquitoes as well as in the fruit fly *Drosophila*, and injected PMP1 was toxic to larvae of *Aedes* and *Anopheles* mosquitoes. However, PMP1 with a mutation in the HExxH motif had no effect, indicating that PMP1 is a metalloprotease and that this activity is essential for its toxicity. PMP1 cleaved mosquito-derived syntaxin, and this also required the HExxH motif. Interestingly, human syntaxin was not cleaved by PMP1 despite a conserved cleavage site, though it appears that nearby nonpolar amino acids prevent PMP1-mediated cleavage.

The crystal structure of the H_C of PMP1 showed that, although the overall fold was similar to that of other clostridial neurotoxins, the binding domain exhibits variation. PMP1 has the ganglioside-binding site motif of SxWY, but it is not in a clear binding pocket. Furthermore, mutations in the SxWY motif did not decrease toxicity, indicating that, unlike in BoNTs, this motif is not essential for toxicity.

6 Recombinantly Engineered BoNTs

As discussed above, recombinant techniques, including site-directed mutations, have been utilized to characterize many of the novel native BoNT and BoNT-like proteins. There are many advantages to using recombinant techniques to express BoNTs in more tractable organisms such as *E. coli*. This can be useful for the new and potential BoNTs identified via bioinformatic methods, in which it may not be known whether the protein is expressed in the native organism. In addition, recombinant overexpression of a BoNT can enable large-scale production for manufacturing purposes. As previously mentioned, some *C. botulinum* strains express more than one toxin serotype, and expression of each independently as a recombinant molecule in *E. coli*, for example, facilitates biological characterization of each toxin.

With the wealth of techniques available in molecular biology, biochemistry, and genetics, a number of novel recombinant BoNTs have been designed for a variety of scientific and clinical applications. Using different LC subtypes, it is possible to customize a BoNT to have specific properties depending on the therapeutic goal. In addition, recombinant engineering of the H_C subunit, or even replacement of the heavy chain with a molecule that binds to a specific receptor, can be used to target different cell types, as will be discussed. A multitude of engineered BoNTs can be constructed to help to fully maximize the therapeutic potential of BoNTs as stand-alone drugs, vaccines, and therapeutic vehicles, several of which will be reviewed below.

6.1 Molecular Design Considerations for Recombinant Production of BoNTs

A number of factors need to be considered in designing novel recombinant BoNTs. Heterologous expression is often employed for ease of production of recombinant BoNTs and also to avoid the hazards of overexpressing toxins in bacteria that can form spores, which are notoriously difficult to eradicate. However, clostridial genes tend to have a high percentage of adenine and thymine nucleotides, which can cause expression problems due to codon usage bias in heterologous hosts. Codon bias is particularly important in prokaryotic expression of heterologous proteins due to the correlation of preferred codons with the concentration of corresponding tRNAs. Codon optimization has been demonstrated to increase heterologous protein expression up to >1,000-fold (Gustafsson et al. 2004). To circumvent this, codon optimization of BoNT genes for the specific host strain using synthetic DNA technologies is a standard approach for efficient production of recombinant BoNTs (Webb 2018).

Another important consideration for BoNT production is the dichain nature of the mature protein. As noted above, native BoNTs are produced as a single-chain protein that is then proteolytically cleaved to produce a dichain comprising an LC protease domain covalently linked to the HC targeting domain via a disulfide bond. While the single-chain BoNT proteins are capable of entering neurons and inhibiting vesicular release to some extent, they are less potent and hence are sometimes employed as a biosafety precaution when studying novel recombinant toxins (Dolly and Wang 2015). The dichain format with an intact disulfide bond is required for full neurotoxicity of BoNT (Schiavo et al. 1990; de Paiva et al. 1993). More recently, it has been demonstrated that the presence of a disulfide bond linking the LC and HC is required for channel formation and delivery of the LC to the cytosol (Simpson et al. 2004; Fischer and Montal 2007; Pirazzini et al. 2011).

When producing recombinant BoNTs as a single-chain molecule, a mechanism of converting it to an activated dichain needs to be considered. While some *C. botulinum* strains are proteolytic (Group I; e.g., A1-8, B1-3, B5-9, F1-5) and can perform this cleavage, other non-proteolytic strains (Group II; e.g., B4, E1-3, E6-11, F6) rely on proteases within the environment or the intoxicated host for activation to the dichain form (Smith 2014; Rummel 2015). As the nicking loops of native BoNTs tend to be rich in lysine and arginine residues, activation of native single-chain BoNTs to the dichain form has been reported with non-clostridial enzymes such as trypsin and the endoprotease Lys-C (Duff et al. 1956; Gerwing et al. 1965; Eklund and Poysky 1972; Ohishi and Sakaguchi 1977; Kozaki et al. 1985; Antharavally and DasGupta 1997). Similarly, activation of recombinant BoNTs has been accomplished with non-specific proteases such as trypsin or Lys-C (Hackett et al. 2018; Chaddock et al. 2002; Bade et al. 2004; Gilmore et al. 2008; Wang et al. 2008, 2012a, b; Elliott et al. 2019). However, even when incubation times and temperatures are carefully controlled, activation to the dichain with non-specific proteases without undesired secondary proteolysis outside of the nicking loop can be a challenge (Hackett et al. 2018). In fact, secondary trypsin proteolysis of BoNT/A outside of the nicking loop has even been exploited for production of BoNT fragments (Chaddock et al. 2002). In an effort to address or limit non-specific proteolysis outside of the targeted nicking loop, nicking loops containing substrate sequences for specific proteases have been engineered. Examples of selective proteases for activation of engineered BoNTs to the dichain include enterokinase (Wang et al. 2012b; Band et al. 2010; Masuyer et al. 2015), thrombin (Wang et al. 2012a, b; Höltje et al. 2013; Weisemann et al. 2015; Kutschenko et al. 2017; López de la Paz et al. 2018), factor Xa (Masuyer et al. 2011, 2015; Somm et al. 2012), and tobacco etch virus protease (Band et al. 2010; Vazquez-Cintron et al. 2014).

In addition to the disulfide bond requirements for a fully functional recombinant BoNT, the protein interface between the LC and HC domains with a large unstructured HC loop that wraps around the LC domain (often referred to as the "belt") needs to be retained for maximal potency (Lacy et al. 1998; Swaminathan and Eswaramoorthy 2000; Kumaran et al. 2009). The belt is a pseudo-substrate that resides in the unusually large substrate-binding cleft of the BoNT protease, blocking substrate binding until the LC is delivered (Chen and Barbieri 2014). It has also been postulated that the belt may serve to protect the LC from proteolysis by host proteases or auto-proteolysis (Chen and Barbieri 2014).

Recombinant BoNTs can be expressed or overexpressed in a variety of systems. The large size of the holotoxin (150 kDa), multiple domains, presence of the linker between the LC and HC, and redox-sensitive disulfide bonding are all considerations for efficient expression of properly folded and functional protein. Whereas all of the BoNTs for commercial use and some for basic science studies are expressed in *C. botulinum* for production (Brin et al. 2014; Pellett et al. 2016), many basic science studies have utilized *E. coli* to express codon-optimized BoNTs. In addition, successful expression of full-length codon-optimized BoNT/A1, /B1, /C1, and /E1 has been achieved in the yeast *Pichia pastoris* (Webb et al. 2009, 2017). Furthermore, the baculovirus insect cell system has been demonstrated to be a viable expression platform for BoNT, enabling production of recombinant molecules that preserve the intraneuronal trafficking properties of native BoNTs (Band et al. 2010; Vazquez-Cintron et al. 2016). Recent strategies, including shuttle vectors, have facilitated the expression of recombinant BoNT from nontoxigenic *C. botulinum* strains, including BoNT/A4 (Bradshaw et al. 2014). In an effort to decipher the protein elements controlling duration of action, neuronal cell entry, and pathology, non-native hybrids of BoNT/A1 and BoNT/A3 have also been produced from engineered strains of *C. botulinum* (Pellett et al. 2018a).

Another consideration in the production of recombinant BoNTs is that the core 150 kDa toxin is typically overexpressed without accessory or complexing proteins. As noted above, the functions of some, but not all, of these accessory proteins are known.

6.2 Vaccines Against BoNT

Engineered BoNTs have played a large role in the development of vaccines. Although it is possible to mass produce functional full-length BoNTs for use as antigens, it is not always preferable due to biosecurity concerns. The first BoNT vaccine, a formalin-inactivated, pentavalent (BoNT/A-E) toxoid, was developed in the 1960s to prevent botulism, particularly against use of BoNT as a biological weapon. This vaccine was in use for more than 30 years before the Centers for Disease Control and Prevention discontinued its use due to decreased immunogenicity (Centers for Disease Control and Prevention (CDC) 2011). A few vaccines have more recently been developed using full-length BoNTs with mutations rendering them nontoxic. An early engineered BoNT was a full-length BoNT/C with three site-directed point mutations in the zinc binding motif of the LC, which rendered it atoxic to mice and unable to cleave syntaxin. Administration of this recombinant toxin or its purified LC to mice resulted in antibody production and protective immunity (Kiyatkin et al. 1997). Several other full-length BoNT/As were rendered nontoxic due to point mutations in the LC and conferred protection to mice immunized against them (Pier et al. 2008; Ravichandran et al. 2016; Przedpelski et al. 2018).

Some vaccines have been developed using only the LC or the HC fragments, but these do not appear to be as effective as the full-length toxin. Catalytically inactive BoNT/A1 in which the histidines and glutamic acid within the conserved LC active site HExxH motif were substituted with nonreactive alanines were shown to confer a more robust immunological response than recombinant BoNT-LC, LC-belt, LC-H_N, and H_C antigens (Webb et al. 2009). This work was later extended to include the full-length BoNT/B1, C1, E1, and F1 subtypes with corresponding alanine substitutions in each HExxH motif (Webb et al. 2017).

6.3 Assembling BoNTs from Individual Subunits

Due to safety precautions and/or technical challenges, a number of approaches have been evaluated to reconstitute functional BoNTs from individual subunits. The earliest reports of this involved creating a modified BoNT holotoxin through reconstituting the LC with the HC (Maisey et al. 1988; Zhou et al. 1995). However, this approach has not been widely implemented, likely due to inefficient reconstitution via a tedious process that requires denaturation and refolding of the LC and HC components. Additionally, this process was not successful when attempting to reconstitute a heterologous LC/A-HC/B molecule (Maisey et al. 1988). More recently, sortase-mediated ligation of adjacent glycine residues was employed to assemble full-length BoNT/X holotoxin from two nontoxic recombinant subunits (Zhang et al. 2017). A "protein stapling" technology has also been developed in which the BoNT/A LC-H_N and H_C domains are expressed separately and joined together via fused SNARE peptides. Animal studies demonstrated that the reassembled toxin (also called BiTox) was able to inhibit nerve function yet was not systemically toxic even at high doses (Darios et al. 2010; Ferrari et al. 2011). Recombinant BoNT engineering has been taken a step further than production of the isolated 150 kDa holotoxin as recombinant variants of the higher-order BoNT/A complex have been produced. An initial report involved recombinant co-expression of an inactive form of BoNT/A and recombinant NTNHA to form the minimal progenitor toxin complex (M-PTC, ca. 300 kDa) (Gu et al. 2012). In that report, acidic residues within BoNT/A that serve as pH sensors were mutated and demonstrated to impact the stability of the M-PTC complex in a pH-dependent manner. Subsequently, reconstitution of a recombinant version of the large progenitor toxin complex (L-PTC, ca. 760 kDa, 14 subunits in the complex) was reported, with BoNT/A, NTNHA, HA70, HA17, and HA33 in a subunit stoichiometry of 1:1:3:3:6, respectively (Lee et al. 2013). This effort provided the first high-resolution structural model of the L-PTC. Additionally, studies with the L-PTC and sub-components of the complex have provided insight into the mechanism of oral toxicity (Lee et al. 2014).

6.4 BoNT as a Delivery Vehicle for Other Proteins to Neurons

Because the H_C of BoNTs efficiently delivers the BoNT protein intracellularly into neurons, this feature can be exploited to use BoNTs as a way to deliver other proteins to neurons. The feasibility of BoNTs as cellular delivery vehicles to neurons was first shown by fusions to the N-terminus of full-length BoNT/D with dihydrofolate reductase, green fluorescent protein, BoNT/A LC, or luciferase (Bade et al. 2004). The creation of atoxic BoNT/A (e.g., inactivated by LC protease mutations E224A and Y366A) has enabled the study of trafficking and neuronal delivery of cargo proteins via BoNT without the dose limitations that are associated with native BoNT/A. These have been shown to bind to but not cleave SNAP-25 (Band et al. 2010; Vazquez-Cintron et al. 2014; Pellett et al. 2011). These constructs also contain a peptide sequence at their N-terminus that allows for the attachment of cargo proteins that could be transported to neurons. An atoxic BoNT/C1 has also been constructed for use as a vehicle to deliver other molecules to neurons (Vazquez-Cintron et al. 2017). A full-length BoNT/B with inactive protease (BoTIM) fused to core streptavidin and linked to biotinylated liposomes has successfully been utilized to deliver liposome-encapsulated cargo into mouse neurons (Edupuganti et al. 2012).

6.5 Tailored BoNTs with Altered Properties

Some recent BoNT engineering efforts have focused on the treatment of pain, either through the use of sensory neuron targeting motifs or through optimization of the pain peptide blockade. For example, a chimera of anthrax toxin and BoNT is being investigated to utilize the anthrax toxin-mediated nociceptor binding combined with the ability of BoNT to cleave SNARE complexes (Yang et al. 2018). In addition, a fusion of an IgG-binding domain from SpA-B (the IgG-binding domain B from *Staphylococcus aureus* virulence factor protein A) to the LC and H_N of BoNT/A, when coupled to anti-tropomyosin kinase A (TrkA) IgG or recombinant fragment crystallizable nerve growth factor, led to targeting to nociceptors expressing TrkA (Nugent et al. 2017). In another example, senrebotase (AGN-214868), a retargeted endopeptidase, has been tested in phase 2 trials for overactive bladder (ClinicalTrials.gov, NCT01157377) and postherpetic neuralgia (ClinicalTrials.gov, NCT01129531). However, this molecule is no longer in development. Finally, a novel BoNT was engineered to maximize the superior ability of the BoNT/E LC to inhibit pain neurotransmitter release with the long duration of LC/A by fusing LC/E to the N-terminus of BoNT/A (Wang et al. 2017). This novel toxin expresses a dual protease fusion and has been demonstrated to be more efficacious than BoNT/A or even systemic pregabalin in neuropathic pain models and has the potential to become a long-acting anti-hyperalgesic for chronic pain conditions (Wang et al. 2017; Nugent et al. 2018).

Engineered BoNTs have also been designed as therapeutics involving non-neuronal cell types. Specifically, one of the treatments recently under investigation for acromegaly is an engineered BoNT targeted to the growth hormone-releasing hormone (GHRH) receptor (Maffezzoni et al. 2016a, b). This targeted secretion inhibitor is composed of a GHRH receptor targeting domain to enable internalization into target cell endosomes, a BoNT/D LC that cleaves VAMP, and a BoNT/D H_N domain that enables transport of the LC into the cytosol. This allows for the inhibition of VAMP-dependent exocytosis of growth hormone vesicles. Intravenous injection in the tail veins of rats resulted in decreases in body weight, body length, growth hormone peaks, production of insulin-like growth factor, and pituitary levels of growth hormone mRNA and protein, indicating specificity for inhibition of hypothalamic GHRH production (Somm et al. 2013). Notably, this inhibitory action was transient, which may have therapeutic value.

Another alteration in BoNT with the goal of trafficking to alternative cell types was a fusion of the cell-penetrating peptide TAT to the LC of BoNT/A. This fusion could enter living cells in vitro and in vivo in a mouse model (Saffarian et al. 2016a, b). However, the utility and safety of such a protein lacking specificity for a certain cell type may prove to be problematic for therapeutic use.

A number of BoNTs have also been engineered to study the general properties of toxins such as their potency and duration of action, some of which may be exploited in the future to generate novel therapeutics. For example, through the creation of recombinant variants, Scheps et al. demonstrated that the C-terminus of the LC plays a role in controlling both the onset and duration of action of BoNT/A1 and that an engineered BoNT with three amino acid changes can have faster onset and shorter duration of action than the wild-type BoNT/A1 (Scheps 2017). In a study designed to address the role of the BoNT/A3 binding and light chain domains in potency and duration, another group created chimeras of BoNT/A1 and BoNT/A3 subtypes that possess unique therapeutic potential while retaining sensitivity to available antiserum (Pellett et al. 2018a). A BoNT/AB chimera, constructed to combine the duration and potency of BoNT/A with the higher neuronal binding affinity of BoNT/B, was shown to have a similar duration as BoNT/A in murine running wheel assay (Kutschenko et al. 2017). With the same aim in mind, a different BoNT/AB chimera was constructed and induced a longer paralysis in mice than BoNT/A (Wang et al. 2012a). Another chimera, of BoNT/F7 in which the activation loop was replaced with that of BoNT/F1 to reduce further proteolysis after posttranslational activation, showed enhanced potency compared with BoNT/F1 (Burgin et al. 2018).

Other recombinant toxins have been engineered to differentially cleave SNARE proteins with the goal of developing BoNT-based therapeutics applicable to non-neuronal cells. Chen and Barbieri were the first group to engineer mutations in the LC of a BoNT molecule, and these conferred the ability to cleave SNAP-23, a non-neuronal SNARE, opening the potential for therapeutics targeting SNAP-23 in diseases involving non-neuronal hypersecretion (Chen and Barbieri 2009). Whereas Chen and Barbieri used BoNT/E as the base molecule to evolve SNARE specificity toward SNAP-23, another group identified SNAP-23-cleaving variants through mutagenesis of BoNT/A SNARE-binding pockets (Sikorra et al. 2016). More recently, a yeast-based assay was utilized to identify BoNT/A variants with enhanced SNAP-23 cleavage activity (Binz et al. 2018). Other work has shown that phage-assisted continuous evolution of the BoNT/F LC could identify variants with novel hydrolytic activities, including some able to cleave VAMP-7. This could lead to the engineering of BoNTs able to cleave different substrates, ultimately for therapeutic use (Foster et al. 2018). In addition, mutants in BoNT/C, the only serotype able to cleave both SNAP-25 and syntaxin, showed decreased ability to cleave SNAP-25, leading to decreased lethality compared with the wild-type BoNT/C. The remaining ability to cleave syntaxin results in notable but incomplete neuromuscular functionality, which may be of clinical use as a toxin with a long duration of action and favorable safety profile (Zanetti et al. 2017). Of note, a BoNT/B1 LC with an enhanced ability to cleave VAMP-1 and VAMP-2 in vitro was not more potent than BoNT/B1 in physiological systems (Elliott et al. 2017).

Some work has been performed to alter the translocation ability of BoNTs. In one study, BoNT/B with a mutated translocation domain showed greater neurotoxicity. The LC was able to reach the cytosol more quickly because membrane translocation could occur at a higher pH (Pirazzini et al. 2013). In addition, chimeras of BoNT/A and BoNT/E in which the H_C were exchanged have been created, revealing differences in translocation time and muscle weakening that may be of use therapeutically (Wang et al. 2008).

In studies of binding domain engineering, recombinant BoNT/B1 proteins with point mutations for increased affinity to the human synaptotagmin II receptor (called BoNT/B_{MY}) were created. Results from hemidiaphragm assays indicate that they were equipotent to BoNT/A, with similar side effect profiles in animal models. Since the clinical efficacy of native BoNT/B is limited by its weaker binding to neuronal receptors, these modified BoNT/B molecules may have clinical use (Elliott et al. 2019; Tao et al. 2017). A chimera was created in which the binding domain from BoNT/B_MY (with enhanced ability to bind to human synaptotagmin II) replaced that of BoNT/X. This chimera appeared to be more potent than BoNT/A or BoNT/B_MY in a cortical neuron SNARE cleavage assay, cleaving both VAMP4 and VAMP2, but was less potent in the mouse hemidiaphragm assay (Beard et al. 2018).

7 Conclusions

In summary, BoNT toxins have powerful therapeutic utility, as shown by the commercially available native forms. The discovery of novel native BoNTs in recent years, as well as the abundance of putative toxins identified through bioinformatic techniques, is exciting because they may lead to new understandings in both biology and medicine. In addition, the modular structure of BoNTs allows for separation of the distinct functions of the domains, which in turn means that the numbers of possible engineered fusions and mutations using molecular techniques are almost limitless. Of course, all recombinant BoNT engineering must be conducted with the utmost consideration of safety.

Acknowledgments The authors thank Dina Anderson, Birgitte Jacky, Mariana Nelson, and Edwin Vazquez-Cintron (Allergan plc) for reviewing the chapter and Maria Rivero (Allergan plc) for the graphics. Medical writing and editorial assistance was provided by Jennifer L. Giel, PhD, on behalf of Evidence Scientific Solutions, Inc, Philadelphia, PA, and was funded by Allergan plc, Dublin, Ireland.

References

Albanese A (2011) Terminology for preparations of botulinum neurotoxins: what a difference a name makes. JAMA Dermatol 305:89–90

Antharavally BS, DasGupta BR (1997) Covalent structure of botulinum neurotoxin type E: location of sulfhydryl groups, and disulfide bridges and identification of C-termini of light and heavy chains. J Protein Chem 16:787–799

Bade S, Rummel A, Reisinger C, Karnath T, Ahnert-Hilger G, Bigalke H et al (2004) Botulinum neurotoxin type D enables cytosolic delivery of enzymatically active cargo proteins to neurones via unfolded translocation intermediates. J Neurochem 91:1461–1472

Band PA, Blais S, Neubert TA, Cardozo TJ, Ichtchenko K (2010) Recombinant derivatives of botulinum neurotoxin A engineered for trafficking studies and neuronal delivery. Protein Expr Purif 71:62–73

Barash JR, Arnon SS (2014) A novel strain of *Clostridium botulinum* that produces type B and type H botulinum toxins. J Infect Dis 209:183–191

Beard M, Burgin D, Hornby F, Donald S, Krupp J, Dong M et al (2018) Recombinant expression and characterization of a botulinum neurotoxin serotype X chimera. Toxicon 156 Suppl 1:S8

Binz T, Beard M, Sikorra S, Schwede S, Elliott M, Donald S et al (2018) Mutations in light chain of botulinum neurotoxin a enable cleavage of human SNAP-23. Toxicon 156 Suppl 1:S10
BOTOX Cosmetic® (onabotulinumtoxinA) [prescribing information] (2018) Irvine: Allergan Pharmaceuticals Ireland, a subsidiary of Allergan, Inc.
BOTOX® (onabotulinumtoxinA) [prescribing information] (2019) Irvine: Allergan Pharmaceuticals Ireland, a subsidiary of Allergan, Inc.
Bradshaw M, Tepp WH, Whitemarsh RC, Pellett S, Johnson EA (2014) Holotoxin activity of botulinum neurotoxin subtype A4 originating from a nontoxigenic *Clostridium botulinum* expression system. Appl Environ Microbiol 80:7415–7422
Brin MF, Lew MF, Adler CH, Comella CL, Factor SA, Jankovic J et al (1999) Safety and efficacy of NeuroBloc (botulinum toxin type B) in type A–resistant cervical dystonia. Neurology 53:1431–1438
Brin MF, James C, Maltman J (2014) Botulinum toxin type A products are not interchangeable: a review of the evidence. Biologics 8:227–241
Brunt J, Carter AT, Stringer SC, Peck MW (2018) Identification of a novel botulinum neurotoxin gene cluster in *Enterococcus*. FEBS Lett 592:310–317
Burgin D, Périer C, Elliot M, Hornby F, Mir I, Beard M (2018) New modified recombinant botulinum neurotoxin type F with enhanced potency. Toxicon 156(Suppl 1):S12–SS3
Burke GS (1919) Notes on bacillus botulinus. J Bacteriol 4:555–70.1
Carruthers J, Solish N, Humphrey S, Rosen N, Muhn C, Bertucci V et al (2017) Injectable daxibotulinumtoxinA for the treatment of glabellar lines: a phase 2, randomized, dose-ranging, double-blind, multicenter comparison with onabotulinumtoxinA and placebo. Dermatol Surg 43:1321–1331
Centers for Disease Control and Prevention (CDC) (2011) Notice of CDC's discontinuation of investigational pentavalent (ABCDE) botulinum toxoid vaccine for workers at risk for occupational exposure to botulinum toxins. MMWR Morb Mortal Wkly Rep 60:1454–1455
Chaddock JA, Herbert MH, Ling RJ, Alexander FC, Fooks SJ, Revell DF et al (2002) Expression and purification of catalytically active, non-toxic endopeptidase derivatives of *Clostridium botulinum* toxin type A. Protein Expr Purif 25:219–228
Chen S, Barbieri JT (2009) Engineering botulinum neurotoxin to extend therapeutic intervention. Proc Natl Acad Sci U S A 106:9180–9184
Chen S, Barbieri JT (2014) Protease activity of the botulinum neurotoxins. In: Foster KA (ed) Molecular aspects of botulinum neurotoxin – current topics in neurotoxicity, vol 4. Springer, New York, pp 171–189
Comella CL, Jankovic J, Shannon KM, Tsui J, Swenson M, Leurgans S et al (2005) Comparison of botulinum toxin serotypes A and B for the treatment of cervical dystonia. Neurology 65:1423–1429
Contreras E, Masuyer G, Qureshi N, Chawla S, Dhillon HS, Lee HL et al (2019) A neurotoxin that specifically targets *Anopheles* mosquitoes. Nat Commun 10:2869
Darios F, Niranjan D, Ferrari E, Zhang F, Soloviev M, Rummel A et al (2010) SNARE tagging allows stepwise assembly of a multimodular medicinal toxin. Proc Natl Acad Sci U S A 107:18197–18201
de Paiva A, Poulain B, Lawrence GW, Shone CC, Tauc L, Dolly JO (1993) A role for the interchain disulfide or its participating thiols in the internalization of botulinum neurotoxin A revealed by a toxin derivative that binds to ecto-acceptors and inhibits transmitter release intracellularly. J Biol Chem 268:20838–20844
Dekleva ML, Dasgupta BR (1990) Purification and characterization of a protease from clostridium botulinum type A that nicks single-chain type A botulinum neurotoxin into the di-chain form. J Bacteriol 172:2498–2503
Dolly O, Wang J (2015) Engineering of botulinum neurotoxins as novel therapeutic tools. In: Alouf J, Ladant D, Popoff MR (eds) The comprehensive sourcebook of bacterial protein toxins, 4th edn. Elsevier Ltd., Amsterdam, pp 995–1015
Dover N, Barash JR, Hill KK, Xie G, Arnon SS (2014) Molecular characterization of a novel botulinum neurotoxin type H gene. J Infect Dis 209:192–202

Dressler D, Truong D, Comella C, Jankovic J, Brashear A, Patel A et al (2018) Immunogenicity of daxibotulinumtoxinA for injection in adults with cervical dystonia from a phase 2 dose-escalation multicenter study. Toxicon 156(Suppl 1):S26–SS7

Duff JT, Wright GG, Yarinsky A (1956) Activation of *Clostridium botulinum* type E toxin by trypsin. J Bacteriol 72:455–460

DYSPORT® (abobotulinumtoxinA) [prescribing information] (2018) Basking Ridge: Ipsen Biopharmaceuticals, Inc.

Edupuganti OP, Ovsepian SV, Wang J, Zurawski TH, Schmidt JJ, Smith L et al (2012) Targeted delivery into motor nerve terminals of inhibitors for SNARE-cleaving proteases via liposomes coupled to an atoxic botulinum neurotoxin. FEBS J 279:2555–2567

Eklund MW, Poysky FT (1972) Activation of a toxic component of *Clostridium botulinum* types C and D by trypsin. Appl Microbiol 24:108–113

Eleopra R, Tugnoli V, Rossetto O, Montecucco C, De Grandis D (1997) Botulinum neurotoxin serotype C: a novel effective botulinum toxin therapy in human. Neurosci Lett 224:91–94

Eleopra R, Tugnoli V, Rossetto O, De Grandis D, Montecucco C (1998) Different time courses of recovery after poisoning with botulinum neurotoxin serotypes A and E in humans. Neurosci Lett 256:135–138

Eleopra R, Tugnoli V, Quatrale R, Rossetto O, Montecucco C, Dressler D (2006) Clinical use of non-A botulinum toxins: botulinum toxin type C and botulinum toxin type F. Neurotox Res 9:127–131

Elliott M, Maignel J, Liu SM, Favre-Guilmard C, Mir I, Farrow P et al (2017) Augmentation of VAMP-catalytic activity of botulinum neurotoxin serotype B does not result in increased potency in physiological systems. PLoS One 12:e0185628

Elliott M, Favre-Guilmard C, Liu SM, Maignel J, Masuyer G, Beard M et al (2019) Engineered botulinum neurotoxin B with improved binding to human receptors has enhanced efficacy in preclinical models. Sci Adv 5:eaau7196

Erbguth FJ (2008) From poison to remedy: the chequered history of botulinum toxin. J Neural Transm 115:559–565

Ferrari E, Maywood ES, Restani L, Caleo M, Pirazzini M, Rossetto O et al (2011) Re-assembled botulinum neurotoxin inhibits CNS functions without systemic toxicity. Toxins 3:345–355

Fischer A, Montal M (2007) Crucial role of the disulfide bridge between botulinum neurotoxin light and heavy chains in protease translocation across membranes. J Biol Chem 282:29604–29611

Foster K, Blum T, Beard M, Chaddock J, Kupinski A, Harding A et al (2018) Phage-assisted continuous evolution of botulinum neurotoxin light chains generates novel light chains with modified snare cleavage specificity. Toxicon 156 Suppl 1:S35

Frevert J, Ahn KY, Park MY, Sunga O (2018) Comparison of botulinum neurotoxin type A formulations in Asia. Clin Cosmet Investig Dermatol 11:327–331

Gerwing J, Dolman CE, Ko A (1965) Mechanism of tryptic activation of *Clostridium botulinum* type E toxin. J Bacteriol 89:1176–1179

Gill DM (1982) Bacterial toxins: a table of lethal amounts. Microbiol Rev 46:86–94

Gilmore MA, Sun S, Ordas JV, Arpawong HT, Okawa Y, Satorius A et al (2008) Fully active recombinant BoNT/E purified from E. coli in high yield. Toxicon 51 Suppl 1:11–12

Giménez DF, Ciccarelli AS (1970) Another type of Clostridium botulinum. Zentralbl Bakteriol Orig 215:221–224

Glogau R, Kane M, Beddingfield F, Somogyi C, Lei X, Caulkins C et al (2012) OnabotulinumtoxinA: a meta-analysis of duration of effect in the treatment of glabellar lines. Dermatol Surg 38:1794–1803

Greene PE, Fahn S (1993) Use of botulinum toxin type F injections to treat torticollis in patients with immunity to botulinum toxin type A. Mov Disord 8:479–483

Gu S, Rumpel S, Zhou J, Strotmeier J, Bigalke H, Perry K et al (2012) Botulinum neurotoxin is shielded by NTNHA in an interlocked complex. Science 335:977–981

Gustafsson C, Govindarajan S, Minshull J (2004) Codon bias and heterologous protein expression. Trends Biotechnol 22:346–353

Gustafsson R, Berntsson RP, Martinez-Carranza M, El Tekle G, Odegrip R, Johnson EA et al (2017) Crystal structures of OrfX2 and P47 from a Botulinum neurotoxin OrfX-type gene cluster. FEBS Lett 591:3781–3792

Hackett G, Moore K, Burgin D, Hornby F, Gray B, Elliott M et al (2018) Purification and characterization of recombinant botulinum neurotoxin serotype FA, also known as serotype H. Toxins 10:E195

Höltje M, Schulze S, Strotmeier J, Mahrhold S, Richter K, Binz T et al (2013) Exchanging the minimal cell binding fragments of tetanus neurotoxin in botulinum neurotoxin A and B impacts their toxicity at the neuromuscular junction and central neurons. Toxicon 75:108–121

Hosomi K, Sakaguchi Y, Kohda T, Gotoh K, Motooka D, Nakamura S et al (2014) Complete nucleotide sequence of a plasmid containing the botulinum neurotoxin gene in *Clostridium botulinum* type B strain 111 isolated from an infant patient in Japan. Mol Gen Genomics 289:1267–1274

Jankovic J, Hunter C, Dolimbek BZ, Dolimbek GS, Adler CH, Brashear A et al (2006) Clinico-immunologic aspects of botulinum toxin type B treatment of cervical dystonia. Neurology 67:2233–2235

Jankovic J, Truong D, Patel AT, Brashear A, Evatt M, Rubio RG et al (2018) Injectable daxibotulinumtoxinaA in cervical dystonia: a phase 2 dose-escalation multicenter study. Mov Disord Clin Pract 5:273–282

JEUVEAU (prabotulinumtoxinA-xvfs) [prescribing information] (2019) Santa Barbara: Evolus Inc.

Kakinuma H, Maruyama H, Takahashi H, Yamakawa K, Nakamura S (1996) The first case of type B infant botulism in Japan. Acta Paediatr Jpn 38:541–543

Kalb SR, Baudys J, Raphael BH, Dykes JK, Lúquez C, Maslanka SE et al (2015) Functional characterization of botulinum neurotoxin serotype H as a hybrid of known serotypes F and A (BoNT F/A). Anal Chem 87:3911–3917

Kauffman JA, Way JF Jr, Siegel LS, Sellin LC (1985) Comparison of the action of types A and F botulinum toxin at the rat neuromuscular junction. Toxicol Appl Pharmacol 79:211–217

Kaufman-Janette JA, Solish N, Liu Y, Rubio RG, Gallagher CJ (2018) Pooled results from 2 phase 3 pivotal studies of daxibotulinumtoxinA for the treatment of glabellar lines. Toxicon 156 Suppl 1:S59

Kennelly M, Dmochowski R, Schulte-Baukloh H, Ethans K, Del Popolo G, Moore C et al (2017) Efficacy and safety of onabotulinumtoxinA therapy are sustained over 4 years of treatment in patients with neurogenic detrusor overactivity: final results of a long-term extension study. Neurourol Urodyn 36:368–375

Kim JE, Song EJ, Choi GS, Lew BL, Sim WY, Kang H (2015) The efficacy and safety of liquid-type botulinum toxin type A for the management of moderate to severe glabellar frown lines. Plast Reconstr Surg 135:732–741

Kiyatkin N, Maksymowych AB, Simpson LL (1997) Induction of an immune response by oral administration of recombinant botulinum toxin. Infect Immun 65:4586–4591

Kozaki S, Oga Y, Kamata Y, Sakaguchi G (1985) Activation of *Clostridium botulinum* type B and E derivative toxins with lysine-specific proteases. FEMS Microbiol Lett 27:149–154

Kukreja RV, Singh BR (2007) Comparative role of neurotoxin-associated proteins in the structural stability and endopeptidase activity of botulinum neurotoxin complex types A and E. Biochemistry 46:14316–14324

Kumaran D, Eswaramoorthy S, Furey W, Navaza J, Sax M, Swaminathan S (2009) Domain organization in *Clostridium botulinum* neurotoxin type E is unique: its implication in faster translocation. J Mol Biol 386:233–245

Kutschenko A, Reinert MC, Krez N, Liebetanz D, Rummel A (2017) BoNT/AB hybrid maintains similar duration of paresis as BoNT/A wild-type in murine running wheel assay. Neurotoxicology 59:1–8

Kutschenko A, Weisemann J, Kollewe K, Fiedler T, Alvermann S, Boselt S et al (2019) Botulinum neurotoxin serotype D – a potential treatment alternative for BoNT/A and B non-responding patients. Clin Neurophysiol 130:1066–1073

Lacy DB, Tepp W, Cohen AC, DasGupta BR, Stevens RC (1998) Crystal structure of botulinum neurotoxin type A and implications for toxicity. Nat Struct Biol 5:898–902

Lee K, Gu S, Jin L, Le TT, Cheng LW, Strotmeier J et al (2013) Structure of a bimodular botulinum neurotoxin complex provides insights into its oral toxicity. PLoS Pathog 9:e1003690

Lee K, Zhong X, Gu S, Kruel AM, Dorner MB, Perry K et al (2014) Molecular basis for disruption of E-cadherin adhesion by botulinum neurotoxin A complex. Science 344:1405–1410

López de la Paz M, Scheps D, Jurk M, Hofmann F, Frevert J (2018) Rational design of botulinum neurotoxin A1 mutants with improved oxidative stability. Toxicon 147:54–57

Ludlow CL, Hallett M, Rhew K, Cole R, Shimizu T, Sakaguchi G et al (1992) Therapeutic use of type F botulinum toxin. N Engl J Med 326:349–350

Maffezzoni F, Formenti AM, Mazziotti G, Frara S, Giustina A (2016a) Current and future medical treatments for patients with acromegaly. Expert Opin Pharmacother 17:1631–1642

Maffezzoni F, Frara S, Doga M, Mazziotti G, Giustina A (2016b) New medical therapies of acromegaly. Growth Hormon IGF Res 30-31:58–63

Maisey EA, Wadsworth JD, Poulain B, Shone CC, Melling J, Gibbs P et al (1988) Involvement of the constituent chains of botulinum neurotoxins A and B in the blockade of neurotransmitter release. Eur J Biochem 177:683–691

Malmirchegini R, Too P, Oliyai C, Joshi A (2018) Revance's novel peptide excipient, RTP004, and its role in stabilizing daxibotulinumtoxinA (DAXI) against aggregation. Toxicon 156:S72–S73

Mansfield MJ, Adams JB, Doxey AC (2015) Botulinum neurotoxin homologs in non-*Clostridium* species. FEBS Lett 589:342–348

Mansfield MJ, Wentz TG, Zhang S, Lee EJ, Dong M, Sharma SK et al (2019) Bioinformatic discovery of a toxin family in *Chryseobacterium piperi* with sequence similarity to botulinum neurotoxins. Sci Rep 9:1634

Maslanka SE, Lúquez C, Dykes JK, Tepp WH, Pier CL, Pellett S et al (2016) A novel botulinum neurotoxin, previously reported as serotype H, has a hybrid-like structure with regions of similarity to the structures of serotypes A and F and is neutralized with serotype A antitoxin. J Infect Dis 213:379–385

Masuyer G, Beard M, Cadd VA, Chaddock JA, Ravi AK (2011) Structure and activity of a functional derivative of *Clostridium botulinum* neurotoxin B. J Struct Biol 174:52–57

Masuyer G, Davies JR, Moore K, Chaddock JA, Ravi AK (2015) Structural analysis of *Clostridium botulinum* neurotoxin type D as a platform for the development of targeted secretion inhibitors. Sci Rep 5:13397

Masuyer G, Zhang S, Barkho S, Shen Y, Henriksson L, Košenina S et al (2018) Structural characterisation of the catalytic domain of botulinum neurotoxin X – high activity and unique substrate specificity. Sci Rep 8:4518

Moriishi K, Koura M, Abe N, Fujii N, Fujinaga Y, Inoue K et al (1996) Mosaic structures of neurotoxins produced from *Clostridium botulinum* types C and D organisms. Biochim Biophys Acta 1307:123–126

MYOBLOC® (rimabotulinumtoxinB) [prescribing information] (2009) South San Francisco: Solstice Neurosciences, Inc

Nugent M, Wang J, Lawrence G, Zurawski T, Geoghegan JA, Dolly JO (2017) Conjugate of an IgG binding domain with botulinum neurotoxin A lacking the acceptor moiety targets its snare protease into TrkA-expressing cells when coupled to anti-TrkA IgG or Fc-betaNGF. Bioconjug Chem 28:1684–1692

Nugent M, Yusef YR, Meng J, Wang J, Dolly JO (2018) A SNAP-25 cleaving chimera of botulinum neurotoxin /A and /E prevents TNFα-induced elevation of the activities of native TRP channels on early postnatal rat dorsal root ganglion neurons. Neuropharmacology 138:257–266

Ohishi I, Sakaguchi G (1977) Activation of botulinum toxins in the absence of nicking. Infect Immun 17:402–407

Pellett S, Tepp WH, Stanker LH, Band PA, Johnson EA, Ichtchenko K (2011) Neuronal targeting, internalization, and biological activity of a recombinant atoxic derivative of botulinum neurotoxin A. Biochem Biophys Res Commun 405:673–677

Pellett S, Tepp WH, Scherf JM, Pier CL, Johnson EA (2015) Activity of botulinum neurotoxin type D (strain 1873) in human neurons. Toxicon 101:63–69

Pellett S, Tepp WH, Bradshaw M, Kalb SR, Dykes JK, Lin G et al (2016) Purification and characterization of botulinum neurotoxin FA from a genetically modified *Clostridium botulinum* strain. mSphere 1:e00100–e00115

Pellett S, Bradshaw M, Tepp WH, Pier CL, Whitemarsh RCM, Chen C et al (2018a) The light chain defines the duration of action of botulinum toxin serotype A subtypes. MBio 9:e00089–e00018

Pellett S, Tepp WH, Lin G, Johnson EA (2018b) Substrate cleavage and duration of action of botulinum neurotoxin type FA ("H, HA"). Toxicon 147:38–46

Pier CL, Tepp WH, Bradshaw M, Johnson EA, Barbieri JT, Baldwin MR (2008) Recombinant holotoxoid vaccine against botulism. Infect Immun 76:437–442

Pirazzini M, Rossetto O, Bolognese P, Shone CC, Montecucco C (2011) Double anchorage to the membrane and intact inter-chain disulfide bond are required for the low pH induced entry of tetanus and botulinum neurotoxins into neurons. Cell Microbiol 13:1731–1743

Pirazzini M, Henke T, Rossetto O, Mahrhold S, Krez N, Rummel A et al (2013) Neutralisation of specific surface carboxylates speeds up translocation of botulinum neurotoxin type B enzymatic domain. FEBS Lett 587:3831–3836

Pirazzini M, Rossetto O, Eleopra R, Montecucco C (2017) Botulinum neurotoxins: biology, pharmacology, and toxicology. Pharmacol Rev 69:200–235

Pons L, Vilain C, Picaut P (2018) Outcomes of the first-in-human study with a recombinant botulinum toxin type E (rBoNT-E): safety and pharmacodynamic profile of rBoNT-E compared with abobotulinumtoxinA (Dysport®). Toxicon. 156(Suppl 1):S93

Przedpelski A, Tepp WH, Zuverink M, Johnson EA, Pellet S, Barbieri JT (2018) Enhancing toxin-based vaccines against botulism. Vaccine 36:827–832

Ravichandran E, Janardhanan P, Patel K, Riding S, Cai S, Singh BR (2016) In vivo toxicity and immunological characterization of detoxified recombinant botulinum neurotoxin type A. Pharm Res 33:639–652

Rummel A (2015) The long journey of botulinum neurotoxins into the synapse. Toxicon 107:9–24

Saffarian P, Peerayeh SN, Amani J, Ebrahimi F, Sedighian H, Halabian R et al (2016a) TAT-BoNT/$A_{(1\text{-}448)}$, a novel fusion protein as a therapeutic agent: analysis of transcutaneous delivery and enzyme activity. Appl Microbiol Biotechnol 100:2785–2795

Saffarian P, Peerayeh SN, Amani J, Ebrahimi F, Sedighianrad H, Halabian R et al (2016b) Expression and purification of recombinant TAT-BoNT/$A_{(1\text{-}448)}$ under denaturing and native conditions. Bioengineered 7:478–483

Sakaguchi Y, Suzuki T, Yamamoto Y, Nishikawa A, Oguma K (2015) Genomics of *Clostridium botulinum* group III strains. Res Microbiol 166:318–325

Scheps D, de la López Paz M, Jurk M, Hofmann F, Frevert J (2017) Design of modified botulinum neurotoxin A1 variants with a shorter persistence of paralysis and duration of action. Toxicon 139:101–108

Schiavo G, Papini E, Genna G, Montecucco C (1990) An intact interchain disulfide bond is required for the neurotoxicity of tetanus toxin. Infect Immun 58:4136–4141

Sikorra S, Litschko C, Müller C, Thiel N, Galli T, Eichner T et al (2016) Identification and characterization of botulinum neurotoxin A substrate binding pockets and their re-engineering for human SNAP-23. J Mol Biol 428:372–384

Simpson LL, Maksymowych AB, Park JB, Bora RS (2004) The role of the interchain disulfide bond in governing the pharmacological actions of botulinum toxin. J Pharmacol Exp Ther 308:857–864

Sloop RR, Cole BA, Escutin RO (1997) Human response to botulinum toxin injection: type B compared with type A. Neurology 49:189–194

Smith TJ (2014) *Clostridium botulinum* genomes and genetic diversity. In: Foster KA (ed) Molecular aspects of botulinum neurotoxin. Springer, New York, pp 207–228

Smith TJ, Hill KK, Foley BT, Detter JC, Munk AC, Bruce DC et al (2007) Analysis of the neurotoxin complex genes in *Clostridium botulinum* A1-A4 and B1 strains: BoNT/A3, /Ba4 and /B1 clusters are located within plasmids. PLoS One 2:e1271

Somm E, Bonnet N, Martinez A, Marks PM, Cadd VA, Elliott M et al (2012) A botulinum toxin-derived targeted secretion inhibitor downregulates the GH/IGF1 axis. J Clin Invest 122:3295–3306

Somm E, Bonnet N, Zizzari P, Tolle V, Toulotte A, Jones R et al (2013) Comparative inhibition of the GH/IGF-I axis obtained with either the targeted secretion inhibitor SXN101959 or the somatostatin analog octreotide in growing male rats. Endocrinology 154:4237–4248

Strahan BL, Failor KC, Batties AM, Hayes PS, Cicconi KM, Mason CT et al (2011) *Chryseobacterium piperi* sp. nov., isolated from a freshwater creek. Int J Syst Evol Microbiol 61:2162–2166

Sugawara Y, Matsumura T, Takegahara Y, Jin Y, Tsukasaki Y, Takeichi M et al (2010) Botulinum hemagglutinin disrupts the intercellular epithelial barrier by directly binding E-cadherin. J Cell Biol 189:691–700

Swaminathan S, Eswaramoorthy S (2000) Structural analysis of the catalytic and binding sites of *Clostridium botulinum* neurotoxin B. Nat Struct Biol 7:693–699

Takeda M, Tsukamoto K, Kohda T, Matsui M, Mukamoto M, Kozaki S (2005) Characterization of the neurotoxin produced by isolates associated with avian botulism. Avian Dis 49:376–381

Tao L, Peng L, Berntsson RP, Liu SM, Park S, Yu F et al (2017) Engineered botulinum neurotoxin B with improved efficacy for targeting human receptors. Nat Commun 8:53

Tehran DA, Pirazzini M (2018) Novel botulinum neurotoxins: exploring underneath the iceberg tip. Toxins 10:E190

Vazquez-Cintron E, Tenezaca L, Angeles C, Syngkon A, Liublinska V, Ichtchenko K et al (2016) Pre-clinical study of a novel recombinant botulinum neurotoxin derivative engineered for improved safety. Sci Rep 6:30429

Vazquez-Cintron EJ, Vakulenko M, Band PA, Stanker LH, Johnson EA, Ichtchenko K (2014) Atoxic derivative of botulinum neurotoxin A as a prototype molecular vehicle for targeted delivery to the neuronal cytoplasm. PLoS One 9:e85517

Vazquez-Cintron EJ, Beske PH, Tenezaca L, Tran BQ, Oyler JM, Glotfelty EJ et al (2017) Engineering botulinum neurotoxin C1 as a molecular vehicle for intra-neuronal drug delivery. Sci Rep 7:42923

Wang J, Meng J, Lawrence GW, Zurawski TH, Sasse A, Bodeker MO et al (2008) Novel chimeras of botulinum neurotoxins A and E unveil contributions from the binding, translocation, and protease domains to their functional characteristics. J Biol Chem 283:16993–17002

Wang J, Zurawski TH, Bodeker MO, Meng J, Boddul S, Aoki KR et al (2012a) Longer-acting and highly potent chimaeric inhibitors of excessive exocytosis created with domains from botulinum neurotoxin A and B. Biochem J 444:59–67

Wang J, Zurawski TH, Meng J, Lawrence GW, Aoki KR, Wheeler L et al (2012b) Novel chimeras of botulinum and tetanus neurotoxins yield insights into their distinct sites of neuroparalysis. FASEB J 26:5035–5048

Wang J, Casals-Diaz L, Zurawski T, Meng J, Moriarty O, Nealon J et al (2017) A novel therapeutic with two SNAP-25 inactivating proteases shows long-lasting anti-hyperalgesic activity in a rat model of neuropathic pain. Neuropharmacology 118:223–232

Webb RP (2018) Engineering of botulinum neurotoxins for biomedical applications. Toxins 10:231

Webb RP, Smith TJ, Wright P, Brown J, Smith LA (2009) Production of catalytically inactive BoNT/A1 holoprotein and comparison with BoNT/A1 subunit vaccines against toxin subtypes A1, A2, and A3. Vaccine 27:4490–4497

Webb RP, Smith TJ, Smith LA, Wright PM, Guernieri RL, Brown JL et al (2017) Recombinant botulinum neurotoxin Hc subunit (BoNT Hc) and catalytically inactive *Clostridium botulinum* holoproteins (ciBoNT HPs) as vaccine candidates for the prevention of botulism. Toxins 9:269

Weisemann J, Krez N, Fiebig U, Worbs S, Skiba M, Endermann T et al (2015) Generation and characterization of six recombinant botulinum neurotoxins as reference material to serve in an international proficiency test. Toxins 7:5035–5054

Wentz TG, Muruvanda T, Lomonaco S, Thirunavukkarasu N, Hoffmann M, Allard MW et al (2017) Closed genome sequence of *Chryseobacterium piperi* strain CTM^{T}/ATCC BAA-1782, a gram-negative bacterium with clostridial neurotoxin-like coding sequences. Genome Announc 5:e01296–e01217

Whitemarsh RC, Tepp WH, Johnson EA, Pellett S (2014) Persistence of botulinum neurotoxin A subtypes 1-5 in primary rat spinal cord cells. PLoS One 9:e90252

Woudstra C, Le Maréchal C, Souillard R, Anniballi F, Auricchio B, Bano L et al (2018) Investigation of *Clostridium botulinum* group III's mobilome content. Anaerobe 49:71–77

Wuchao L, Ya F, Junhui S, Zheng Z, Lizhen P, Libin Z et al (2018) An experimental study about neurotoxic potency of a novel botulinum toxin type A free of neurotoxin-associated proteins based on muscle strength measurement. Toxicon 156 Suppl 1:S117

XEOMIN® (incobotulinumtoxinA) [prescribing information] (2019) Raleigh: Merz Pharmaceuticals, LLC

Yang N, Liu SM, Marlin S, Lewandowska A, Heyang M, Voisin T et al (2018) Chimeras of anthrax toxin and botulinum neurotoxin as novel analgesic proteins. Toxicon 156 Suppl 1:S118

Yoelin SG, Dhawan SS, Vitarella D, Ahmad W, Hasan F, Abushakra S (2018) Safety and efficacy of EB-001, a novel type E botulinum toxin, in subjects with glabellar frown lines: results of a phase 2, randomized, placebo-controlled, ascending-dose study. Plast Reconstr Surg 142:847e–855e

Zanetti G, Sikorra S, Rummel A, Krez N, Duregotti E, Negro S et al (2017) Botulinum neurotoxin C mutants reveal different effects of syntaxin or SNAP-25 proteolysis on neuromuscular transmission. PLoS Pathog 13:e1006567

Zhang S, Masuyer G, Zhang J, Shen Y, Lundin D, Henriksson L et al (2017) Identification and characterization of a novel botulinum neurotoxin. Nat Commun 8:14130

Zhang S, Lebreton F, Mansfield MJ, Miyashita SI, Zhang J, Schwartzman JA et al (2018) Identification of a botulinum neurotoxin-like toxin in a commensal strain of *Enterococcus faecium*. Cell Host Microbe 23:169–176.e6

Zhou L, de Paiva A, Liu D, Aoki R, Dolly JO (1995) Expression and purification of the light chain of botulinum neurotoxin A: a single mutation abolishes its cleavage of SNAP-25 and neurotoxicity after reconstitution with the heavy chain. Biochemistry 34:15175–15181

Zornetta I, Azarnia Tehran D, Arrigoni G, Anniballi F, Bano L, Leka O et al (2016) The first non Clostridial botulinum-like toxin cleaves VAMP within the juxtamembrane domain. Sci Rep 6:30257

Part II

Clinical Practice

Clinical Pharmacology of Botulinum Toxin Drugs

Dirk Dressler

Contents

Abstract

Botulinum toxin (BT) has changed from a deadly poison to a novel therapeutic principle for a large number of disorders in many medical areas.

BT drugs are special in many ways: they are biologicals, their active ingredient BT is not patentable, their spectrum of clinical applications is extremely broad, their dose range is enormous, their mode of action is local and their life cycles are special.

This review covers BT's therapeutic mode of action, time course of action, target tissues, pharmacological profile, adverse effects, interactions, potency labelling and antigenicity as well as BT's therapeutic preparations.

D. Dressler (✉)
Movement Disorders Section, Department of Neurology, Hannover Medical School, Hannover, Germany
e-mail: dressler.dirk@mh-hannover.de

S. M. Whitcup, M. Hallett (eds.), *Botulinum Toxin Therapy*,
Handbook of Experimental Pharmacology 263, https://doi.org/10.1007/164_2019_273

Keywords
Adverse effects · Antigenicity · Botulinum toxin · Clinical pharmacology · Interactions · Mode of action · Pharmacological profile · Potency labelling · Target tissues · Therapeutic preparations · Time course

1 Introduction

Botulinum toxin (BT) has made one of the most remarkable transitions in the history of mankind: once infamous as a food safety hazard and a means of biological warfare it is now a drug for a large number of disorders for many of which it has revolutionised their treatment.

BT drugs are special in many ways: As biologicals they are not only specified by their chemicophysical properties alone, but also by their steric confirmation which is heavily influenced by the manufacturing process and handling conditions. Comparing biologicals is challenging and has to consider many parameters. Nevertheless, also biologicals can and need to be compared with respect to efficacy, adverse effects and economics. BT as a natural compound cannot be protected by patents. Intellectual property protection is mainly gained by the BT drug's registration status. This generates very special drug life cycles and the option of alternative registration pathways through biosimilarity approval. This may dramatically influence the future development of BT drugs. The broad spectrum of clinical applications generates enormous dose ranges leading to serious pricing issues. As a strictly local agent BT drugs require specific registration pathways.

This review covers BT's therapeutic mode of action, time course of action, target tissues, pharmacological profile, adverse effects, interactions, potency labelling and antigenicity as well as BT's therapeutic preparations.

2 Therapeutic Mode of Action

Botulinum neurotoxin (BNT) is blocking the SNARE protein-mediated excretion process of acetylcholine from cholinergic neurons. BT's binding, internalisation and intracellular action involve highly complex and specific molecular mechanisms (Dressler and Foster 2018). The BT-induced cholinergic blockade is long-lasting, but temporary and fully reversible and does not produce any structural damage even in delicate tissues and after prolonged application. It reduces the activity of the cholinergic neuromuscular junction and of the cholinergic autonomic junction through interaction with the cholinergic nerve terminal, i.e. the peripheral nervous system. Additional effects on pain perception and pain processing have been described and a registration to treat chronic migraine has subsequently been granted (Aurora et al. 2010; Diener et al. 2010). Interactions with calcitonin-gene-related peptide (CGRP) and other transmitters have been proposed. Exact mechanisms involved, however, are largely unknown, but would – most likely – involve direct central nervous system interactions. It has long been known that BT does not only interact with the SNARE proteins in the nerve terminals but can also be transported to the alpha motoneuron's soma by retrograde axonal transport (Wiegand et al. 1976). Recent evidence suggests that BT – like the structurally closely related tetanus toxin – might be able to transgress to secondary neurons (Antonucci et al. 2008). This would require an externalisation process equally complex as BT's internalisation process. So far, nothing is known about such a process. Indirect central nervous system effects are not surprisingly numerous as BT's peripheral effect on the human body is considerable. As BT was demonstrated to interact not only with striate or smooth muscle fibres but also with the intrafusal muscle fibres of the Golgi muscle tendon organ (Filippi et al. 1993; Dressler et al. 1993; Rosales et al. 1996), additional therapeutic effects on muscle hyperactivity syndromes beyond the well-described peripheral paresis were discussed. Recently, antidepressive effects of BT application have been described (Wollmer et al. 2012). Even more so than in analgesia, underlying mechanisms remain unclear and central as well as peripheral nervous system effects have been discussed.

3 Time Course of Action

BT has a prolonged biological effect. This is one of the key prerequisites for its therapeutic use. BT drug effects follow a distinct *time course* (Dressler and Benecke 2007). With a delay of 2–5 days, their therapeutic effect builds up (build-up phase), stays on a plateau for 6–10 weeks (plateau phase) and then gradually declines (wearing-off phase). To maintain a steady clinical improvement, reinjections become necessary. This is usually the case after 8–14 weeks, but may be delayed in reality by many considerations including the physician's availability, travel logistics, economic considerations and the patient's personal choices.

Different BT types have different *durations of action*. Exact data on the duration of action of BT in humans is sparse, as in most treatment studies patients would not accept reinjection delays once they enter the wearing-off phase. In humans, BT-A has the longest duration of action, whereas BT-B may have a slightly shorter one and BT type C and BT type E seem to have a particularly short duration of action (Dressler and Foster 2018). BT's duration of action may also depend on the particular target tissue chosen. BT therapy of hyperhidrosis, where it is applied to the sweat glands, may produce somewhat longer effects than BT therapy of muscle hyperactivity syndromes (Naumann and Jost 2004). Comparing the therapeutic duration of action requires identical endpoint definitions and identical treatment parameters, so that valid data on the duration of action of different BT types or different BT drugs can only be generated by direct head-to-head comparison studies.

The *build-up phase* is therapeutically less relevant as it only occurs in full at the initiation of BT therapy and as it is very short in relation to the plateau phase. Different BT types do not seem to have much different build-up phases. The same seems to be true for the *wearing-off phase*. Long-term data show that the temporal profile of BT's action is remarkably stable over even decade-long applications (Dressler et al. 2015b). This lack of enzymatic induction and lack of tachyphylaxis are other key prerequisites for BT's therapeutic use.

BT's effect follows a dose-effect correlation and a dose-duration correlation. With the *dose-effect correlation* (Dressler and Rothwell 2000), it is possible to adjust therapeutic BT doses to the effect size necessary. Dose-effect curves can also be used to detect BT antibodies (Dressler et al. 2000) and to compare BT drug potencies (Dressler et al. 2018). The *dose-duration correlation* is weak. With usual BT doses applied, the duration effect is in most cases saturated.

4 Target Tissues

BT drugs can be applied to all tissues with *cholinergic innervation* including muscle tissue (striate and smooth), exocrine glandular tissue (sweat glands, saliva gland, lacrimal glands) and pain relevant structures. The *target tissue affinity* seems to be the same for BT drugs of the same BT type, but is different between BT drugs of different BT types. Whereas BT-A has a relatively strong affinity to muscle tissue and a relatively weak one to autonomic tissue, this relationship is reversed in BT-B (Dressler and Benecke 2003). Potentially different durations of action in different target tissues have been discussed above.

5 General Pharmacological Profile

With all features described above, BT drugs can be characterised as long-term and long-lasting, but temporary and fully reversible, well controllable, local and – even in very high doses – surprisingly safe drugs for muscle relaxation, exocrine gland secretion suppression and analgesia.

6 Adverse Effects

The adverse effect profile of BT is remarkably benign, as BT remains at its injection site and does not participate in the body's general metabolism thus sparing critical absorption and secretion organs. Adverse effects of BT therapy include unwanted paresis and unwanted exocrine gland suppression. All adverse effects are acute only. They may be divided in local and systemic ones. *Local adverse effects* are caused by BT spread from the target tissue into adjacent tissues. *Systemic adverse effects* occur when relevant amounts of BT are distributed with the bloodstream producing effects which cannot be explained by local BT spread. As the fraction of BT which is not bound to the target tissue is very low (Takamizawa et al. 1986), systemic adverse effects would require application of substantial doses. Recently, BT *high-dose therapy* was introduced (Dressler et al. 2015a, b) and subsequently reconfirmed (Wissel et al. 2017). With high-dose therapy total incobotulinumtoxinA doses of up to 1250MU reconstituted with 2.5 ml of 0.9% NaCl/H_2O, distributed over a large number of target muscles and using several injection sites per target muscle are applied. Acute and long-term follow-up demonstrated not only safety with respect to systemic toxicity but also safety with respect of BT antibody formation (Dressler et al. 2015a). With this, the therapeutic dose range of BT is now considerable and allows treatment not only of focal muscle hyperactivity disorders but also of segmental and generalised ones, thus advancing BT therapy considerably (Dressler et al. 2016b). Repeated BT therapy – often for decades – does not produce additional adverse effects indicating also an exceptional *long-term safety* (Dressler et al. 2013).

The adverse effect profile of BT-B is much different from that of BT-A drugs. Even at moderate BT-B doses, patients may experience *anticholinergic adverse effects* including dryness of mouth, dryness of eyes and accommodation difficulties (Dressler and Benecke 2004).

7 Interactions

Registration documents warn not to perform BT therapy in the presence of anticoagulation. However, it was recently shown that BT therapy can be performed safely as long as thin injection needles are used (Schrader et al. 2018). It is strongly advised not to interrupt coagulation in preparation of BT therapy as this may greatly increase the risk of thrombosis or haemorrhage. Underlying neuromuscular transmission disorders including Lambert-Eaton myasthenic syndrome and myasthenia gravis may increase the sensitivity to BT therapy (Dressler 2010; Erbguth et al. 1993), but are no contraindications as long as BT doses are adjusted accordingly. In case of unusual hypersensitivity towards BT therapy, hitherto undetected underlying neuromuscular transmission disorders should be considered.

8 Potency Labelling

According to the European Pharmacopoeia, the biological potency of BT drugs is measured by a standardised LD50 assay and expressed in mouse units (European Pharmacopoeia 2008a, b). However, clinical practise suggests that the potency labelling of different BT type A drugs is not identical. Between the potency labelling of Botox® and Dysport® conversion factors from 1:5 to 1:2.41 have been reported in clinical studies (Brin and Blitzer 1993; Marion et al. 1995; Marsden 1993; Van den Berg et al. 1996; Ranoux et al. 2002). In LD50 assays conversion factors of 1:2.89 (Pickett and Hambleton 1994), 1:2.86 (Van den Berg et al. 1996) and 1:1.9 (First et al. 1994) were determined. Between the potency labelling of Xeomin® and Botox® a conversion factor of 1:1 is established (Dressler 2009; Dressler et al. 2012, 2018; Scaglione 2016). The conversion factor between the BT-A drug Botox® and the BT-B drug Myobloc® seems to be 1:40. Reasons for the contradictory potency labelling are unclear, but may include differences in the potency assays (for BT-A drug differences) and different species susceptibilities (for BT-A and BT-B drug differences). Clinical studies to determine the conversion factors between different BT drugs are usually of limited validity, especially when clinical models with limited sensitivity and low adverse effect frequencies as in blepharospasm and small sample sizes are used. To obtain more precise results, full dose-effect curves need to be compared.

9 Antigenicity

The current understanding of BT antigenicity was recently reviewed (Dressler and Bigalke 2017a). As BNT is a *protein*, antigenicity of BT drugs has always been a concern. After it was believed that BNT amounts applied were too small to induce immune responses, it became clear in the early 1990s that even the minute therapeutic BNT doses can induce BT antibody formation reducing BNT's therapeutic effects and adverse effects. Although the actual frequency of complete antibody-induced therapy failure is – as will be subsequently described – low, measures to prevent it include reducing BT dosages, and interinjection intervals are considerably limiting the real potential of BT therapy. Reducing antigenicity in BT therapy should, therefore, be an important development goal.

BT Antibodies BT antibodies can be neutralising, i.e. blocking BNT's mode of action, and they can be non-neutralising, i.e. targeting non-functional BNT epitopes, indicating high BT antibody specificity. BT antibodies may occur in different *titres* (Dressler et al. 2002). Those titres may be very low and may not reduce BT's therapeutic effects, thus making them *therapeutically irrelevant.* When titres are intermediate, they may produce *partial therapy failure.* High titres elicit *complete therapy failure*. As also the amount of BNT applied is relevant (Dressler et al. 2002), we described the interaction between BNT and BT antibodies as a *balance* indicating the importance of BT antibody titre determination.

BT Antibody Detection Detection of BT antibodies bears numerous risks of misinterpretation concerning test system sensitivity and specificity and the underlying balance model. BT antibodies can be detected by *structural tests* using ELISA arrays (Dressler et al. 2014). They are not able to distinguish between neutralising and non-neutralising antibodies. Antibodies can also be detected by *functional tests* detecting only neutralising ones. In principle, all biological BNT effects can be used in a test system. Usually, lethality is used in animal tests. The mouse diaphragm assay is an advanced and animal friendly ex vivo test with an elaborate quality assessment (Goeschel et al. 1997). In humans usually paretic effects are used as in the EDB test (Kessler and Benecke 1997) or the SCM test (Dressler et al. 2000). However, also sweating may be a test parameter.

Interpretation of BT Antibody Measurements Results from BT antibody tests need to be interpreted carefully. Demonstrating the shear presence of BT antibodies is usually not helpful, as they may present false-positive results, results from hypersensitive test systems and as they may not be correlated to clinical nonresponsiveness. Only quantitative measurements of BT antibody titres generate data for exact clinical interpretation.

Risk Factors for BT Antibody Formation Classical risk factors are the single dose, i.e. the amount of BNT applied at each injection series; the interinjection interval, i.e. the time between two subsequent injection series; and the application of booster injections, i.e. two BT injection series with an interinjection interval of less than 2 weeks. More recently, it became clear that also the immunological quality of the BT drug used as described by the specific biological potency may be a risk factor (Dressler and Bigalke 2017a). Sex and age of the patients treated, the cumulative BT dose applied and the treatment duration do not seem to be risk factors. Also, so far, there is no indication that the particular target tissue injected may change the risk of BT antibody formation.

Occurrence of BT Antibodies There are no exact data on the frequency of BT antibody formation available, as they would require prospective monitoring of large patient groups over prolonged periods of time. Interestingly, BT antibody formation seems to occur in a time window early in the treatment (Dressler 2004). After several years of BT therapy, the risk for BT antibody formation actually seems to drop (Dressler 2004). Estimates suggest that complete antibody-induced therapy failure for low- to intermediate-dose indications has a frequency of 1–5% when onabotulinumtoxinA and abobotulinumtoxinA are used. When rimabotulinumtoxinA is used this frequency may go up to 40% (Dressler and Bigalke 2005). For incobotulinumtoxinA there have not been any reports on complete antibody-induced therapy failure, although this BT drug has been worldwide available since 2005. This low – or even non-existent – antigenicity was the basis to improve the BT treatment algorithms by introducing the *short interval therapy* (Dressler and Adib Saberi 2017a) and the *high-dose therapy* (Dressler et al. 2015a, b), thus improving BT therapy considerably. The particular aetiology of the muscle hyperactivity syndrome treated and the target tissue type do not seem to matter.

10 Therapeutic Preparations

BT drugs are complex mixtures of compounds. Their various features are shown in Table 1. BT drugs consist of BNT, complexing proteins (CP) and excipients.

BNT BNT is the therapeutically active ingredient. It exists in seven different *subtypes* named type A to type G. BNT used in BT drugs is either BT type A (BT-A) or BT type B (BT-B). BT types E, C, D and F have only experimentally been used in humans. Therapeutically relevant parameters of different BT types may differ considerably, whereas BT drugs of the same BT type produce very similar effects as they are based on virtually identical BNT. As described above, BT subtypes are different with respect to target tissue affinity, antigenicity and time course of action.

Table 1 Features of botulinum toxin drugs

Botulinum neurotoxin	Subtype Target tissue affinity Antigenicity Time course of action Duration of action Latency of onset
Complexing proteins	Presence or removal Antigenicity
Excipients	Human serum albumin, gelatine, polysorbate Risk of HIV, BSE, anaphylaxis Others pH value
Manufacturing process	Production continuity Aliquotation, continuous production Purification Crystallisation, dialysis, chromatography, Precipitation, High-Pure technology Activation Specific biological activity Degree of BNT inactivation during purification Degree of knicking during activation Potency testing Animal-based assays, cell-based assays Potency consistency Stabilisation Lyophilisation (freeze drying), vacuum drying, pH Reduction Potency stability Unreconstituted, reconstituted drug Potency labelling
Manufacturer's support	Product documentation Product support Reliability of drug supply Counterfeit protection Handling safety Differentiability of vials with different potencies Denomination of packaging size Potency per vial Packages per over-pack Competitive pricing per adjusted mouse unit

CP CP are a residue of the natural development process and are not necessary for BT's therapeutic action. Their role in BT's antigenicity is unclear. It was suggested they may indirectly increase BT's antigenicity by attracting leucocytes into the injection area (Lee et al. 2005). Based on these considerations, CP have been removed from the BT drug incobotulinumtoxinA without changing its therapeutic and adverse effect profile, but potentially contributing to its particularly low antigenicity.

Excipients Excipients are added during the manufacturing process to stabilise BT drugs. Usually, they contain human serum albumin, but lanbotulinumtoxinA contains bovine gelatine instead and Coretox® uses polysorbate only and, thus, is free of any biological additives. Sugars including maltose, lactose, sucrose and dextran may also be added as may be NaCl and methionine. In liquid preparations buffer systems may be used to reduce the pH value to stabilise the solution. Reduced pH values, however, are increasing injection site pain (Dressler et al. 2016a).

Manufacturing The manufacturing of BT drugs is a complex and closely controlled process as it directly influences core features of the final drug.

Manufacturing differs with respect of *production continuity*. BT drugs may be produced by aliquotation of a single masterbatch licensed by the registration authorities. They may also be produced by a controlled and licensed continuous production process generating batches continuously. Originally, Botox® was based on the masterbatch 79/11 acquired from the Alan Scott's Oculinum Company. 1998 the manufacturing process was changed to a continuous one to increase SBA and to reduce antigenicity (Jankovic et al. 2003). All major BT drugs are now manufactured continuously. *Purification* of the bacterial broth is an important manufacturing process. It should generate high drug purity together with low degree of BNT degradation. BNT *activation* ('knicking') should yield a high degree of activated BNT. The *specific biological activity* (SBA), i.e. the potency per total BNT content, was introduced some time ago as a parameter to predict the antigenicity of BT drugs (Dressler and Hallett 2006). A low SBA indicates high risks for BT antibody formation. Low SBA may be the product of suboptimal purification and incomplete knicking causing SBA differences between BT drugs of the same BT type. SBA may also be different amongst different BT types. BT-A has the highest SBA, and BT-B has a particularly low one. Whether, in the case of the only available BT-B drug Myobloc®, the low SBA is a general property of BT-B or whether it reflects its particular manufacturing remains open. Repeated *potency testing* during the manufacturing process of BT drugs was based on animal testing sacrificing large numbers of mice. State of the art potency testing is now performed by cell-based assays. Although Allergan, Ipsen and Merz are applying this technique, still a large percentage of their BT drugs have to be produced with conventional animal potency tests as many registration authorities are not yet prepared to accept cell-based assays. In the future entry of new BT drugs to the North American and European markets will most likely only be granted when the manufacturing uses cell-based assays. Optimal manufacturing excels with a high inter-batch *potency consistency*. *Stabilisation* of BT drugs may be achieved by various processes including lyophilisation (freeze drying), vacuum drying and pH reduction. *Potency stability* of the BT drug is an important feature as it affects logistics (temperature control, shelf life) as well as economics of use (Dressler and Bigalke 2017b). *Potency labelling* should be directly comparable between BT drugs.

Manufacturer's Support The BT drug itself should be backed up by a reliable manufacturer providing sufficient product documentation, product support, reliability of drug supply, counterfeit protection, handling safety by differentiability of vials with different potencies (Dressler and Adib Saberi 2017b) and reasonable denomination of packaging size (potency per vial, packages per overpack) and – last but not least – a competitive pricing per adjusted mouse unit.

Botulinum Toxin Drugs BT drugs have *unique features*: they are not patentable as such, they have an enormous spectrum of indications, their therapeutic doses range spread with a factor of 1,000, they represent a completely new therapeutic principle, they require highly individualised injection schemes developed by trained injectors and they have remarkably long commercial life cycles. This profile demonstrates the enormous therapeutic potential, but, at the same time, generates numerous challenges including qualitative and quantitative off-label use and the necessity to offer user training and to prevent therapy failure due to suboptimal application techniques. Therapeutic BT preparations currently available or under development are shown in Table 2.

In 1989 the US drug Botox® (onabotulinumtoxinA) was the worldwide first BT drug registered. In 1991 came the UK drug Dysport® (abobotulinumtoxinA), in 2000 another US drug Myobloc® (rimabotulinumtoxinB) and in 2005 the German drug Xeomin® (incobotulinumtoxinA). These drugs and their aesthetic analogues are currently the only BT drugs available in North America and Europe. In 1997 Hengli was first registered in the People's Republic of China, but, so far, has not reached core foreign markets. In the meantime, many other BT drugs are being developed in the USA, the Republic of Korea, the Russian Federation, the UK, Iran and India. Table 2 gives an overview. Most of these drugs are commercially targeting aesthetic uses in fringe markets, but also therapeutic registrations in North America and Europe are being prepared. The most active BT drug development place is currently the Republic of Korea with companies including Medytox, Daewoong and Hugel.

Further Development There are several directions of further BT drug development. Firstly, all companies are continuously trying to expand the *indication spectrum* of their BT drugs and to search for novel indications. Whether the antidepressive effects described will, indeed, establish a new class of action of BT drugs remains unclear. Also unclear is the perspective of developing potential anti-inflammatory effects. Some companies are trying to prolong or to reduce BT's *duration of action*. Whilst prolonging the duration of action may provide some benefit for patients, it is hard to imagine indications for BT drugs with reduced duration of action. Most advanced is a project by Revance of the USA to prolong BT's duration of action by co-administration of a proprietary protein. So far, proof of concept is still lacking and potential risks to the motoneuron pool by long-term exposure to a foreign protein have not been sufficiently addressed. Another development goal are *liquid preparations* of existing BT drugs. Most advanced is Ipsen.

Table 2 Therapeutic botulinum toxin preparations

Generic name	Trade name	Manufacturers and partners (past and present)	Country	Specifics
OnabotulinumtoxinA	Botox Botox® Cosmetics® Vistabel®	Allergan-AbbVie	USA/ Ireland	BT-A
AbobotulinumtoxinA	Dysport® Azzalure® Reloxin®	Ipsen/Medicis	UK/France/ USA	BT-A
IncobotulinumtoxinA	Xeomin® Xeomin Cosmetics® Bocouture®	Merz Pharmaceuticals	Germany	BT-A, no complexing proteins
RimabotulinumtoxinB	NeuroBloc® Myobloc® NerBloc®	US WorldMeds/Eisai/ Sloan/Elan/Solstice	USA	BT-B, liquid preparation
LanbotulinumtoxinA	Hengli® Lantox® Lanzox® CBTX-A® Prosigne® Redux® Liftox® Dituroxal®	Lanzhou Institute of Biological Products/Hugh Source	P.R. China	BT-A, Botox® analogon
	Neuronox® Meditoxin® Botulift® Cunox®	Medytox	R. Korea	BT-A, Botox® analogon
	Coretox®	Medytox	R. Korea	BT-A, Xeomin® analogon, no complexing proteins, no biological excipients
	Innotox®	Medytox/Allergan	R. Korea/ USA	BT-A, liquid preparation
	Botulax® Zentox® Regenox®	Hugel	R. Korea	
PrabotulinumtoxinA	Nabota® Jeuveau® Evosyal®	Daewoong/Evolus-Alphaeon	R. Korea/ USA	BT-A, Botox® analogon
DaxibotulinumtoxinA	RTT150	Revance	USA	BT-A, protein additive
		Revance/Mylan	USA/ Netherlands	BT-A, Botox® analogon, biosimilar approach
	Relatox®	Microgen	Russia	BT-A, Botox® analogon
	Botulax®	Hugel	R. Korea	BT-A, Botox® analogon
	Masport®	Masoundarou	I.R. Iran	BT-A, Dysport® analogon
	CosmeTox®	Transdermal	USA	BT-A, cream
	BTXA®	Intas	India	BT-A, Botox® analogon
	Botogenie	BioMed	India	BT-A, Botox® analogon
	EB-001	Bonti/Allergan	USA	BT-E
	MCL005	Malvern Cosmeceuticals	UK	BT-A, topic gel
	ANT-1207	Anterios/Allergan	USA	BT-A, lotion

However, the registration process seems to falter, probably due to potency differences between both Dysport® preparations. Another liquid preparation is Innotox® provided by Medytox. In the future, most new BT drugs will be second-generation BT drugs *lacking CP*. Several companies are trying to develop BT drugs without *biological excipients* to reduce the potential risk of HIV or BSE transmission. Medytoxin's Innotox® is the first of this kind of BT drugs. Mainly for aesthetic indications, transdermal BT application avoiding injection site pain seems attractive. Revance apparently has stopped their project using their proprietary protein for this purpose. All new BT drugs should have improved *antigenicity* to advance the potential of BT therapy. High antigenicity, unexpected by lack of SBA calculations and undetected by insufficient registration trials, may lead to failure of the drug in the market as seen with the registration of Myobloc®. Trying to provide data to apply for drug registration based on a *biosimilarity approach* is challenging, but may have huge potential marketing implications. Most new BT drugs, however, are based on a business model using mainly Botox® analogons and trying to get a market share by *price reduction*.

References

Antonucci F, Rossi C, Gianfranceschi L, Rossetto O, Caleo M (2008) Long-distance retrograde effects of botulinum neurotoxin A. J Neurosci 28:3689–3696

Aurora SK, Dodick DW, Turkel CC, De Gryse RE, Silberstein SD, Lipton RB, Diener HC, Brin MF, PREEMPT 1 Chronic Migraine Study Group (2010) OnabotulinumtoxinA for treatment of chronic migraine: results from the double-blind, randomized, placebo-controlled phase of the PREEMPT 1 trial. Cephalalgia 30:793–803

Brin MF, Blitzer A (1993) Botulinum toxin: dangerous terminology errors. J R Soc Med 86:493–494

Diener HC, Dodick DW, Aurora SK, Turkel CC, RE DG, Lipton RB, Silberstein SD, Brin MF, PREEMPT 2 Chronic Migraine Study Group (2010) OnabotulinumtoxinA for treatment of chronic migraine: results from the double-blind, randomized, placebo-controlled phase of the PREEMPT 2 trial. Cephalalgia 30:804–814

Dressler D (2004) Clinical presentation and management of antibody-induced failure of botulinum toxin therapy. Mov Disord 19(Suppl 8):S92–S100

Dressler D (2009) Routine use of Xeomin® in patients pre-treated with Botox®. Eur J Neurol 16 (Suppl 2):2–5

Dressler D (2010) Subclinical myasthenia gravis causing increased sensitivity to botulinum toxin therapy. J Neural Transm 117:1293–1294

Dressler D, Adib Saberi F (2017a) Safety of botulinum toxin short interval therapy using incobotulinumtoxin A. J Neural Transm 124:437–440

Dressler D, Adib Saberi F (2017b) Economics of botulinum toxin therapy: influence of the abobotulinumtoxinA package size on the costs of botulinum toxin therapy. J Clin Mov Disord 4:6

Dressler D, Benecke R (2003) Autonomic side effects of botulinum toxin type B treatment of cervical dystonia and hyperhidrosis. Eur Neurol 49:34–38

Dressler D, Benecke R (2004) Autonomic side effects of botulinum toxin type B therapy. Adv Neurol 94:315–320

Dressler D, Benecke R (2007) Pharmacology of therapeutic botulinum toxin preparations. Disabil Rehabil 29:1761–1768

Dressler D, Bigalke H (2005) Botulinum toxin type B de novo therapy of cervical dystonia: frequency of antibody induced therapy failure. J Neurol 252:904–907

Dressler D, Bigalke H (2017a) Immunological aspects of botulinum toxin therapy. Expert Rev Neurother 17:487–494
Dressler D, Bigalke H (2017b) Long-term stability of reconstituted incobotulinumtoxinA: how can we reduce costs of botulinum toxin therapy? J Neural Transm 124:1223–1225
Dressler D, Foster K (2018) Pharmacology of botulinum toxins. In: Dressler D, Altenmüller E, Krauss JK (eds) Treatment of dystonia. Cambridge University Press, Cambridge
Dressler D, Hallett M (2006) Immunological aspects of Botox, Dysport and Myobloc/NeuroBloc. Eur J Neurol 13(Suppl 1):11–15
Dressler D, Rothwell JC (2000) Electromyographic quantification of the paralysing effect of botulinum toxin in the sternocleidomastoid muscle. Eur Neurol 43:13–16
Dressler D, Eckert J, Kukowski B, Meyer BU (1993) Somatosensorisch evozierte potentiale bei schreibkrampf: normalisierung pathologischer befunde unter Botulinum Toxin Therapie. Z EEG EMG 24:191
Dressler D, Bigalke H, Rothwell JC (2000) The sternocleidomastoid test: an in vivo assay to investigate botulinum toxin antibody formation in humans. J Neurol 247:630–632
Dressler D, Münchau A, Bhatia KP, Quinn NP, Bigalke H (2002) Antibody-induced botulinum toxin therapy failure: can it be overcome by increased botulinum toxin doses? Eur Neurol 47:118–121
Dressler D, Mander G, Fink K (2012) Measuring the potency labelling of onabotulinumtoxinA (Botox®) and Xeomin® in an LD50 assay. J Neural Transm 119:13–15
Dressler D, Paus S, Seitzinger A, Gebhardt B, Kupsch A (2013) Long-term efficacy and safety of incobotulinumtoxinA injections in patients with cervical dystonia. J Neurol Neurosurg Psychiatry 84:1014–1019
Dressler D, Gessler F, Tacik P, Bigalke H (2014) An enzyme-linked immunosorbent assay for detection of botulinum toxin-antibodies. Mov Disord 29:1322–1324
Dressler D, Adib Saberi F, Kollewe K, Schrader C (2015a) Safety aspects of incobotulinumtoxinA high-dose therapy. J Neural Transm 122:327–333
Dressler D, Tacik P, Adib Saberi F (2015b) Botulinum toxin therapy of cervical dystonia: duration of therapeutic effects. J Neural Transm 122:297–300
Dressler D, Adib Saberi F, Bigalke H (2016a) Botulinum toxin therapy: reduction of injection site pain by pH normalisation. J Neural Transm 123:527–531
Dressler D, Altenmueller E, Bhidayasiri R, Bohlega S, Chana P, Chung TM, Frucht S, Garcia-Ruiz PJ, Kaelin A, Kaji R, Kanovsky P, Laskawi R, Micheli F, Orlova O, Relja M, Rosales R, Slawek J, Timerbaeva S, Warner TT, Adib Saberi F (2016b) Strategies for treatment of dystonia. J Neural Transm 123:251–258
Dressler D, Pan L, Bigalke H (2018) Comparing incobotulinumtoxinA (Xeomin®) and onabotulinumtoxinA (Botox®): identical potency labelling in the hemidiaphragm assay. J Neural Transm 125:1351–1354
Erbguth F, Claus D, Engelhardt A, Dressler D (1993) Systemic effect of local botulinum toxin injections unmasks subclinical Lambert-Eaton myasthenic syndrome. J Neurol Neurosurg Psychiatry 56:1235–1236
European Pharmacopoeia 6.0 (2008a) Botulinum toxin type A for injection. pp 1327–1329
European Pharmacopoeia 6.0 (2008b) Statistical analysis of biological assays and tests. 5.3, pp 571–600
Filippi GM, Errico P, Santarelli R, Bagolini B, Manni E (1993) Botulinum A toxin effects on rat jaw muscle spindles. Acta Otolaryngol 113:400–404
First ER, Pearce LB, Borodic GE (1994) Dose standardisation of botulinum toxin. Lancet 343:1035
Goeschel H, Wohlfahrt K, Frevert J, Dengler R, Bigalke H (1997) Botulinum A toxin: neutralizing and nonneutralizing antibodies – therapeutic consequences. Exp Neurol 147:96–102
Jankovic J, Vuong KD, Ahsan J (2003) Comparison of efficacy and immunogenicity of original versus current botulinum toxin in cervical dystonia. Neurology 60:1186–1188
Kessler KR, Benecke R (1997) The EBD test--a clinical test for the detection of antibodies to botulinum toxin type A. Mov Disord 12:95–99

Lee JC, Yokota K, Arimitsu H, Hwang HJ, Sakaguchi Y, Cui J, Takeshi K, Watanabe T, Ohyama T, Oguma K (2005) Production of anti-neurotoxin antibody is enhanced by two subcomponents, HA1 and HA3b, of Clostridium botulinum type B 16S toxin-haemagglutinin. Microbiology 151:3739–3747

Marion MH, Sheehy M, Sangla S, Soulayrol S (1995) Dose standardisation of botulinum toxin. J Neurol Neurosurg Psychiatry 59:102–103

Marsden CD (1993) Botulinum toxin: dangerous terminology errors. J R Soc Med 86:494

Naumann M, Jost W (2004) Botulinum toxin treatment of secretory disorders. Mov Disord 19 (Suppl 8):S137–S141

Pickett AM, Hambleton P (1994) Dose standardisation of botulinum toxin. Lancet 344:474–475

Ranoux D, Gury C, Fondarai J, Mas JL, Zuber M (2002) Respective potencies of Botox and Dysport: a double blind, randomised, crossover study in cervical dystonia. J Neurol Neurosurg Psychiatry 72:459–462

Rosales RL, Arimura K, Takenaga S, Osame M (1996) Extrafusal and intrafusal muscle effects in experimental botulinum toxin-A injection. Muscle Nerve 19:488–496

Scaglione F (2016) Conversion ratio between Botox®, Dysport®, and Xeomin® in clinical practice. Toxins 8:E65. https://doi.org/10.3390/toxins8030065

Schrader C, Ebke M, Adib Saberi F, Dressler D (2018) Botulinum toxin therapy in patients with oral anticoagulation: is it safe? J Neural Transm 125:173–176

Takamizawa K, Iwamori M, Kozaki S, Sakaguchi G, Tanaka R, Takayama H, Nagai Y (1986) TLC immunostaining characterization of Clostridium botulinum type A neurotoxin binding to gangliosides and free fatty acids. FEBS Lett 201:229–232

Van den Bergh P, Lison D (1996) Dose standardisation of BTX. 3rd international dystonia symposium, October 9–11, 1996, Miami, Florida. Affiliated National Dystonia Associations, Chicago, p 30

Wiegand H, Erdmann G, Wellhöner HH (1976) 125I-labelled botulinum A neurotoxin: pharmacokinetics in cats after intramuscular injection. Naunyn Schmiedebergs Arch Pharmacol 292:161–165

Wissel J, Bensmail D, Ferreira JJ, Molteni F, Satkunam L, Moraleda S, Rekand T, McGuire J, Scheschonka A, Flatau-Baqué B, Simon O, Rochford ET, Dressler D, Simpson DM, TOWER Study Investigators (2017) Safety and efficacy of incobotulinumtoxinA doses up to 800 U in limb spasticity: the TOWER study. Neurology 88:1321–1328

Wollmer MA, de Boer C, Kalak N, Beck J, Götz T, Schmidt T, Hodzic M, Bayer U, Kollmann T, Kollewe K, Sönmez D, Duntsch K, Haug MD, Schedlowski M, Hatzinger M, Dressler D, Brand S, Holsboer-Trachsler E, Kruger TH (2012) Facing depression with botulinum toxin: a randomized controlled trial. J Psychiatr Res 46:574–581

The Use of Botulinum Toxin for Treatment of the Dystonias

Alfredo Berardelli and Antonella Conte

Contents

Abstract

Dystonias are characterized by involuntary muscle contractions, twisting movements, abnormal postures, and often tremor in various body regions. However, in the last decade several studies have demonstrated that dystonias are also characterized by sensory abnormalities. While botulinum toxin is the gold standard therapy for focal dystonia, exactly how it improves this disorder is not entirely understood. Neurophysiological studies in animals and humans have clearly demonstrated that botulinum toxin improves dystonic motor manifestations by inducing chemodenervation, therefore weakening the injected muscles. In addition, neurophysiological and neuroimaging evidence also suggests that botulinum toxin modulates the activity of various neural structures in the CNS distant from the injected site, particularly cortical motor and sensory areas. Concordantly, recent studies have shown that in patients with

The original version of this chapter was revised. A correction to this chapter can be found at https://doi.org/10.1007/164_2020_414

A. Berardelli (✉) · A. Conte
Department of Human Neurosciences, Sapienza University of Rome, Rome, Italy

IRCCS Neuromed, Pozzilli, IS, Italy
e-mail: alfredo.berardelli@uniroma1.it

S. M. Whitcup, M. Hallett (eds.), *Botulinum Toxin Therapy*,
Handbook of Experimental Pharmacology 263, https://doi.org/10.1007/164_2019_339

focal dystonias botulinum toxin ameliorates sensory disturbances, including reduced spatial discrimination acuity and pain. Overall, these observations suggest that in these patients botulinum toxin-induced effects encompass complex mechanisms beyond chemodenervation of the injected muscles.

Keywords
Blepharospasm · Botulinum toxin · Cervical dystonia · Focal dystonia · Mechanisms of action · Upper limb dystonia

Dystonia is currently defined as "sustained or intermittent muscle contractions causing abnormal, often repetitive, movements, postures, or both" (Albanese et al. 2013; Jinnah et al. 2013). Tremor can also be part of the motor phenomenology of dystonia and may be present in the body part affected by dystonia (dystonic tremor) or in a body part unaffected by dystonia (tremor associated with dystonia) (Defazio et al. 2015).

Dystonia can be classified as focal (only one region of the body is affected), segmental (adjacent regions of the body are affected), or generalized (several body parts are affected) (Jinnah et al. 2013; Albanese et al. 2013). Focal dystonias are comprised of cranial, cervical, and limb dystonias. Cranial dystonia includes blepharospasm characterized by involuntary eyelid closure (Defazio et al. 2017) and oromandibular dystonia (OMD) characterized by movements in the lower part of the face (Albanese et al. 2013). Cervical dystonia is characterized by neck muscle involvement. The most frequent form of cervical dystonia is torticollis (head turning to one side), but other forms also exist, including laterocollis (neck tilting to the side), retrocollis (neck extension), and anterocollis (neck flexion). Limb dystonia usually consists of involuntary contractions of the arms or legs associated with abnormal posturing, repetitive movements, and functional impairment. Upper limb dystonia is more common and usually has an adult onset, whereas lower limb dystonia is rare and usually affects people under 26 years of age. There are also forms of focal dystonia that appear only during the execution of specific movements, classified as task-specific dystonias. Task-specific dystonias typically affect the upper limb and are commonly seen in writers, piano players, typists, golfers, hairdressers, and in others who perform activities involving prolonged repetitive and stereotyped movements.

During the course of disease, dystonia can spread to adjacent body parts (Martino et al. 2012; Norris et al. 2016). The different types of dystonia variably tend to spread depending on age at onset and body distribution. In the last few decades, the non-motor features of dystonic patients have also attracted considerable attention and several studies have reported that patients with dystonia often complain of sensory symptoms, such as pain that often precedes or is associated with dystonic symptoms, as well as psychiatric and cognitive disorders (Heiman et al. 2004;

Lencer et al. 2009; Fabbrini et al. 2010, 2011; Berardelli et al. 2015, 2019; Conte et al. 2016; Norris et al. 2016; Berman et al. 2017; Jahanshahi and Rothwell 2017; Ferrazzano et al. 2019).

1 Pathophysiology of Dystonia

The anatomical basis for dystonia is still debated. Early studies in humans showed that basal ganglia lesions (Marsden et al. 1985) determined hemidystonia (Pettigrew and Jankovic 1985), cervical dystonia (LeDoux and Brady 2003), blepharospasm (Khooshnoodi et al. 2013), and upper limb dystonia (Liuzzi et al. 2016), suggesting that basal ganglia may play a prominent role in the pathophysiology of dystonia. In addition, structural lesions in the brainstem and cerebellum have been reported as determinants of cervical dystonia, oromandibular dystonia, and, less frequently, upper limb dystonia (Krauss et al. 1997; Tan et al. 2005; Agrawal et al. 2009; Kojovic et al. 2013; Ogawa et al. 2018; Corp et al. 2019), thus implying that both structures may be involved in dystonia.

Early neurophysiological studies investigating motor symptoms and reflexes evaluated inhibitory functions at various levels of the central nervous system (cortical motor areas, brainstem, and spinal cord) and reported loss of inhibition in patients with dystonia (Rothwell et al. 1983; Berardelli et al. 1985, 1998; Sohn and Hallett 2004; Beck et al. 2008; Hallett 2011; Quartarone and Hallett 2013; Jinnah et al. 2013; Balint et al. 2018) (Table 1). Loss of inhibition seems to be more evident in circuits whose functional role is relevant for the body part affected by dystonia (reduced cortical inhibitory mechanisms, including altered surround inhibition, and reduced spinal inhibition in patients with upper limb dystonia, reduced inhibitory mechanisms in the brainstem in patients with cranial and cervical dystonia). However, loss of inhibition may also alter other cortical and subcortical functions, including plasticity and sensorimotor integration (Conte et al. 2019).

Besides clinical motor symptoms, several abnormalities in sensory processing have also been identified, including altered spatial and temporal discrimination of tactile stimuli and distorted body representation in the primary sensory cortex, whose role in determining motor symptoms is still unclear (Stamelou et al. 2012; Patel et al. 2014; Hutchinson et al. 2018; Conte et al. 2019). Sensory deficits may cause disordered mechanisms of sensorimotor integration which are necessary for the execution of accurate movements. For example, sensory information can be attenuated ("gating") or prioritized during movement and consequently the motor output can be remodulated on the basis of unexpected salient sensory inputs that might disrupt the action. Abnormal sensory processing can therefore determine defective gating, as reported in studies with somatosensory-evoked potentials and somatosensory temporal discrimination thresholds (Murase et al. 2000; Macerollo et al. 2016, Conte et al. 2018).

Table 1 Neurophysiological abnormalities in dystonia

	Function	Neurophysiological techniques	Findings
Primary motor cortex	Inhibitory interneuron activity	SICI	Reduced intracortical inhibition
Primary motor cortex	Surround inhibition	Motor surround inhibition	Reduced inhibition of the surround muscles during hand muscle activation
Primary motor cortex	Cortical plasticity	PAS	Abnormally increased cortical plasticity
Brainstem	Trigemino-facial circuit excitability	Blink reflex recovery cycle	Increased excitability due to reduced descending inhibitory control
Spinal cord	Presynaptic inhibition of Ia afferents	Reciprocal inhibition	Reduced spinal reciprocal inhibition
Cerebellum	Pavlovian learning protocol	EBCC	Reduced conditioning of blinking
Cerebellum	Adaptive learning	Anticipatory adaptation	Normal adaptive learning or, if altered, related to the presence of tremor
Somatosensory system	Lateral inhibition	GOT	Increased spatial discrimination threshold
Somatosensory system	Inhibitory interneurons	STDT	Increased somatosensory temporal discrimination threshold
Somatosensory system	Nociceptive pathways	Laser-evoked potential	Normal latency and amplitude of laser evoked potentials

EBCC eyeblink classical conditioning, *GOT* grating orientation task, *PAS* paired associative plasticity, *SICI* short interval intracortical inhibition, *STDT* somatosensory temporal discrimination threshold

2 Therapeutic Strategies for the Various Forms of Dystonia

Notwithstanding several studies that have investigated the pathophysiology of focal dystonia, the role of various neurophysiological abnormalities in causing motor symptoms is still unclear. Therefore, treatment of focal dystonia is still symptomatic and usually chosen depending on distribution and severity of symptoms. The gold standard therapy for focal and segmental dystonias is botulinum toxin injected in the muscle affected by dystonia. Since the mid-1980s, several clinical trials have been published that highlight the positive short-term results of treatment with botulinum toxin for the various forms of focal dystonia. Two botulinum toxin serotypes, serotypes A and B, have been approved by the FDA and the European Medicines Agency (EMA). Botulinum toxin is administered by local injection to the target muscle and then it is distributed in the muscle belly by convection (Hallett 2015).

Botulinum toxin may also diffuse from the initial site by Brownian motion depending on the concentration gradient and molecular size (Hallett 2015). After injection, botulinum toxin is rapidly taken up into the presynaptic nerve terminals and exerts its paralytic action by inhibiting the release of acetylcholine, thereby blocking neuromuscular transmission. This results in temporary weakening of the muscle tissue that lasts up to 12–14 weeks. The most common adverse effects of botulinum toxin are transient and consist of excessive weakness of the injected muscle (for example, ptosis in patients with blepharospasm or neck weakness and dysphagia in patients with cervical dystonia).

3 Evidence-Based Guidelines for the Use of Botulinum Toxin in Dystonia

Practice guidelines for the use of botulinum toxin have variable levels of recommendations, but all guidelines agree that botulinum toxin is the preferred treatment for most patients with focal or segmental dystonia. In particular, for blepharospasm onabotulinumtoxinA and incobotulinumtoxinA injections should be considered Level B, and abobotulinumtoxinA may be considered Level C (Girlanda et al. 1996; Hsiung et al. 2002; Roggenkämper et al. 2006; Truong et al. 2008; Bentivoglio et al. 2009; Gill and Kraft 2010; Jankovic et al. 2011; Wabbels et al. 2011; Simpson et al. 2016) (Table 2). All three type-A toxins appear to have similar efficacy and are efficacious over long periods (Hsiung et al. 2002; Bentivoglio et al. 2009; Gill and Kraft 2010; Simpson et al. 2016). OMD is a rarer condition and the published literature is scarce. The literature review provides evidence to support only a Level C recommendation for the use of botulinum toxin type A for the treatment of OMD (Hallett et al. 2013; Comella 2018).

For the treatment of cervical dystonia, abobotulinumtoxinA, onabotulinumtoxinA, and rimabotulinumtoxinB should be offered (Level A), and incobotulinumtoxinA should be considered (Level B) (Contarino et al. 2017). However, although evidence levels may differ across botulinum toxin serotypes and brands, studies indicate similar efficacy for rimabotulinumtoxinB and onabotulinumtoxinA, and for abobotulinumtoxinA and onabotulinumtoxinA for the treatment of cervical dystonia (Geenen et al. 1996; Lew et al. 1997; Poewe et al. 1998; Brashear et al. 1999; Brin et al. 1999; Truong et al. 2005; Comella et al.

Table 2 Evidence-based guidelines for the use of botulinum toxins in the various forms of focal dystonia

Forms of dystonia	AbobotulinumtoxinA	OnabotulinumtoxinA	IncobotulinumtoxinA	RimabotulinumtoxinB
Blepharospasm	C	B	B	–
Cervical dystonia	A	A	B	A
Upper limb dystonia	B	B	–	–

A = strongly recommended; B = recommended; C = option

2005, 2011; Gill and Kraft 2010; Camargo et al. 2011; Simpson et al. 2016; Poewe et al. 2016). In a comparative study, Ranoux et al. (2002) reported that the effect duration was longer with the abobotulinumtoxinA dose regimen than with the onabotulinumtoxinA dose regimen (mean 114 days vs. 89.3 days for abobotulinumtoxinA and onabotulinumtoxinA, respectively). However, the longer clinical efficacy of abobotulinumtoxinA was associated with a greater frequency of adverse events (abobotulinumtoxinA: 36.0% vs. onabotulinumtoxinA: 17.6%). Jochim et al. (2019) analyzed the long-term efficacy and safety of onabotulinumtoxinA and abobotulinumtoxinA treatment in patients with cervical dystonia (2,592 onabotulinumtoxinA treatment sessions in 135 patients and 6,660 abobotulinumtoxinA treatment sessions in 209 patients) and found stable mean dose and injection intervals for both formulations, thus implying that therapy is safe and effective even over a long treatment duration.

As regards the long-term effect of botulinum toxin on patients with cervical dystonia, several studies (Brin et al. 2008; Gill and Kraft 2010; Camargo et al. 2011; Ramirez-Castaneda and Jankovic 2014; Bentivoglio et al. 2017; Colosimo et al. 2019) have documented the long-term efficacy of botulinum toxin and the study with the longest follow-up demonstrated that the efficacy of botulinum toxin persists for over 20 years (Mejia et al. 2005).

For limb dystonia, evidence supports a Level B recommendation for both abobotulinumtoxinA and onabotulinumtoxinA (Tsui et al. 1993; Cole et al. 1995; Kruisdijk et al. 2007; Contarino et al. 2007) (Table 2). No studies have been performed to test the efficacy of incobotulinumtoxinA or rimabotulinumtoxinB for limb dystonia, and therefore treatment efficacy is unproven for these two formulations. Given the complexity of the hand and the variability of patients, each patient will have to be individually considered for dose optimization.

As regards treatment response, it has been shown that a lack of response to botulinum toxin type A treatment in patients with dystonia may occur due to inadequate dosage, inappropriate muscle selection due to the non-use of EMG/ultrasound guide, a change in the pattern of dystonic muscle contractions, or even the development of neutralizing antibodies. Neutralizing antibodies may be the cause of an initially good clinical response followed by the therapeutic failure of subsequent injections (secondary non-response). It is known that the development of neutralizing antibodies may occur within the first years of botulinum toxin treatment and may be dose dependent (Jankovic and Schwartz 1995; Papapetropoulos and Singer 2006). A higher dose per session and frequent injections are associated with an increased risk of developing neutralizing antibodies and booster injections are discouraged (Fabbri et al. 2016). The protein load linked to the specific brand of botulinum toxin may also account for the development of neutralizing antibodies (Atassi 2015; Ferreira et al. 2015; Kutschenko et al. 2019). However, in current products the protein load is much lower (Brin et al. 2008) and in a recent study by Jochim et al. (2019) only two patients were positive for the antibody against botulinum toxin.

Furthermore, a new type of botulinum toxin, botulinum toxin type D, has been proposed for patients with neutralizing antibodies towards botulinum toxin types A and B. Botulinum toxin type D injections may represent a therapeutic alternative for non-responding patients with high-dose indications such as cervical dystonia due to their high sequence divergence antigenicity (Kutschenko et al. 2019). In a neurophysiological study in humans, Eleopra et al. (2013), however, reported that botulinum toxin type D is less effective than expected (Eleopra et al. 2013). Only limited evidence is available for the use of botulinum toxin serotypes C, E, and F in humans (Eleopra et al. 1998, 2006; Chen et al. 1998). Whereas serotype C showed a temporal profile similar to that of A and a positive clinical outcome, botulinum toxin serotypes E and F were both associated with a shorter duration of effects in comparison to serotype A (Eleopra et al. 1998).

In addition to response differences due to technical aspects related to dose and injection site, it has also been observed that patients with anxiety or depression respond less effectively to treatment. Thus, these symptoms may represent a psychopathological basis for response differences (Müller et al. 2002; Fabbrini et al. 2010).

4 Botulinum Toxin Mechanisms of Action in Dystonia

After local injection into muscles, botulinum toxin inhibits the vesicular release of acetylcholine. Botulinum toxin binds and cleaves SNARE (soluble N-ethylmaleimide-sensitive factor attachment receptor) complex proteins, thus blocking acetylcholine release at the neuromuscular junction (Brin 1997). The resulting effect is a transient denervation and a decrease in muscle strength and contraction that makes the injected muscles less active. The ability of botulinum toxin to remain localized at the injection site reflects the remarkable safety of this therapy (Ramirez-Castaneda et al. 2013).

Botulinum toxin can modify synaptic transmission related to reflex activity involving gamma motor neurons and intrafusal muscle fibers (Priori et al. 1995; Rosales et al. 1996; Gilio et al. 2000; Trompetto et al. 2006; Currà and Berardelli 2009). Trompetto et al. (2006) showed that botulinum toxin acts differently on extra and intrafusal muscle spindles as measured by changes in M-wave and maximal voluntary contraction as well as by changes in the tonic vibration reflex. Interestingly, the authors also found that botulinum toxin induced a persistent clinical benefit even though indicators of extrafusal chemodenervation had fully recovered in patients with focal hand dystonia (Trompetto et al. 2006). There is also evidence that changes in afferent input determined by botulinum toxin injection modulate spinal cord excitability. Indeed, Wohlfarth et al. (2001) reported that F-waves from distant, non-injected muscles of patients with cervical dystonia mildly changed following botulinum toxin injection, possibly due to reduced spinal motoneuronal excitability. Moreover, when investigating recurrent inhibition in non-injected remote muscles of patients treated for lower limb spasticity, Marchand-Pauvert et al. (2013) demonstrated a decreased recurrent inhibition from soleus motoneurons (injected with botulinum toxin) to quadricep muscle motoneurons and suggested that botulinum toxin affects spinal synaptic transmission by acting on the cholinergic synapses of Renshaw cells in humans.

Several early animal studies demonstrated a retrograde transport of botulinum toxin (Habermann and Erdmann 1978; Montecucco and Schiavo 1994), but whether this is also true in humans remains highly debated. In cats, intramuscular injection of radiolabeled botulinum toxin in the gastrocnemius muscle has been radioactively observed in the sciatic nerve, the ipsilateral spinal ventral roots, and the spinal cord with a distal-proximal gradient (Wiegand et al. 1976). In addition, Antonucci et al. (2008) reported that botulinum toxin was found in facial nucleus neurons after injection in the whisker muscles. In rats, functional evidence for bilateral muscle relaxation was observed after unilateral injection of commercially used botulinum toxin in the paw (Akaike et al. 2013). However, the dose used in this experiment was higher than that used for clinical purposes. In a recent study on rats, Caleo and Restani (2018) demonstrated that botulinum toxin type A is retrogradely trafficked to brainstem motoneurons, released within the facial nucleus to enter upstream neurons, and preferentially targets central cholinergic synapses. These observations were replicated using both high and low therapeutic toxin doses and in both the absence and presence of neurotoxin-associated proteins. However, it is still unclear whether this type of botulinum toxin diffusion may also account for central effects in humans.

Conversely, neurophysiological studies investigating the central effects of botulinum toxin on the excitability of supraspinal neural structures have yielded controversial findings in humans. In patients with spasmodic dysphonia, botulinum toxin reduced the muscle activity of the injected and non-injected thyroarytenoid muscle, and this finding was interpreted as excitability changes at the brainstem level (Bielamowicz and Ludlow 2000). Using a neurophysiological protocol to test brainstem circuit excitability and plasticity in patients with blepharospasm, Quartarone et al. (2006) reported that the enhanced facilitation of the R2 response of the blink reflex, a neurophysiological hallmark of blepharospasm reflecting brainstem hyperexcitability, was normalized by botulinum toxin. However, enhanced plasticity and its normalization following botulinum toxin injection were not confirmed in a subsequent study (Zeuner et al. 2010), and other studies have not revealed any modulation of hyperexcitable brainstem pathways in patients with blepharospasm following botulinum toxin injection (Valls-Sole et al. 1994; Grandas et al. 1998; Conte et al. 2010). Finally, botulinum toxin did not affect brainstem auditory-evoked potentials in patients with craniocervical dystonia and hemifacial spasm (Ce 2000).

Studies on cortical excitability and plasticity revealed different results for the sensory and motor cortices. Kanovský et al. (1998) first showed that intramuscular injections of botulinum toxin normalized the abnormally enhanced somatosensory-evoked potential amplitudes in patients with cervical dystonia, thus implying that botulinum toxin-induced effects on the primary sensory cortex are possibly due to changes in afferent input processing. As regards motor areas, botulinum toxin changed the topography of the upper limb representation in the primary motor cortex in patients with focal hand dystonia, cervical dystonia, and primary hand tremor (Byrnes et al. 1998, 2005; Thickbroom et al. 2003). Besides motor maps, an earlier

Table 3 Effects of botulinum toxin injection on neurophysiological abnormalities

Structure tested	Neurophysiological techniques	Effects of botulinum toxin
Primary motor cortex	SICI	Normalized or failed to modify intracortical inhibition
Primary motor cortex	PAS	Reduced excessive cortical plasticity
Brainstem	Blink reflex recovery cycle	Left the increased blink reflex excitability unchanged
Spinal cord	Reciprocal inhibition	Restored spinal reciprocal inhibition
Primary somatosensory cortex	GOT	Reduced spatial discrimination threshold
Primary somatosensory cortex	STDT	Left somatosensory temporal discrimination threshold unchanged

GOT grating orientation task, *PAS* paired associative plasticity, *SICI* short interval intracortical inhibition, *STDT* somatosensory temporal discrimination threshold

study (Gilio et al. 2000) reported that botulinum toxin normalized increased motor cortical excitability and reduced intracortical inhibition. However, these botulinum toxin-induced effects on cortical excitability were not confirmed by subsequent studies (Boroojerdi et al. 2003; Allam et al. 2005) (Table 3).

Results from studies with paired associative stimulation may give further insights on this topic. Paired associative stimulation combines repetitive electrical stimulation of an upper limb peripheral nerve with subsequent transcranial magnetic stimulation of the contralateral motor cortex (Stefan et al. 2004). Abnormally enhanced changes in motor cortical excitability following this paired associative stimulation protocol in dystonia (Weise et al. 2006, 2011) were modulated by botulinum toxin injections in patients with cervical dystonia (Kojovic et al. 2011) (Table 3). Overall, neurophysiological investigations on the effects of botulinum toxin at the cortical level may suggest that changes in afferent input may modulate cortical activity of the somatosensory cortex, which in turn influences motor cortex activity.

Finally, results from some neurophysiological investigations on autonomic function (cardiovascular reflexes) using single-fiber electromyography (SFEMG) suggested that botulinum toxin central effects were the consequence of hematogenic spread. Some authors indeed showed decreased heart rate variability and altered baroreflex sensitivity in patients treated with botulinum toxin for focal dystonia (Meischner and Reichel 2005; Tiple et al. 2008). In addition, there is evidence that after botulinum toxin injection in the neck muscles, jitter tested with SFEMG increased in the extensor digitorum muscles in patients with cervical dystonia (Lange et al. 1987; Girlanda et al. 1992; Erdal et al. 1999). Finally, clinical reports of a flu-like syndrome in some patients injected with botulinum toxin may be considered a sign of hematogenous spread. However, evidence exists that the above reported autonomic and neuromuscular effects are subclinical, thus contradicting the hypothesized role of hematogenous spread in determining remote changes in the CNS. Moreover, it is known that botulinum toxin does not cross the blood–brain barrier. It is therefore unlikely that hematogenous spread explains the reported central effects of botulinum toxin.

Neuroimaging studies indicate distinct functional and structural changes in several brain regions induced by botulinum toxin injections (Weise et al. 2019). In patients with cervical dystonia, botulinum toxin treatment was associated with widespread changes in functional magnetic resonance imaging (fMRI)-measured activation within several brain regions, including the bilateral primary and secondary somatosensory cortex, the bilateral supplementary motor area, the contralateral primary motor cortex, and the cerebellum (Nevrlý et al. 2018). Several previous studies investigating cranial and cervical dystonia (Dresel et al. 2006, 2011; Opavský et al. 2011, 2012) showed significant changes within the sensorimotor network in patients receiving long-term treatment with botulinum toxin in comparison with healthy controls. Modulation of cortical activity induced by botulinum toxin was even found on fMRI 1 month after injection in patients with spasmodic dysphonia, thus implying both a short- and long-term central effect of botulinum toxin in focal dystonia. Patients with different types of focal dystonia also have impaired resting-state fMRI connectivity within the sensorimotor and basal ganglia network (Dresel et al. 2006, 2011, 2014; Mohammadi et al. 2012; Delnooz et al. 2012, 2015a, b; Haslinger et al. 2017; Jochim et al. 2018) and botulinum toxin at least in part modulated disease-related altered functional connectivity patterns (Delnooz et al. 2013, 2015a, b; Jochim et al. 2018). Furthermore, in patients with cervical dystonia Brodoehl et al. (2019) recently reported a synchronous activation of the putamen, thalamus, and motor cortex, reflecting a hyperactive direct striatal-thalamic-cortical pathway, and a botulinum toxin-induced reduction in connectivity between the putamen, sensorimotor cortex, thalamus, and subthalamic nucleus.

A few studies have investigated the structural changes induced by botulinum toxin in focal hand and cervical dystonia using diffusion tensor imaging (DTI) and reported that botulinum toxin normalized structural abnormalities in the thalamus and sensorimotor cortex 1 month after injection (Colosimo et al. 2005; Blood et al. 2006). Using voxel-based morphometry, Delnooz and colleagues (2015a, b) reported an increase in gray matter volume (GMV) exclusively within the right precentral sulcus following botulinum toxin treatment in patients with cervical dystonia and suggested that these findings reflect central consequences of modified peripheral sensory input. More recently, Blood et al. (2019) found that left/right asymmetry in brain white matter microstructure medial to the globus pallidus interna in patients with cervical dystonia was reduced 4 weeks after peripheral botulinum toxin injection, and that there was a linear relationship between the magnitude of white matter changes and clinical response to treatment. Finally, Weise et al. (2019) recently compared drug-naïve and drug-treated patients with cervical dystonia in order to differentiate disease and therapy-specific gray matter changes. Since the two groups differed in bilateral mesiotemporal gray matter volume, the authors concluded that their findings reflected long-term effects of continuous botulinum toxin therapy.

In conclusion, results from neurophysiological and neuroimaging investigations in humans support the hypothesis that the central effects induced by botulinum toxin in patients with focal dystonia are more likely secondary manifestations of changes in afferent input processing rather than the result of retrograde/hematogenous diffusion of botulinum to the CNS.

5 Botulinum Toxin Effects on Sensory Abnormalities in Focal Dystonia

Patients with focal dystonia have altered spatial and temporal discrimination thresholds and these sensory abnormalities likely involve defective lateral and feedforward inhibition in the somatosensory pathways (Stamelou et al. 2012; Conte et al. 2019). In a study by Walsh and Hutchinson (2007) that tested the effects of botulinum toxin injection on the spatial discrimination threshold, which reflects mechanisms of lateral inhibition in the primary sensory cortex, the authors found a significant improvement after botulinum toxin injection, thus implying cortical reorganization of sensory areas. Conversely, botulinum toxin failed to restore the altered temporal discrimination threshold (Scontrini et al. 2011). This evidence is in line with the current hypothesis that an abnormal temporal discrimination threshold reflects feedforward inhibition mechanisms that cause increased susceptibility to dystonia but are not directly linked to the emergence of motor symptoms (Conte et al. 2019).

Besides tactile information processing, pain is another non-motor disturbance frequently reported by patients with focal dystonia. Pain is often the reason patients with cervical dystonia seek treatment (Avenali et al. 2018; Marciniec et al. 2019; Novaretti et al. 2019). A multicenter study showed that up to 90% of subjects reported pain associated with cervical dystonia (Charles et al. 2014), though it is not clear whether pain contributes directly to, or is a consequence of, increased severity of the condition. The pain perception threshold is decreased in patients with cervical dystonia (Paracka et al. 2017), although laser-evoked potentials have been found to be normal in cervical dystonia patients both with and without pain, thus implying that cutaneous nociceptive pathway function is normal and muscle pain is not due to any central sensitization of nociceptive inputs in either painful or non-painful body areas (Tinazzi et al. 2012).

The higher prevalence of pain in affected and adjacent regions and the pain relief after local treatment with botulinum toxin may suggest that pain arises due to local effects induced by dystonic activity (Kutvonen et al. 1997). Botulinum toxin may improve pain by reducing the tone and volume of dystonic muscles and increasing tissue perfusion and oxygenation, thus improving muscle metabolism and eliminating sensitization (Wissel et al. 2001). However, one-third of patients do not report pain and patients with pain do not relate it to the intensity of motor symptoms (Camargo et al. 2015). Furthermore, pain relief without improvement in motor symptoms in patients with bilateral globus pallidus interna deep brain stimulation and the presence of pain in areas far from the dystonia suggest other central mechanisms of pain generation (Kulisevsky et al. 2000).

In line with the hypothesis that pain arises from complex mechanisms and not simply from muscle conditions per se, evidence suggests that the time course of pain relief differs from that of botulinum toxin-induced muscle weakness in cervical dystonia (Marciniec et al. 2019). Pain improvement often occurs before motor improvement and pain relief lasts longer than muscle weakness (Relja and Miletić 2017). In a recent study, Tinazzi et al. (2019) reported a normal baseline pain rating and laser-evoked potentials in patients with cervical dystonia, therefore confirming that there is no overactivity of the ascending pain pathways. Conversely, this same study also reported that these patients had a reduced response to a conditioned pain modulation protocol as compared to healthy subjects and patients with cranial dystonia. The authors thus suggested that the endogenous inhibitory pain system may be defective in patients with cervical dystonia. However, since the authors did not test the effect of botulinum toxin on this abnormality it remains to be determined whether pain relief after botulinum toxin is due to the modulation of the endogenous inhibitory pain system.

6 Botulinum Toxin Effects on Tremor in Dystonia

Several neurophysiological studies in humans have investigated the pathophysiological mechanisms of tremor in dystonia and concurred that tremor arises from the entrainment of cerebellar output in the dysfunctional network connectivity typical of dystonia (Sadnicka et al. 2014, 2015, 2018; Defazio et al. 2015; Antelmi et al. 2016; Avanzino et al. 2018). The use of botulinum toxin injections to improve tremor in dystonia is still based on anecdotal observations (Mittal et al. 2019). For example, Niemann and Jankovic (2018) used onabotulinumtoxinA to treat dystonic hand tremor in 31 patients. Papapetropoulos and Singer (2006) injected botulinum toxin in four patients with primary writing tremor and all patients reported moderate motor improvement. Finally, botulinum toxin injections have been employed in patients with head tremor (Borodic et al. 1991; Boghen and Flanders 1993) and tremor improved in about 50–65% of these patients (Borodic et al. 1991; Boghen and Flanders 1993).

Since the muscle afferents influence the activity of motor network structures including the motor cortex, thalamus, and cerebellum, a botulinum toxin-induced decrease in afferent input may dampen peripheral oscillation as well as the afferent input arriving at the central structures, thus leading to a decrease in central oscillatory activity and consequently in tremor.

References

Agrawal A, Cincu R, Joharapurkar SR, Bhake A, Hiwale KM (2009) Hemorrhage in brainstem cavernoma presenting with torticollis. Pediatr Neurosurg 45(1):49–52

Akaike N, Shin MC, Wakita M, Torii Y, Harakawa T, Ginnaga A, Kato K, Kaji R, Kozaki S (2013) Transsynaptic inhibition of spinal transmission by A2 botulinum toxin. J Physiol 591 (4):1031–1043
Albanese A, Bathia K, Bressman SB et al (2013) Phenomenology and classification of dystonia: a consensus update. Mov Disord 28:863–873
Allam N, Fonte-Boa PM, Tomaz CA, Brasil-Neto JP (2005) Lack of effect of botulinum toxin on cortical excitability in patients with cranial dystonia. Clin Neuropharmacol 28(1):1–5
Antelmi E, Di Stasio F, Rocchi L et al (2016) Impaired eye blink classical conditioning distinguishes dystonic patients with and without tremor. Parkinsonism Relat Disord 31:23–27
Antonucci F, Rossi C, Gianfranceschi L, Rossetto O, Caleo M (2008) Long-distance retrograde effects of botulinum neurotoxin A. J Neurosci 28(14):3689–3696
Atassi MZ (2015) Molecular basis of immunogenicity to botulinum neurotoxins and uses of the defined antigenic regions. Toxicon 107(Pt A):50–58
Avanzino L, Ravaschio A, Lagravinese G, Bonassi G, Abbruzzese G, Pelosin E (2018) Adaptation of feedforward movement control is abnormal in patients with cervical dystonia and tremor. Clin Neurophysiol 129:319–326
Avenali M, De Icco R, Tinazzi M, Defazio G, Tronconi L, Sandrini G, Tassorelli C (2018) Pain in focal dystonias – a focused review to address an important component of the disease. Parkinsonism Relat Disord 54:17–24
Balint B, Mencacci NE, Valente EM et al (2018) Dystonia. Nat Rev Dis Primers 4:25
Beck S, Richardson SP, Shamim EA, Dang N, Schubert M, Hallett M (2008) Short intracortical and surround inhibition are selectively reduced during movement initiation in focal hand dystonia. J Neurosci 28:10363–10369
Bentivoglio AR, Fasano A, Ialongo T, Soleti F, Lo Fermo S, Albanese A (2009) Fifteen-year experience in treating blepharospasm with Botox or Dysport: same toxin, two drugs. Neurotox Res 15(3):224–231
Bentivoglio AR, Di Stasio E, Mulas D, Cerbarano ML, Ialongo T, Laurienzo A, Petracca M (2017) Long-term abobotulinumtoxin A treatment of cervical dystonia. Neurotox Res 32(2):291–300
Berardelli A, Rothwell JC, Day BL, Marsden CD (1985) Pathophysiology of blepharospasm and oromandibular dystonia. Brain 108:593–608
Berardelli A, Rothwell JC, Hallett M, Thompson PD, Manfredi M, Marsden CD (1998) The pathophysiology of primary dystonia. Brain 121:1195–1212
Berardelli I, Ferrazzano G, Pasquini M, Biondi M, Berardelli A, Fabbrini G (2015) Clinical course of psychiatric disorders in patients with cervical dystonia. Psychiatry Res 229(1-2):583–585
Berardelli I, Pasquini M, Conte A, Bologna M, Berardelli A, Fabbrini G (2019) Treatment of psychiatric disturbances in common hyperkinetic movement disorders. Expert Rev Neurother 19(1):55–65
Berman BD, Junker J, Shelton E et al (2017) Psychiatric associations of adult-onset focal dystonia phenotypes. J Neurol Neurosurg Psychiatry 88:595–602
Bielamowicz S, Ludlow CL (2000) Effects of botulinum toxin on pathophysiology in spasmodic dysphonia. Ann Otol Rhinol Laryngol 109(2):194–203
Blood AJ, Tuch DS, Makris N, Makhlouf ML, Sudarsky LR, Sharma N (2006) White matter abnormalities in dystonia normalize after botulinum toxin treatment. Neuroreport 17:1251–1255
Blood AJ, Kuster JK, Waugh JL, Levenstein JM, Multhaupt-Buell TJ, Sudarsky LR, Breiter HC, Sharma N (2019) White matter changes in cervical dystonia relate to clinical effectiveness of botulinum toxin treatment. Front Neurol 10:265
Boghen D, Flanders M (1993) Effectiveness of botulinum toxin in the treatment of spasmodic torticollis. Eur Neurol 33(3):199–203
Borodic GE, Mills L, Joseph M (1991) Botulinum A toxin for the treatment of adult-onset spasmodic torticollis. Plast Reconstr Surg 87(2):285–289
Boroojerdi B, Cohen LG, Hallett M (2003) Effects of botulinum toxin on motor system excitability in patients with writer's cramp. Neurology 61(11):1546–1550
Brashear A, Lew MF, Dykstra DD, Comella CL, Factor SA, Rodnitzky RL, Trosch R, Singer C, Brin MF, Murray JJ, Wallace JD, Willmer-Hulme A, Koller M (1999) Safety and efficacy of

NeuroBloc (botulinum toxin type B) in type A-responsive cervical dystonia. Neurology 53 (7):1439–1446

Brin MF (1997) Botulinum toxin: chemistry, pharmacology, toxicity, and immunology. Muscle Nerve Suppl 6:S146–S168

Brin MF, Lew MF, Adler CH, Comella CL, Factor SA, Jankovic J, O'Brien C, Murray JJ, Wallace JD, Willmer-Hulme A, Koller M (1999) Safety and efficacy of NeuroBloc (botulinum toxin type B) in type A-resistant cervical dystonia. Neurology 53(7):1431–1438

Brin MF, Comella CL, Jankovic J, Lai F, Naumann M, CD-017 BoNTA Study Group (2008) Long-term treatment with botulinum toxin type A in cervical dystonia has low immunogenicity by mouse protection assay. Mov Disord 23(10):1353–1360

Brodoehl S, Wagner F, Prell T, Klingner C, Witte OW, Günther A (2019) Cause or effect: altered brain and network activity in cervical dystonia is partially normalized by botulinum toxin treatment. Neuroimage Clin 22:101792

Byrnes ML, Thickbroom GW, Wilson SA, Sacco P, Shipman JM, Stell R, Mastaglia FL (1998) The corticomotor representation of upper limb muscles in writer's cramp and changes following botulinum toxin injection. Brain 121(Pt 5):977–988

Caleo M, Restani L (2018) Direct central nervous system effects of botulinum neurotoxin. Toxicon 147:68–72

Camargo CH, Teive HA, Becker N, Munhoz RP, Werneck LC (2011) Botulinum toxin type A and cervical dystonia: a seven-year follow-up. Arq Neuropsiquiatr 69(5):745–750

Camargo CH, Cattai L, Teive HA (2015) Pain relief in cervical dystonia with botulinum toxin treatment. Toxins (Basel) 7(6):2321–2335

Ce P (2000) Central effects of botulinum toxin: study of brainstem auditory evoked potentials. Eur J Neurol 7(6):747

Charles PD, Adler CH, Stacy M, Comella C, Jankovic J, Manack Adams A, Schwartz M, Brin MF (2014) Cervical dystonia and pain: characteristics and treatment patterns from CD PROBE (cervical dystonia patient registry for observation of onabotulinumtoxinA efficacy). J Neurol 261:1309–1319

Chen R, Karp BI, Hallett M (1998) Botulinum toxin type F for treatment of dystonia: long-term experience. Neurology 51:1494–1496

Cole R, Hallett M, Cohen LG (1995) Double-blind trial of botulinum toxin for treatment of focal hand dystonia. Mov Disord 10(4):466–471

Colosimo C, Pantano P, Calistri V, Totaro P, Fabbrini G, Berardelli A (2005) Diffusion tensor imaging in primary cervical dystonia. J Neurol Neurosurg Psychiatry 76:1591–1593

Colosimo C, Charles D, Misra VP, Maisonobe P, Om S, INTEREST IN CD2 Study Group (2019) How satisfied are cervical dystonia patients after 3 years of botulinum toxin type A treatment? Results from a prospective, long-term observational study. J Neurol. https://doi.org/10.1007/s00415-019-09527-2

Comella CL (2018) Systematic review of botulinum toxin treatment for oromandibular dystonia. Toxicon 147:96–99

Comella CL, Jankovic J, Shannon KM, Tsui J, Swenson M, Leurgans S, Fan W, Dystonia Study Group (2005) Comparison of botulinum toxin serotypes A and B for the treatment of cervical dystonia. Neurology 65(9):1423–1429

Comella CL, Jankovic J, Truong DD, Hanschmann A, Grafe S, U.S. XEOMIN Cervical Dystonia Study Group (2011) Efficacy and safety of incobotulinumtoxinA (NT 201, XEOMIN®, botulinum neurotoxin type A, without accessory proteins) in patients with cervical dystonia. J Neurol Sci 308(1–2):103-109

Contarino MF, Kruisdijk JJ, Koster L, Ongerboer de Visser BW, Speelman JD, Koelman JH (2007) Sensory integration in writer's cramp: comparison with controls and evaluation of botulinum toxin effect. Clin Neurophysiol 118(10):2195–2206

Contarino MF, Van Den Dool J, Balash Y, Bhatia K, Giladi N, Koelman JH, Lokkegaard A, Marti MJ, Postma M, Relja M, Skorvanek M, Speelman JD, Zoons E, Ferreira JJ, Vidailhet M,

Abanese A, Tijssen MA (2017) Clinical practice: evidence-based recommendations for the treatment of cervical dystonia with botulinum toxin. Front Neurol 8:35

Conte A, Fabbrini G, Belvisi D, Marsili L, Di Stasio F, Berardelli A (2010) Electrical activation of the orbicularis oculi muscle does not increase the effectiveness of botulinum toxin type A in patients with blepharospasm. Eur J Neurol 17:449–455

Conte A, Berardelli I, Ferrazzano G, Pasquini M, Berardelli A, Fabbrini G (2016) Non-motor symptoms in patients with adult-onset focal dystonia: sensory and psychiatric disturbances. Parkinsonism Relat Disord 22:S111–S114

Conte A, Belvisi D, De Bartolo MI, Manzo N, Cortese FN, Tartaglia M, Ferrazzano G, Fabbrini G, Berardelli A (2018) Abnormal sensory gating in patients with different types of focal dystonias. Mov Disord 33(12):1910–1917

Conte A, Defazio G, Hallett M, Fabbrini G, Berardelli A (2019) The role of sensory information in the pathophysiology of focal dystonias. Nat Rev Neurol 15:224–233

Corp DT, Joutsa J, Darby RR, Delnooz CCS, van de Warrenburg BPC, Cooke D, Prudente CN, Ren J, Reich MM, Batla A, Bhatia KP, Jinnah HA, Liu H, Fox MD (2019) Network localization of cervical dystonia based on causal brain lesions. Brain 142(6):1660–1674

Currà A, Berardelli A (2009) Do the unintended actions of botulinum toxin at distant sites have clinical implications? Neurology 72(12):1095–1099

Defazio G, Conte A, Gigante AF, Fabbrini G, Berardelli A (2015) Is tremor in dystonia a phenotypic feature of dystonia? Neurology 84:1053–1059

Defazio G, Hallett M, Jinnah HA, Conte A, Berardelli A (2017) Blepharospasm 40 years later. Mov Disord 32:498–509

Delnooz CCS, Helmich RC, Toni I, van de Warrenburg BPC (2012) Reduced parietal connectivity with a premotor writing area in writer's cramp. Mov Disord 27:1425–1431

Delnooz CCS, Pasman JW, Beckmann CF, van de Warrenburg BPC (2013) Task-free functional MRI in cervical dystonia reveals multi-network changes that partially normalize with botulinum toxin. PLoS One 8:e62877

Delnooz CCS, Pasman JW, Beckmann CF, van de Warrenburg BPC (2015a) Altered striatal and pallidal connectivity in cervical dystonia. Brain Struct Funct 220:513–523

Delnooz CCS, Pasman JW, van de Warrenburg BPC (2015b) Dynamic cortical gray matter volume changes after botulinum toxin in cervical dystonia. Neurobiol Dis 73:327–333

Dresel C, Haslinger B, Castrop F, Wohlschlaeger AM, Ceballos-Baumann AO (2006) Silent event-related fMRI reveals deficient motor and enhanced somatosensory activation in orofacial dystonia. Brain 129:36–46

Dresel C, Bayer F, Castrop F, Rimpau C, Zimmer C, Haslinger B (2011) Botulinum toxin modulates basal ganglia but not deficient somatosensory activation in orofacial dystonia. Mov Disord 26:1496–1502

Dresel C, Li Y, Wilzeck V, Castrop F, Zimmer C, Haslinger B (2014) Multiple changes of functional connectivity between sensorimotor areas in focal hand dystonia. J Neurol Neurosurg Psychiatry 85:1245–1252

Eleopra R, Tugnoli V, Rossetto O, De Grandis D, Montecucco C (1998) Different time courses of recovery after poisoning with botulinum neurotoxin serotypes A and E in humans. Neurosci Lett 256:135–138

Eleopra R, Tugnoli V, Quatrale R, Rossetto O, Montecucco C, Dressler D (2006) Clinical use of non-A botulinum toxins: botulinum toxin type C and botulinum toxin type F. Neurotox Res 9:127–131

Eleopra R, Montecucco C, Devigili G, Lettieri C, Rinaldo S, Verriello L, Pirazzini M, Caccin P, Rossetto O (2013) Botulinum neurotoxin serotype D is poorly effective in humans: an in vivo electrophysiological study. Clin Neurophysiol 124:999–1004

Erdal J, Ostergaard L, Fuglsang-Frederiksen A, Werdelin L, Dalager T, Sjö O, Regeur L (1999) Long-term botulinum toxin treatment of cervical dystonia – EMG changes in injected and noninjected muscles. Clin Neurophysiol 110:1650–1654

Fabbri M, Leodori G, Fernandes RM, Bhidayasiri R, Marti MJ, Colosimo C, Ferreira JJ (2016) Neutralizing antibody and botulinum toxin therapy: a systematic review and meta-analysis. Neurotox Res 29(1):105–117

Fabbrini G, Berardelli I, Moretti G et al (2010) Psychiatric disorders in adult-onset focal dystonia: a case-control study. Mov Disord 25:459–465

Fabbrini G, Berardelli I, Moretti G, Pasquini M, Colosimo C, Berardelli A (2011) Nonmotor symptoms in adult-onset focal dystonia: psychiatric abnormalities. Mov Disord 26:1765–1766

Ferrazzano G, Berardelli I, Conte A, Baione V, Concolato C, Belvisi D, Fabbrini G, Defazio G, Berardelli A (2019) Motor and non-motor symptoms in blepharospasm: clinical and pathophysiological implications. J Neurol. https://doi.org/10.1007/s00415-019-09484-w

Ferreira JJ, Colosimo C, Bhidayasiri R, Marti MJ, Maisonobe P, Om S (2015) Factors influencing secondary non-response to botulinum toxin type A injections in cervical dystonia. Parkinsonism Relat Disord 21(2):111–115

Geenen C, Consky E, Ashby P (1996) Localizing muscles for botulinum toxin treatment of focal hand dystonia. Can J Neurol Sci 23(3):194–197

Gilio F, Currà A, Lorenzano C, Modugno N, Manfredi M, Berardelli A (2000) Effects of botulinum toxin type A on intracortical inhibition in patients with dystonia. Ann Neurol 48(1):20–26

Gill HS, Kraft SP (2010) Long-term efficacy of botulinum a toxin for blepharospasm and hemifacial spasm. Can J Neurol Sci 37(5):631–636

Girlanda P, Vita G, Nicolosi C, Milone S, Messina C (1992) Botulinum toxin therapy: distant effects on neuromuscular transmission and autonomic nervous system. J Neurol Neurosurg Psychiatry 55(9):844–845

Girlanda P, Quartarone A, Sinicropi S, Nicolosi C, Messina C (1996) Unilateral injection of botulinum toxin in blepharospasm: single fiber electromyography and blink reflex study. Mov Disord 11(1):27–31

Grandas F, Traba A, Alonso F, Esteban A (1998) Blink reflex recovery cycle in patients with blepharospasm unilaterally treated with botulinum toxin. Clin Neuropharmacol 21:307–311

Habermann E, Erdmann G (1978) Pharmacokinetic and histoautoradiographic evidence for the intraaxonal movement of toxin in the pathogenesis of tetanus. Toxicon 16(6):611–623

Hallett M (2011) Neurophysiology of dystonia: the role of inhibition. Neurobiol Dis 42:177–184

Hallett M (2015) Explanation of timing of botulinum neurotoxin effects, onset and duration, and clinical ways of influencing them. Toxicon 107:64–67

Hallett M, Albanese A, Dressler D, Segal KR, Simpson DM, Truong D, Jankovic J (2013) Evidence-based review and assessment of botulinum neurotoxin for the treatment of movement disorders. Toxicon 67:94–114

Haslinger B, Noé J, Altenmüller E, Riedl V, Zimmer C, Mantel T, Dresel C (2017) Changes in resting-state connectivity in musicians with embouchure dystonia. Mov Disord 32:450–458

Heiman GA, Ottman R, Saunders-Pullman RJ, Ozelius LJ, Risch NJ, Bressman SB (2004) Increased risk for recurrent major depression in DYT1 dystonia mutation carriers. Neurology 63:631–637

Hsiung GY, Das SK, Ranawaya R, Lafontaine AL, Suchowersky O (2002) Long-term efficacy of botulinum toxin A in treatment of various movement disorders over a 10-year period. Mov Disord 17(6):1288–1293

Hutchinson M, McGovern EM, Narasimham S et al (2018) The premotor syndrome of cervical dystonia: disordered processing of salient environmental stimuli. Mov Disord 33:232–237

Jahanshahi M, Rothwell JC (2017) Inhibitory dysfunction contributes to some of the motor and non-motor symptoms of movement disorders and psychiatric disorders. Philos Trans R Soc Lond B Biol Sci 372:1718

Jankovic J, Schwartz K (1995) Response and immunoresistance to botulinum toxin injections. Neurology 45(9):1743–1746

Jankovic J, Comella C, Hanschmann A, Grafe S (2011) Efficacy and safety of incobotulinumtoxinA (NT 201, Xeomin) in the treatment of blepharospasm-a randomized trial. Mov Disord 26 (8):1521–1528

Jinnah HA, Berardelli A, Comella C, Defazio G, Delong MR, Factor S, Galpern WR, Hallett M, Ludlow CL, Perlmutter JS, Rosen AR, Dystonia Coalition Investigators (2013) The focal dystonias: current views and challenges for future research. Mov Disord 28(7):926–943

Jochim A, Li Y, Gora-Stahlberg G, Mantel T, Berndt M, Castrop F, Dresel C, Haslinger B (2018) Altered functional connectivity in blepharospasm/orofacial dystonia. Brain Behav 8:e00894

Jochim A, Meindl T, Mantel T, Zwirner S, Zech M, Castrop F, Haslinger B (2019) Treatment of cervical dystonia with abo- and onabotulinumtoxinA: long-term safety and efficacy in daily clinical practice. J Neurol 266(8):1879–1886

Kanovský P, Streitová H, Dufek J, Znojil V, Daniel P, Rektor I (1998) Change in lateralization of the P22/N30 cortical component of median nerve somatosensory evoked potentials in patients with cervical dystonia after successful treatment with botulinum toxin A. Mov Disord 13 (1):108–117

Khooshnoodi MA, Factor SA, Jinnah HA (2013) Secondary blepharospasm associated with structural lesions of the brain. J Neurol Sci 331:98–101

Kojovic M, Caronni A, Bologna M, Rothwell JC, Bhatia KP, Edwards MJ (2011) Botulinum toxin injections reduce associative plasticity in patients with primary dystonia. Mov Disord 26:1282–1289

Kojovic M, Parees I, Kassavetis P et al (2013) Secondary and primary dystonia: pathophysiological differences. Brain 136:2038–2049

Krauss JK, Seeger W, Jankovic J (1997) Cervical dystonia associated with tumors of the posterior fossa. Mov Disord 12(3):443–447

Kruisdijk JJ, Koelman JH, Ongerboer de Visser BW, de Haan RJ, Speelman JD (2007) Botulinum toxin for writer's cramp: a randomised, placebo-controlled trial and 1-year follow-up. J Neurol Neurosurg Psychiatry 78(3):264–270

Kulisevsky J, Lleó A, Gironell A, Molet J, Pascual-Sedano B, Parés P (2000) Bilateral pallidal stimulation for cervical dystonia: dissociated pain and motor improvement. Neurology 55 (11):1754–1755

Kutschenko A, Weisemann J, Kollewe K, Fiedler T, Alvermann S, Böselt S, Escher C, Garde N, Gingele S, Kaehler SB, Karatschai R, Krüger THC, Sikorra S, Tacik P, Wegner F, Wollmann J, Bigalke H, Wohlfarth K, Rummel A (2019) Botulinum neurotoxin serotype D – a potential treatment alternative for BoNT/A and B non-responding patients. Clin Neurophysiol 130 (6):1066–1073

Kutvonen O, Dastidar P, Nurmikko T (1997) Pain in spasmodic torticollis. Pain 69(3):279–286

Lange DJ, Brin MF, Warner CL, Fahn S, Lovelace RE (1987) Distant effects of local injection of botulinum toxin. Muscle Nerve 10:552–555

LeDoux MS, Brady KA (2003) Secondary cervical dystonia associated with structural lesions of the central nervous system. Mov Disord 18:60–69

Lencer R, Steinlechner S, Stahlberg J et al (2009) Primary focal dystonia: evidence for distinct neuropsychiatric and personality profiles. J Neurol Neurosurg Psychiatry 80:1176–1179

Lew MF, Adornato BT, Duane DD, Dykstra DD, Factor SA, Massey JM, Brin MF, Jankovic J, Rodnitzky RL, Singer C, Swenson MR, Tarsy D, Murray JJ, Koller M, Wallace JD (1997) Botulinum toxin type B: a double-blind, placebo-controlled, safety and efficacy study in cervical dystonia. Neurology 49(3):701–707

Liuzzi D, Gigante AF, Leo A, Defazio G (2016) The anatomical basis of upper limb dystonia: lesson from secondary cases. Neurol Sci 37:1393–1398

Macerollo A, Chen JC, Parees I, Sadnicka A, Kassavetis P, Bhatia KP, Kilner JM, Rothwell JC, Edwards MJ (2016) Abnormal movement-related suppression of sensory evoked potentials in upper limb dystonia. Eur J Neurol 23(3):562–568

Marchand-Pauvert V, Aymard C, Giboin LS, Dominici F, Rossi A, Mazzocchio R (2013) Beyond muscular effects: depression of spinal recurrent inhibition after botulinum neurotoxin A. J Physiol 591(4):1017–1029

Marciniec M, Szczepańska-Szerej A, Kulczyński M, Sapko K, Popek-Marciniec S, Rejdak K (2019) Pain in cervical dystonia and the antinociceptive effects of botulinum toxin: what is currently known? Rev Neurosci. https://doi.org/10.1515/revneuro-2018-0119

Marsden CD, Obeso JA, Zarranz JJ, Lang AE (1985) The anatomical basis of symptomatic hemidystonia. Brain 108:463–483

Martino D, Berardelli A, Abbruzzese G, Bentivoglio AR, Esposito M, Fabbrini G, Guidubaldi A, Girlanda P, Liguori R, Marinelli L, Morgante F, Santoro L, Defazio G (2012) Age at onset and symptom spread in primary adult-onset blepharospasm and cervical dystonia. Mov Disord 27 (11):1447–1450

Mejia NI, Vuong KD, Jankovic J (2005) Long-term botulinum toxin efficacy, safety, and immunogenicity. Mov Disord 20(5):592–597

Mittal SO, Lenka A, Jankovic J (2019) Botulinum toxin for the treatment of tremor. Parkinsonism Relat Disord 63:31–41

Mohammadi B, Kollewe K, Samii A, Beckmann CF, Dengler R, Münte TF (2012) Changes in resting-state brain networks in writer's cramp. Hum Brain Mapp 33:840–848

Montecucco C, Schiavo G (1994) Mechanism of action of tetanus and botulinum neurotoxins. Mol Microbiol 13(1):1–8

Müller J, Kemmler G, Wissel J, Schneider A, Voller B, Grossmann J, Diez J, Homann N, Wenning GK, Schnider P, Poewe W, Austrian Botulinum toxin and Dystonia Study Group (2002) The impact of blepharospasm and cervical dystonia on health-related quality of life and depression. J Neurol 249(7):842–846

Murase N, Kaji R, Shimazu H, Katayama-Hirota M, Ikeda A, Kohara N, Kimura J, Shibasaki H, Rothwell JC (2000) Abnormal premovement gating of somatosensory input in writer's cramp. Brain 123(Pt 9):1813–1829

Nevrlý M, Hluštík P, Hok P, Otruba P, Tüdös Z, Kaňovský P (2018) Changes in sensorimotor network activation after botulinum toxin type A injections in patients with cervical dystonia: a functional MRI study. Exp Brain Res 236(10):2627–2637

Niemann N, Jankovic J (2018) Botulinum toxin for the treatment of hand tremor. Toxins (Basel) 10(7)

Norris SA, Jinnah HA, Espay AJ, Klein C, Brüggemann N, Barbano RL, Malaty IA, Rodriguez RL, Vidailhet M, Roze E, Reich SG, Berman BD, LeDoux MS, Richardson SP, Agarwal P, Mari Z, Ondo WG, Shih LC, Fox SH, Berardelli A, Testa CM, Cheng FC, Truong D, Nahab FB, Xie T, Hallett M, Rosen AR, Wright LJ, Perlmutter JS (2016) Clinical and demographic characteristics related to onset site and spread of cervical dystonia. Mov Disord 31(12):1874–1882

Novaretti N, Cunha ALN, Bezerra TC, Pena Pereira MA, de Oliveira DS, Macruz Brito MMC, Pimentel AV, Brozinga TR, Foss MP, Tumas V (2019) The prevalence and correlation of non-motor symptoms in adult patients with idiopathic focal or segmental dystonia. Tremor Other Hyperkinet Mov (N Y) 9:596

Ogawa T, Shojima Y, Kuroki T, Eguchi H, Hattori N, Miwa H (2018) Cervico-shoulder dystonia following lateral medullary infarction: a case report and review of the literature. J Med Case Reports 12(1):34

Opavský R, Hluštík P, Otruba P, Kaňovský P (2011) Sensorimotor network in cervical dystonia and the effect of botulinum toxin treatment: a functional MRI study. J Neurol Sci 306:71–75

Opavský R, Hluštík P, Otruba P, Kaňovský P (2012) Somatosensory cortical activation in cervical dystonia and its modulation with botulinum toxin: an fMRI study. Int J Neurosci 122:45–52

Papapetropoulos S, Singer C (2006) Treatment of primary writing tremor with botulinum toxin type a injections: report of a case series. Clin Neuropharmacol 29(6):364–367

Paracka L, Wegner F, Blahak C, Abdallat M, Saryyeva A, Dressler D, Karst M, Krauss JK (2017) Sensory alterations in patients with isolated idiopathic dystonia: an exploratory quantitative sensory testing analysis. Front Neurol 8:553

Patel N, Jankovic J, Hallett M (2014) Sensory aspects of movement disorders. Lancet Neurol 13:100–112

Pettigrew LC, Jankovic J (1985) Hemidystonia: a report of 22 patients and a review of the literature. J Neurol Neurosurg Psychiatry 48:650–657

Poewe W, Deuschl G, Nebe A, Feifel E, Wissel J, Benecke R, Kessler KR, Ceballos-Baumann AO, Ohly A, Oertel W, Künig G (1998) What is the optimal dose of botulinum toxin A in the treatment of cervical dystonia? Results of a double blind, placebo controlled, dose ranging study using Dysport. German Dystonia Study Group. J Neurol Neurosurg Psychiatry 64(1):13–17

Poewe W, Burbaud P, Castelnovo G, Jost WH, Ceballos-Baumann AO, Banach M, Potulska-Chromik A, Ferreira JJ, Bihari K, Ehler E, Bares M, Dzyak LA, Belova AN, Pham E, Liu WJ, Picaut P (2016) Efficacy and safety of abobotulinumtoxinA liquid formulation in cervical dystonia: a randomized-controlled trial. Mov Disord 31(11):1649–1657

Priori A, Berardelli A, Mercuri B, Manfredi M (1995) Physiological effects produced by botulinum toxin treatment of upper limb dystonia. Changes in reciprocal inhibition between forearm muscles. Brain 118:801–807

Quartarone A, Hallett M (2013) Emerging concepts in the physiological basis of dystonia. Mov Disord 28:958–967

Quartarone A, Sant'Angelo A, Battaglia F, Bagnato S, Rizzo V, Morgante F, Rothwell JC, Siebner HR, Girlanda P (2006) Enhanced long-term potentiation-like plasticity of the trigeminal blink reflex circuit in blepharospasm. J Neurosci 26(2):716–721

Ramirez-Castaneda J, Jankovic J (2014) Long-term efficacy, safety, and side effect profile of botulinum toxin in dystonia: a 20-year follow-up. Toxicon 90:344–348

Ramirez-Castaneda J, Jankovic J, Comella C, Dashtipour K, Fernandez HH, Mari Z (2013) Diffusion, spread, and migration of botulinum toxin. Mov Disord 28(13):1775–1783

Ranoux D, Gury C, Fondarai J, Mas JL, Zuber M (2002) Respective potencies of botox and Dysport: a double blind, randomised, crossover study in cervical dystonia. J Neurol Neurosurg Psychiatry 72(4):459–462

Relja M, Miletić V (2017) When movement disorders hurt: addressing pain in hyperkinetic disorders. Parkinsonism Relat Disord 44:110–113

Roggenkämper P, Jost WH, Bihari K, Comes G, Grafe S, Blepharospasm Study Team (2006) Efficacy and safety of a new Botulinum toxin type A free of complexing proteins in the treatment of blepharospasm. J Neural Transm (Vienna) 113(3):303–312

Rosales RL, Arimura K, Takenaga S, Osame M (1996) Extrafusal and intrafusal muscle effects in experimental botulinum toxin-A injection. Muscle Nerve 19(4):488–496

Rothwell JC, Obeso JA, Day BL, Marsden CD (1983) Pathophysiology of dystonias. Adv Neurol 39:851–863

Sadnicka A, Patani B, Saifee TA et al (2014) Normal motor adaptation in cervical dystonia: a fundamental cerebellar computation is intact. Cerebellum 13:558–567

Sadnicka A, Teo JT, Kojovic M et al (2015) All in the blink of an eye: new insight into cerebellar and brainstem function in DYT1 and DYT6 dystonia. Eur J Neurol 22:762–767

Sadnicka A, Stevenson A, Bhatia KP, Rothwell JC, Edwards MJ, Galea JM (2018) High motor variability in DYT1 dystonia is associated with impaired visuomotor adaptation. Sci Rep 8:3653

Scontrini A, Conte A, Fabbrini G, Colosimo C, Di Stasio F, Ferrazzano G, Berardelli A (2011) Somatosensory temporal discrimination tested in patients receiving botulinum toxin injection for cervical dystonia. Mov Disord 26:742–746

Simpson DM, Hallett M, Ashman EJ, Comella CL, Green MW, Gronseth GS, Armstrong MJ, Gloss D, Potrebic S, Jankovic J, Karp BP, Naumann M, So YT, Yablon SA (2016) Practice guideline update summary: botulinum neurotoxin for the treatment of blepharospasm, cervical dystonia, adult spasticity, and headache: report of the guideline development Subcommittee of the American Academy of Neurology. Neurology 86(19):1818–1826

Sohn YH, Hallett M (2004) Disturbed surround inhibition in focal hand dystonia. Ann Neurol 56:595–599

Stamelou M, Edwards MJ, Hallett M, Bhatia KP (2012) The non-motor syndrome of primary dystonia: clinical and pathophysiological implications. Brain 135:1668–1681

Stefan K, Wycislo M, Classen J (2004) Modulation of associative human motor cortical plasticity by attention. J Neurophysiol 92(1):66–72

Tan EK, Chan LL, Auchus AP (2005) Hemidystonia precipitated by acute pontine infarct. J Neurol Sci 234(1-2):109–111

Thickbroom GW, Byrnes ML, Stell R, Mastaglia FL (2003) Reversible reorganisation of the motor cortical representation of the hand in cervical dystonia. Mov Disord 18(4):395–402

Tinazzi M, Valeriani M, Squintani G, Corrà F, Recchia S, Defazio G, Berardelli A (2012) Nociceptive pathway function is normal in cervical dystonia: a study using laser-evoked potentials. J Neurol 259(10):2060–2066

Tinazzi M, Squintani GM, Bhatia KP, Segatti A, Donato F, Valeriani M, Erro R (2019) Pain in cervical dystonia: evidence of abnormal inhibitory control. Parkinsonism Relat Disord. https://doi.org/10.1016/j.parkreldis.2019.06.009

Tiple D, Strano S, Colosimo C, Fabbrini G, Calcagnini G, Prencipe M, Berardelli A (2008) Autonomic cardiovascular function and baroreflex sensitivity in patients with cervical dystonia receiving treatment with botulinum toxin type A. J Neurol 255(6):843–847

Trompetto C, Currà A, Buccolieri A, Suppa A, Abbruzzese G, Berardelli A (2006) Botulinum toxin changes intrafusal feedback in dystonia: a study with the tonic vibration reflex. Mov Disord 21 (6):777–782

Truong D, Duane DD, Jankovic J, Singer C, Seeberger LC, Comella CL, Lew MF, Rodnitzky RL, Danisi FO, Sutton JP, Charles PD, Hauser RA, Sheean GL (2005) Efficacy and safety of botulinum type A toxin (Dysport) in cervical dystonia: results of the first US randomized, double-blind, placebo-controlled study. Mov Disord 20(7):783–791

Truong D, Comella C, Fernandez HH, Ondo WG, Dysport Benign Essential Blepharospasm Study Group (2008) Efficacy and safety of purified botulinum toxin type A (Dysport) for the treatment of benign essential blepharospasm: a randomized, placebo-controlled, phase II trial. Parkinsonism Relat Disord 14(5):407–414

Tsui JK, Bhatt M, Calne S, Calne DB (1993) Botulinum toxin in the treatment of writer's cramp: a double-blind study. Neurology 43(1):183–185

Valls-Sole J, Tolosa ES, Marti MJ, Allam N (1994) Treatment with botulinum toxin injections does not change brainstem interneuronal excitability in patients with cervical dystonia. Clin Neuropharmacol 17(3):229–235

Wabbels B, Reichel G, Fulford-Smith A, Wright N, Roggenkämper P (2011) Double-blind, randomised, parallel group pilot study comparing two botulinum toxin type A products for the treatment of blepharospasm. J Neural Transm (Vienna) 118(2):233–239

Walsh R, Hutchinson M (2007) Molding the sensory cortex: spatial acuity improves after botulinum toxin treatment for cervical dystonia. Mov Disord 22:2443–2446

Weise D, Schramm A, Stefan K, Wolters A, Reiners K, Naumann M, Classen J (2006) The two sides of associative plasticity in writer's cramp. Brain 129(Pt10):2709–2721

Weise D, Schramm A, Beck M, Reiners K, Classen J (2011) Loss of topographic specificity of LTD-like plasticity is a trait marker in focal dystonia. Neurobiol Dis 42(2):171–176

Weise D, Weise CM, Naumann M (2019) Central effects of botulinum neurotoxin-evidence from human studies. Toxins (Basel) 11(1):6

Wiegand H, Erdmann G, Wellhöner HH (1976) 125I-labelled botulinum A neurotoxin: pharmacokinetics in cats after intramuscular injection. Naunyn Schmiedebergs Arch Pharmacol 292 (2):161–165

Wissel J, Kanovsky P, Ruzicka E, Bares M, Hortova H, Streitova H, Jech R, Roth J, Brenneis C, Müller J, Schnider P, Auff E, Richardson A, Poewe W (2001) Efficacy and safety of a standardised 500 unit dose of Dysport (clostridium botulinum toxin type A haemaglutinin complex) in a heterogeneous cervical dystonia population: results of a prospective, multicentre, randomised, double-blind, placebo-controlled, parallel group study. J Neurol 248 (12):1073–1078

Wohlfarth K, Schubert M, Rothe B, Elek J, Dengler R (2001) Remote F-wave changes after local botulinum toxin application. Clin Neurophysiol 112(4):636–640

Zeuner KE, Knutzen A, Al-Ali A, Hallett M, Deuschl G, Bergmann TO, Siebner HR (2010) Associative stimulation of the supraorbital nerve fails to induce timing-specific plasticity in the human blink reflex. PLoS One 5(10):e13602

The Use of Botulinum Toxin for Treatment of Spasticity

Sheng Li and Gerard E. Francisco

Contents

Abstract

Spasticity is one component of the upper motor neuron (UMN) syndrome resulting from a multitude of neurologic conditions, such as stroke, brain injury, spinal cord injury, multiple sclerosis, and cerebral palsy. It is clinically recognized as a phenomenon of velocity-dependent increase in resistance, i.e., hypertonia. Recent advances in the pathophysiology of spasticity improve our understanding of mechanisms underlying this complex phenomenon and its relations to other components of UMN syndrome (weakness and disordered motor control), as well as the resultant clinical problems. This theoretical framework provides a foundation to set up treatment goals and to guide goal-oriented clinical assessment and treatment. Among a spectrum of treatment options, botulinum toxin (BoNT) therapy is the preferred treatment for focal spasticity. The evidence is very robust that BoNT therapy effectively reduces spasticity;

The original version of this chapter was revised. A correction to this chapter can be found at https://doi.org/10.1007/164_2020_412

S. Li (✉) · G. E. Francisco
Department of Physical Medicine and Rehabilitation, McGovern Medical School, University of Texas Health Science Center Houston (UTHealth), Houston, TX, USA

TIRR Memorial Hermann Hospital, Houston, TX, USA
e-mail: sheng.li@uth.tmc.edu; gerard.e.francisco@uth.tmc.edu

S. M. Whitcup, M. Hallett (eds.), *Botulinum Toxin Therapy*,
Handbook of Experimental Pharmacology 263, https://doi.org/10.1007/164_2019_315

however, it does not improve voluntary movement. In this chapter, we highlight a few issues on how to achieve the best clinical outcomes of BoNT therapy, such as dosing, dilution, guidance techniques, adjunctive therapies, early treatment, repeated injections, and central effects, as well as the ways to improve motor function in selected subgroups of patients with spasticity. We also discuss the reasons of poor responses to BoNT therapy and when not to use BoNT therapy.

Keywords
Botulinum toxin · Brain injury · Human · Motor recovery · Rehabilitation · Spasticity · Spinal cord injury · Stroke

1 Introduction

Spasticity is one component of the upper motor neuron (UMN) syndrome resulting from a multitude of neurologic conditions. Clinically, spasticity is easily recognized as a phenomenon of velocity-dependent increase in tonic stretch reflexes ("muscle tone") with exaggerated tendon jerks, resulting from hyperexcitability of the stretch reflex (Lance 1980). Estimates of spasticity incidence and prevalence vary, due to the lack of a strict definition and clinical measurement of spasticity. It is estimated to occur in around 80% of persons with multiple sclerosis (Patejdl and Zettl 2017) and 65–78% in those with spinal cord injuries (Maynard et al. 1990). Prevalence in stroke is about 20–40% (Zorowitz et al. 2013). Within the first year of stroke, spasticity was found in 38% of survivors (Watkins et al. 2002). However, spasticity is present in 97% of chronic stroke survivors with moderate to severe motor impairments (Pundik et al. 2018). Presence of spasticity in persons with traumatic brain injuries (TBI) depends on the severity of injury. Spasticity can exist in up to 40% of those with severe brainstem involvement (Wedekind and Lippert-Grüner 2005).

Spasticity is significant because it not only causes problems directly, such as pain, distorted joint position, and posture and hygiene difficulties, but it also predisposes to other complications, such as joint contractures and permanent deformities. Furthermore, spasticity interacts with and amplifies the effects of other impairments, such as weakness, exaggerated stretch reflexes, clonus, impaired coordination, and motor control and planning, thus contributing to limitations in activity and participation (Mayer and Esquenazi 2003). These numerous abnormalities and impairments intersect and evolve over time, thus producing a dynamic picture of varying clinical presentations after an UMN lesion (Gracies 2005a, b). These interactions often result in abnormal joint postures, disordered motor control, and functional limitations, such as difficulty in grasping, reaching, walking, transferring, and performing hygiene, dressing, self-care, and other activities of daily living. In addition, spasticity-related stiffness and discomfort can interfere with these physical activities and contribute to psychological consequences on mood and self-esteem (Thompson et al. 2005). Collectively, these motor impairments limit their vocational and social participation in more than half of stroke survivors at age 65 and over (Murphy and Carmine 2012; Benjamin et al. 2017).

2 Pathophysiology and Clinical Presentations

The underlying mechanisms of spasticity are still poorly understood. This partly makes it a challenge for clinicians to understand the clinical presentations and problems and to develop a plan of care. Here we first briefly summarize current understandings of poststroke spasticity and its relation to clinical presentations and problems (Brown 1994; Gracies 2005b; Nielsen et al. 2007; Mukherjee and Chakravarty 2010; Burke et al. 2013; Stecco et al. 2014; Li and Francisco 2015) (Fig. 1). A stroke often damages the motor cortex and its descending corticospinal tract (CST), immediately causing muscle weakness (usually unilateral), subsequently resulting in incoordination and often joint immobilization. On the other hand, neuroplastic changes occur after stroke as well. Due to lesions of corticobulbar pathways accompanied with lesion of motor cortices and/or descending CST, bulbospinal hyperexcitability gradually develops due to loss of cortical inhibition. This is mainly a phenomenon of disinhibition or unmasking effects. Potential candidates include reticulospinal, vestibulospinal, and rubrospinal projections (Miller et al. 2014; Li and Francisco 2015; Owen et al. 2017). Medial reticulospinal (RS) hyperexcitability appears to be the most likely mechanism (Brown 1994; Li and Francisco 2015). RS hyperexcitability provides unopposed excitatory descending inputs to spinal stretch reflex circuits, resulting in elevated excitability of spinal motor neurons and hyperreflexia. This adaptive change can account for most clinical findings, for example, exaggerated stretch reflex, velocity-dependent resistance to stretch, muscle overactivity, or spontaneous firings of motor units. Such muscle overactivity in a joint position at a shortened muscle length facilitates limb immobilization, development of muscle and tendon contractures, and accumulation of extracellular matrix deposits (Stecco et al. 2014; Raghavan et al. 2016). Muscle fiber shortening and fibrosis secondary to limb immobilization increase mechanical muscle stiffness. Hyaluronan is the primary component in the extracellular matrix (Fraser et al. 1997). Accumulation and crowding of hyaluronan decrease lubrication between different layers of collagen and muscle fibers, thus perceived as increased stiffness (Stecco et al. 2013). Though not adequately distinguished in clinical (Vattanasilp et al. 2000) or laboratory examinations (Malhotra et al. 2009), these components collectively contribute to increased resistance or spastic hypertonia.

Understanding the different mechanisms of weakness and spasticity and the various components of spastic hypertonia provides a useful theoretical framework to understand the clinical presentations and problems related to spasticity and, subsequently, to develop treatment plans for an effective motor rehabilitation program. Clinical presentations of spasticity vary widely across individuals within and across patient populations. Common postural patterns, including elbow flexion, finger flexion, and equinovarus, are shown in Fig. 1. It is of clinical significance to understand that abnormal postures are almost always manifestations of imbalance of weakness and hypertonia. For example, a flexed elbow posture is not necessarily due to flexor muscle group hypertonia solely, but may be a combination of hypertonic flexors and weak extensors; or it could also be that both flexor and extensor muscle groups are both hypertonic, but the former predominates. It is important to point out

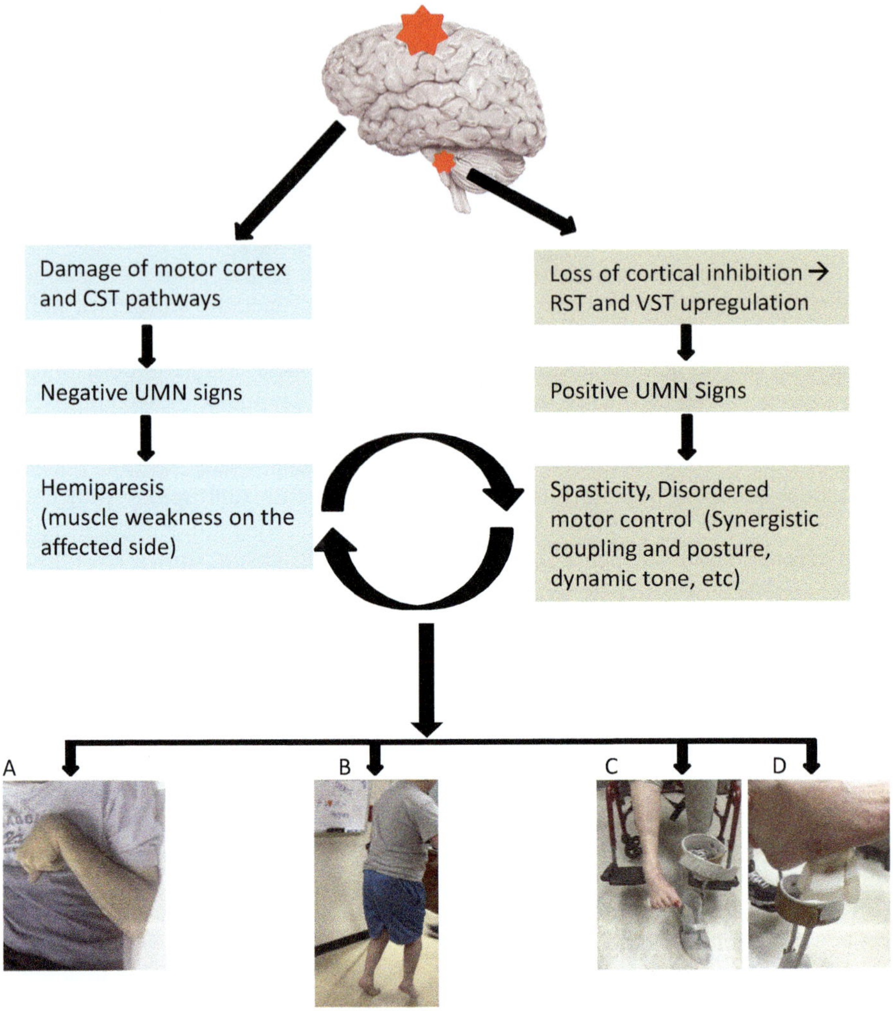

Fig. 1 Pathophysiology of spasticity and its relations to clinical problems. *UMN* upper motor neuron, *CST* corticospinal tract, *RST* reticulospinal tract, *VST* vestibulospinal tract, *MN* motor neuron (**a**) Abnormal posture leading to difficulty with hygiene and dressing; (**b**) Abnormal gait; (**c**) Spastic equinovarus; (**d**) Pressure sore

that clinical problems of spasticity are not the abnormal joint postures caused by spasticity; instead, consequences of abnormal joint postures are usually the problems. As shown in Fig. 1, difficulty to clean the clenched fist and armpit is more problematic than the non-movable clenched fist and shoulder joint, because it may lead to skin maceration and infection. Similarly, the problem of a spastic equinovarus is mainly manifested by the pressure sore developed during constant abnormal pressure during walking. Impaired motor control of spastic muscle is another example of clinical problems associated with spasticity. Sustained activation of spastic calf muscles during weight-bearing can cause spastic foot drop and abnormal gait pattern (Fig. 1). This theoretical framework can also guide the development of treatment plans. These are detailed in the management section.

3 Goal-Setting and Goal-Oriented Clinical Assessment

It is clear that spasticity is only one component of the clinical problems, as mentioned above. The problems are usually associated with consequences of spasticity or disordered motor control that a limb could not be moved, or the resultant functional limitations, such as the inability to release a grasped object or difficulty with walking due to an in-turned foot. Spastic muscles should be treated only if they are causing or predisposing to other problems. However, it is not uncommon for patients to desire goals of regaining normal function, but since this is usually not achievable, a discussion regarding goal-setting prior to initiating treatment can help manage expectations of treatment outcomes.

Patient-centered goal-setting should be the key driver of management decision-making. Treatment goals should be mutually agreed upon by the patient (or caregiver) and clinician. All factors should be considered, including findings from focused medical history, functional history, the patient's realistic expectations, inputs from care-provider(s) and therapists, and social support system. For example, a medical cause of a transient increase in the severity of spasticity, such as urinary tract infection or pressure sores, should be considered and treated prior to setting the treatment goal. It is therefore important to obtain a thorough, yet focused, medical and functional history to guide the examination and to formulate treatment goals and plans. A systematic approach to history-taking and clinical assessment of spasticity is proposed in Tables 1 and 2. It can be modified for different clinical scenarios.

Spasticity of individual muscles and muscle groups is often assessed by clinical scales. The commonly used scales include the Ashworth scale (AS), the modified Ashworth scale (MAS), and the Tardieu scale (Tables 3 and 4). The Tardieu scale has advantages over the MAS because it not only quantifies the muscles' reaction to stretch, but it controls for the velocity of the stretch and measures the angle at which the catch, or clonus, occurs. However, neither scale has shown to be more reliable than the other. In addition, a limitation of both Tardieu and AS/MAS scales is the fact that they are performed at rest, whereas spasticity may be bothersome during active function when the person is upright and attempting to move or perform an activity. Thus these clinical assessments do not correspond with the treatment goal.

Quantitative measures, such as biomechanical and electrophysiological tests, are desirable because of their inherent objectivity and reliability. Unfortunately, many of the devices are not available to a typical clinician, or the tests are too time-consuming. On the other hand, clinical problems are the consequences associated with spasticity, rather than the spasticity itself in most situations. Clinical scales are often sufficient to guide the treatment.

Table 1 Some important medical and functional history in spasticity assessment

Medical history
• Is there any new medical condition? (e.g., urinary tract infection, other infections)
• What are the current medications? (e.g., any spasmolytic agents? And dose?)
• Is there any change of medications? (e.g., addition of neurostimulants)
• Was there a recent increase in tightness (that may warrant further diagnostic testing to rule out a new neurologic or medical problem)?
• What treatments for muscle tightness have been tried previously and their outcome?
Functional history
• Is the limb tight all the time or only at certain times?
• Does a particular position or movement trigger tightness?
• Is the tightness related to spasms?
• Does the tightness cause pain?
• Have there been episodes of skin compromise due to tightness or spasm?
• Does the tightness result in difficulty with cleaning?
• Does the tightness result in difficulty donning splints?
• Does the tightness limit the ability to move limbs, reach for objects, and use the hands?
• Does the tightness of the lower limbs result in problems with transferring from one surface to another or with walking?

Table 2 Practical clinical examination sequence

Tasks	What to look for	What can be gleaned
Observation	Observe limb posture at rest and how they change with position	Abnormal posture at rest – sustained muscle contraction (dystonia), contracture, pressure sore, wound (that would worsen spasticity)
Voluntary and functional activities[a]	How limbs move and how much active range is available Gait characteristics and associated upper limb and trunk postural abnormalities	Functional strength, coordination, spastic co-contraction, contractures, presence of other movement disorders, synkinesis, or associated reactions Position-dependent postural changes – dynamic tone Pain and discomfort during voluntary and functional movements
Passive (MAS, AS, Tardieu)	Passive range of motion, strength, muscle tone, velocity-dependent "angle of catch," clonus	Spasticity Rigidity Contracture Clonus Pain and discomfort during passive stretch

[a]Voluntary movements, such as sit to stand, transfer, ambulation, and other functional activities

Table 3 Modified Ashworth scale

0	No increase in muscle tone
1	Slight increase in tone, manifested by a catch and release at the end of range of motion (ROM)
1+	Slight increase in tone, manifested by a catch, followed by minimal resistance throughout the remainder (less than half) of the ROM (catch in the first half of ROM)
2	Marked increase in tone through most of the ROM, still easily moved
3	Considerable increase in tone, passive movement difficult
4	Affected part(s) rigid in flexion or extension

Table 4 Tardieu scale

Quality of muscle reaction
0. No resistance
1. Slight resistance
2. Catch followed by a release
3. Fatigable clonus (<10 s)
4. Continuous clonus (>10 s)
Angle of muscle reaction at different velocities of stretch
V1. As slow as possible
V2. Speed of limb falling under gravity
V3. As fast as possible

4 Botulinum Toxin (BoNT) Therapy for Spasticity Management and Related Clinical Issues

There are a number of treatment options for management of spasticity, including physical modalities; oral medications; chemodenervation with botulinum toxins (BoNT), phenol, or alcohol; intrathecal baclofen therapy; and surgical interventions. There are different approaches to utilize these treatment options, e.g., the sequential approach from the least invasive treatment to surgical procedure and combined therapy with both "invasive" and "noninvasive" treatments. Selection of treatment options is discussed in more detail elsewhere (Francisco and Li 2015).

Nevertheless, chemodenervation with BoNT has become a widely used spasticity treatment. It is preferred for the management of focal spasticity or when the treatment plan targets a particular muscle (Simpson et al. 2016). Botulinum toxin exerts its effect through inhibition of acetylcholine release at the neuromuscular junction via a complex process (see other chapters for details) (Wheeler and Smith 2013; Jankovic 2017; Pirazzini et al. 2017). Currently, serotypes A and B of *Clostridium botulinum* are utilized clinically: *abobotulinumtoxinA*, *incobotulinumtoxinA*, *onabotulinumtoxinA*, and *rimabotulinumtoxinB*. They all inhibit acetylcholine release and the muscle paralysis they produce is reversible. The clinical effects of BoNT do not manifest until several days following an injection. The clinical effects last about 3 months, and recurrence of spasticity is likely due to functional repair of the neuromuscular junctions previously paralyzed by the toxin (de Paiva et al. 1999). Usually, patients require repeated BoNT injections every 3–4 months (Moeini-Naghani et al. 2016; Simpson et al. 2016). However, majority survey of treating physicians and patients found that a majority prefer more frequent injections to achieve better clinical outcome (Bensmail et al. 2014). A new injectable BoNT, daxibotulinumtoxinA (an investigational BoNT, RT002), may offer a more prolonged duration of treatment effect (Jankovic et al. 2018). Though still under investigation (Fonfria et al. 2018; Webb 2018), engineered BoNT appears able to enhance receptor binding and thus increase the efficacy of BoNT (Tao et al. 2017). Advantages of BoNT treatment over oral medications are target specificity and a more favorable adverse event profile. Drowsiness and sedation are practically nonexistent with BoNT treatment.

Over three decades, overwhelming evidence demonstrates that BoNT therapy results in significant improvement at the body function and structure level (Bakheit et al. 2000; Burridge et al. 2005; Rosales and Chua-Yap 2008; Simpson et al. 2008; Wissel et al. 2009; Bensmail et al. 2010; Sheean et al. 2010; Shaw et al. 2011; Rosales et al. 2012; Lampire et al. 2013; Holman Barden et al. 2014; Tenniglo et al. 2014). In a recent meta-analysis study that included 40 trials (Andringa et al. 2019), the authors reported robust evidence of BoNT on reducing resistance to passive movement and on self-care, as measured with the (modified) Ashworth scale, and improving self-care ability for the affected side after intervention and at follow-up. Similarly, evidence of the absence of the effect on the "arm-hand capacity" at follow-up was also robust. BoNT significantly reduced "involuntary movements," "spasticity-related pain," and "carer burden" and improved "passive range of motion," while no evidence was found for "arm and hand use" after the intervention.

The main clinical issue is how to achieve the best outcome with BoNT therapy. The relevant issues are (1) medication related (dosing, dilution, molecular manipulation, and immunoresistance), (2) injection related (injection guidance and motor innervation zone), (3) use of adjunct therapy, (4) relation to motor recovery (therapeutic weakness and central mechanisms), and (5) alterative treatment options.

Dosing Clinical experience, regulatory and insurance coverage restrictions, and manufacturers' recommendations based on a few studies largely dictate the choice of doses of the various botulinum toxins. There are a handful of dose-ranging studies that define dose-related therapeutic and adverse effects in spasticity (Bhakta et al. 1996; Simpson et al. 1996, 2016; Hyman et al. 2000; Baker et al. 2002; Childers et al. 2004; Gracies et al. 2014). Dosages that are used in current practice recommended by consensus statements (Wissel et al. 2009) are higher than doses used in published randomized controlled studies. The use of escalating doses of botulinum toxins was becoming a common practice until safety concerns were raised and fueled by mandates from the US Food and Drug Administration (FDA). Responding to reports suggestive of systemic toxicity of botulinum toxins, in 2009 the FDA required new label warnings and a risk mitigation strategy that requires clinicians to discuss the risks and provide written material that details the warnings. The current experience of many clinicians is that using dosages of inco- and onabotulinumtoxinA as high as 600–800 units (U) is effective and safe (Santamato et al. 2013; Wissel et al. 2013). Two comprehensive reviews concluded that higher doses of botulinum toxin type A appeared to be efficacious in reducing spasticity of the upper and lower limbs after stroke, with minimal adverse effects (Santamato et al. 2015; Baricich et al. 2018). Recently, Wissel et al. (2017) reported on the safety and efficacy of escalating doses of incobotulinumtoxinA up to 800 units. The resistance to passive movement scale improved significantly. The proportion of subjects achieving at least three of four pre-identified treatment goals increased with higher doses of the toxin. No neutralizing antibody was detected.

Dilution It is believed that increasing the volume of botulinum toxin solution injected magnifies its therapeutic effects by facilitating the toxin's ability to reach more motor endplates. This has been demonstrated in animal studies (Shaari and Sanders 1993; Kim et al. 2003) where muscle paralysis and atrophy were greater when a more dilute preparation, i.e., higher volume relative to dose, or lower concentration, of botulinum toxin is injected. Human studies are equivocal in demonstrating superiority of higher volumes of botulinum toxin injections (Francisco 2004; Lee et al. 2004) largely due to methodological limitations of studies although some investigation have found that high-volume or endplate-targeted botulinum toxin injections result in more profound neuromuscular blockade and spasticity and co-contraction reduction, as compared to low-volume, non-endplate-targeted injections (Gracies et al. 2009). As much as high-volume injections appear attractive, it may be a double-edged sword in that it may facilitate distant spread of the toxin. Cases have been reported wherein patients with poststroke spasticity who receive large dilution volumes in proximal upper limb muscles developed transient weakness in the non-injected contralateral upper limb. Based on electrophysiologic abnormalities documented following the injection, weakness was attributed to neuromuscular blockade.

Techniques to Enhance BoNT Effectiveness There is a lot of interest in techniques to enhance the clinical effects of botulinum toxin, without concomitantly increasing the risk for adverse events. According to the mechanism of action of BoNT – blockade of acetylcholine release at the neuromuscular junction, different techniques have been tried. Injections at multiple sites within a muscle and using a higher-volume/more dilute toxin solution (already discussed above) are regarded as ways to reach more neuromuscular junctions, than to increase effectiveness. Other techniques used to attempt enhancement of toxin effectiveness include guided injection by listening to EMG activity, motor point identification through electrical stimulation (ES), or visualizing target sites by sonography. The superiority of one guidance technique over another is yet to be established, but consistently studies have demonstrated that EMG, ES, or sonography is better than anatomic localization through muscle palpation (Schnitzler et al. 2012; Picelli et al. 2014a, b, c; Ploumis et al. 2014).

A novel neuroengineering technique can provide information of accurate localization of neuromuscular junctions of a muscle (Barbero et al. 2012). Using surface EMG recording with a high-density EMG electrode, neuromuscular junctions can be determined from visual inspection or analysis of surface EMG signals. This surface projection is called innervation zone (IZ). In healthy subjects, it has been shown that the effect of BoNT decreases by 46% if BoNT is injected by 1 cm away from the innervation zone (Lapatki et al. 2011). Due to secondary and adaptive changes, IZ location changes in the spastic muscles after stroke. The difference in IZ locations between spastic biceps muscle and the contralateral biceps muscles was up to 3 cm (Bhadane et al. 2016). Through advanced computational algorithms, the information of the depth of IZ locations within the biceps muscles is obtained and validated, i.e., IZ location in three-dimensional space within a muscle (Zhang et al. 2017, 2019). Comparisons of clinical outcomes of BoNT therapy between IZ-guided injection and conventional methods are ongoing in our lab. As expected, our preliminary data showed better reduction in spasticity after IZ-guided injection.

Adjunctive Therapies to Enhance the Effect of BoNT Therapy When used alone or in combination of BoNT therapy, physical modalities have been shown to be effective in reducing spasticity and increasing range of motion. Splinting and casting are often used in the acute setting for sustained stretching to prevent contracture and reduce spasticity (Booth et al. 1983; Preissner 2002; Mortenson and Eng 2003; Pohl et al. 2003; Bovend'Eerdt et al. 2008). Casting alone seems sufficient to prevent contracture and reduce spasticity if the intervention is initiated early after severe brain injury. However, a systematic review on the use of upper extremity casting found high variability in casting protocols which indicates no consensus in technique (Lannin et al. 2007). Casting can enhance the effect of onabotulinumtoxinA (Farina et al. 2008), as prolonged stretching of spastic muscles after BoNT injections affords long-lasting therapeutic benefit. Another promising technique to magnify the clinical effect of BoNT therapy is pairing it with superficial electrical stimulation, which influences activity of synaptobrevin-2 receptors that facilitate neuronal binding and subsequent uptake of BoNT (Hesse et al. 1998; Bayram et al. 2006; Mayer et al. 2008; Wilkenfeld 2013). More recently, extracorporeal shockwave therapy (ESWT) has been shown to have a greater magnitude of BoNT enhancement than electrical stimulation, most likely through modulation of muscle rheology and neurotransmission (Santamato et al. 2013; Wilkenfeld 2013). For a more in-depth discussion of this topic, Mills et al. (2016) conducted an excellent systematic review of how adjunct therapies improve outcomes of botulinum toxin injections for spasticity.

Early Treatment: When to Start BoNT Therapy? There is no standard in how early BoNT can be safely and effectively administered. A few studies reported that treatment as early as 3–6 months of disease onset effectively manages muscle hypertonia and decreases risk of later complications, such as contracture development (Hesse et al. 2012; Fietzek et al. 2014). Results of an exploratory, double-blind, randomized, placebo-controlled trial (Rosales et al. 2018) using abobotulinumtoxinA 500 U in subjects with upper limb spasticity within 2 to 12 weeks poststroke suggested that early treatment significantly delayed time to reach reinjection criteria when compared with placebo.

Repeated Injections: Are Repeated Injections "Safe"? In clinical practice many patients receive multiple injections over a period of many years, sometimes decades, while the long-term effects are not systematically documented. The fact that patients continue to receive BoNT therapy over a long period of time implies that the patients continue to benefit from it. Most studies involving the use of botulinum toxin for spasticity involve only a few cycles of injection. A rare few have reported safety and sustained efficacy up to five injection cycles over a few years (Lagalla et al. 2000; Gordon et al. 2004; Elovic et al. 2008; Santamato et al. 2017). Although the few studies claimed that repeated injections were safe, concerns remain about the long-term effect of BoNT on muscles. An animal study concluded that the contractile properties of target and nontarget muscles did not fully recover within 6 months of BoNT injections (Fortuna et al. 2013). The same investigators also found that following repeated BoNT injections muscle atrophy sets in and contractile material is replaced by fat (Fortuna et al. 2011). Recognition of BoNT's effects on muscle length and force (Turkoglu et al. 2014) is also emerging, although how this translates clinically is still unclear. These concerning findings need to be investigated further in clinical studies emphasizing muscle changes in recovery after BoNT injections.

Poor Responses to BoNT Therapy: What Are the Reasons? The effectiveness of BoNT therapy in spasticity reduction is well documented, as discussed above. However, the response varies from person to person. Poor response is defined as the treatment goals are not met. There are many potential reasons for poor responses to BoNT therapy in spasticity management (Table 5). One of the most common reasons is unrealistic expectations from patients and family members and/or caregivers. It is important to set the treatment goals prior to BoNT therapy, and the goals need to be agreed upon between the patient and the treating physician. What needs to bear in mind is that not all increased resistance (hypertonia) is caused by spasticity, as discussed in the Pathophysiology and the Clinical Presentations. Adaptive muscular changes are likely to occur, such as hyaluronan accumulation, muscle fiber shortening, and fibrosis. Hypertonia caused by these changes is not expected to respond to BoNT therapy. On the other hand, alternative treatment options should be used. Hyaluronidase is an enzyme that hydrolyzes hyaluronan. It is reported that hypertonia was significantly reduced after hyaluronidase injection (Raghavan et al. 2016).

Table 5 Potential reasons for poor outcome of botulinum toxin therapy

Patient-related	Injector-related	Drug-related
Unrealistic expectations Disease conditions Concurrent medications that interact with spasmolytic drugs or alter muscle tone Immunoresistance Secondary muscle changes, fibrosis, etc.	Incorrect diagnosis Incorrect muscle selection Improper injection technique (far away from the neuromuscular junction)	Incorrect dose (over- or underdose) Incorrect preparation Inactive medication

Immunoresistance is a potential factor that causes suboptimal or no responses to BoNT therapy. Bioassay of neutralizing antibodies (NABs) to BoNT is considered the gold standard in confirming immunoresistance. Based on early reports in the cervical dystonia population, high doses and frequent injections of BoNT were identified as risk factors for immunoresistance (Zuber et al. 1993; Greene et al. 1994).This also provided support for the practice of allowing no less than 90 days in between exposures to BoNT. A much higher incidence of antibody formation has been associated with cervical dystonia than in spasticity (Naumann et al. 2010). There is a growing interest in incobotulinumtoxinA, which is free of excipient proteins and, as such, may have a lower propensity to induce an immunogenic response relative to the other botulinum toxin preparations with complexing proteins (Albrecht et al. 2019). A meta-analysis of 16 clinical trials involving a total of 3,006 subjects with various diagnoses found that neutralizing antibodies determined by mouse protection assay appeared in 1.28% of cervical dystonia, as opposed to only 0.32% poststroke subjects. In another pooled analysis involving three 12- to 42-week clinical poststroke spasticity studies, the formation of neutralizing antibodies was found to be 0.5% (1/191 subjects) (Yablon et al. 2007). However, there are heterogeneous reports. In children with cerebral palsy, high neutralizing antibody (NAB) frequencies of up to 30% have been described (Herrmann et al. 2004). The most significant risk factors for antibody formation were frequent treatment and high dose per treatment in this study. In a more recent cohort of patients with different neurological impairments, 83 of 596 patients (13.9%) had measureable NAB (Albrecht et al. 2019). The probability of developing antibodies increased with repeated treatment and was influenced by the BoNT/A formulation. The NAB rates were similar for aboBoNT/A and ona-BoNT/A (6% and 7%, respectively), while no NABs were observed in patients treated exclusively with inco-BoNT/A. The difference in NAB rates is likely related to the amount of 150 kDa BoNT/A neurotoxin. It was found that, at current FDA-approved doses, abobotulinumtoxinA contains greater amounts of active neurotoxin as compared to other BoNT/A products (Field et al. 2018). Disease entity and treatment duration had no additional influence (Albrecht et al. 2019). In the same study (Albrecht et al. 2019), those patients with positive NABs still responded to BoNT therapy (at least partially), while NAB was positive in only 57% (20 out of 35) in patients with spasticity who failed BoNT therapy. Overall, the prevalence of NAB has dropped from 10% in the past (Jankovic et al. 2003) to the current level about 1% (Mathevon et al. 2019), with no NABs in patients treated with inco-BoNT/A (Albrecht et al. 2019). Therefore, it is important to note that it is extremely rare that NAB is the cause of non-responders (Jankovic 2017; Mathevon et al. 2019).

Recovery of Motor Function: Can BoNT Therapy Help Recover Motor Function in a Subgroup of Patients? The evidence is robust for the effect of BoNT therapy in spasticity, while it is also robust that BoNT therapy does not improve voluntary movement (Andringa et al. 2019). However, there are unusual cases when the outcome of BoNT injections surpasses this expectation and results in an increase in functional abilities of the hand in chronic stroke survivors (Fridman et al. 2010;

Chang et al. 2012; Mas et al. 2017). In a case study (Chang et al. 2012), the patient was a 53-year-old female, who sustained a hemorrhagic right middle cerebral artery stroke 3 years earlier. She had finger flexor spasticity and residual weak finger/wrist extension. She received 50 units of onabotulinumtoxinA injection to each of the left flexor digitorum superficialis and flexor digitorum profundus, respectively. As expected, BoNT injection led to weakness and spasticity reduction in the spastic finger flexors. However, she was able to open her hand faster due to improved grip release time. This was accompanied by shortened finger flexor EMG activity during hand and finger opening. Similarly, another chronic stroke survivor regained the ability to open the hand 4 years poststroke after several BoNT injections to finger flexors (Mas et al. 2017). In these cases, natural motor recovery is not likely after 3 to 4 years after stroke. Regardless of underlying mechanisms, these reports suggest that late motor recovery is possible in selected chronic patients when motor recovery is presumed plateaued.

Advancement in understanding the pathophysiology of spasticity helps understand the phenomenon of late motor recovery after BoNT injections. As illustrated in Fig. 1, spasticity and weakness are mediated by different mechanisms secondary to neural plasticity (Li 2017). When finger flexor spasticity is addressed by BoNT injections, with concomitant reduction in spastic co-contraction in finger flexors during finger extension attempts (Chang et al. 2012), weak finger extensors became functional, and motor function of the hand improved. In this regard, weakness produced by BoNT injection is therapeutic, i.e., therapeutic weakness (Francisco and Li 2015). It follows that interventions to strengthen the finger extensors after BoNT injections to the spastic finger flexors are expected to better improve motor recovery. The expected results were confirmed in a recent study (Lee et al. 2018).

Central Effects: Does the BoNT Effect Go Beyond the Injected Muscles? As described earlier, the therapeutic effects of BoNT therapy on spasticity reduction is widely accepted to be realized via blockade of acetylcholine release presynaptically at neuromuscular junctions of the targeted muscles. Neuromuscular blockade affects both extrafusal and intrafusal muscle fibers (Filippi et al. 1993; Rosales et al. 1996). It is estimated that BoNT injection results in a decrease in activity of intrafusal muscle fibers, i.e., afferent input by 33%. Such decrease was found to be greatest at 2 weeks, and tapered off at 12 weeks postinjection, and correlated with spasticity reduction (Phadke et al. 2013). BoNT-related blockade at intrafusal fibers decreases spindle inflow to spinal stretch reflex circuits, thus contributing to spasticity reduction. Furthermore, decreased afferent inputs via intrafusal blockade can further alter spinal motor neuron excitability and sensorimotor integration. Trompetto et al. reported that suppression of tonic vibration reflex was still observed at 7 months after BoNT injection when muscle strength and the magnitude of maximal M-wave have fully recovered (Trompetto et al. 2006). In another study, recurrent inhibition from soleus motor axons to motor neurons supplying the quadriceps muscle was suppressed after BoNT injection in the soleus muscle (Marchand-Pauvert et al. 2013). This suppression was considered to be induced by BoNT through axonal transport and blockade of the cholinergic synapses of

Renshaw cells. Accumulated evidence from animal and human studies have shown that intramuscularly injected BoNT could reach further to brainstem and cortical levels indirectly through hematogenic spread, retrograde transport of BoNT, and plastic reorganization of the central nervous system due to altered afferent inputs (see reviews Mas et al. 2017; Caleo and Restani 2018; Weise et al. 2019). Collectively, substantial evidence demonstrates that intramuscularly injected BoNT is able to reach and modulate excitability of motor neurons at the spinal and supraspinal levels.

Alternative Treatment Options: When Not to Use BoNT Therapy for Spasticity Management? BoNT therapy is the preferred treatment option and is widely used for spasticity management. However, there are a number of other treatment options. To know when not to use BoNT therapy for spasticity management is also very important, since different treatment options have their advantages and indications as well, and BoNT therapy also has potential adverse effects, even if rare. For example, how many BoNT injections need to be done to address a spastic-dystonic "clenched fist" before surgical release of the finger and thumb flexor tendons should be entertained? How many times should a person with severe spastic paraplegia receive BoNT injections before intrathecal baclofen therapy is considered? The economic impact of these clinical decisions will also need to be weighed to better appreciate the cost-effectiveness of spasticity interventions. An alternative, such as hyaluronidase, may be considered to address different components of spastic hypertonia when there is suboptimal response to BoNT injections (Raghavan et al. 2016). Phenol neurolysis is likely to be a better choice to address moderate to severe spasticity for an inpatient where its immediate effects on spasticity reduction would be highly appreciated (Karri et al. 2017). Other emerging adjunctive therapies, such as noninvasive brain (Kumru et al. 2010; Wu et al. 2013; Barros Galvao et al. 2014; Gunduz et al. 2014), spinal (Pinter et al. 2000), and transcutaneous nerve (Hofstoetter et al. 2014; Oo 2014) stimulation, may be considered to enhance the effect of BoNT therapy.

References

Albrecht P, Jansen A, Lee JI et al (2019) High prevalence of neutralizing antibodies after long-term botulinum neurotoxin therapy. Neurology 92:e48–e54

Andringa A, van de Port I, van Wegen E, Ket J, Meskers C, Kwakkel G (2019) Effectiveness of botulinum toxin treatment for upper limb spasticity after stroke over different ICF domains: a systematic review and meta-analysis. Arch Phys Med Rehabil 100:1703

Baker R, Jasinski M, Maciag-Tymecka I et al (2002) Botulinum toxin treatment of spasticity in diplegic cerebral palsy: a randomized, double-blind, placebo-controlled, dose-ranging study. Dev Med Child Neurol 44:666–675

Bakheit AMO, Thilmann AF, Ward AB, Poewe W, Wissel J, Muller J, Neumann C (2000) A randomized, double-blind, placebo-controlled, dose-ranging study to compare the efficacy and safety of three doses of botulinum toxin type A (Dysport) with placebo in upper limb spasticity after stroke. Stroke 31:2402–2406

Barbero M, Merletti R, Rainoldi A (2012) Atlas of muscle innervation zones: understanding surface electromyography and its applications. Springer, Berlin

Baricich A, Picelli A, Santamato A et al (2018) Safety profile of high-dose botulinum toxin type A in post-stroke spasticity treatment. Clin Drug Investig 38:991–1000

Barros Galvao SC, Borba Costa dos Santos R, Borba dos Santos P, Cabral ME, Monte-Silva K (2014) Efficacy of coupling repetitive transcranial magnetic stimulation and physical therapy to reduce upper-limb spasticity in patients with stroke: a randomized controlled trial. Arch Phys Med Rehabil 95:222–229

Bayram S, Sivrioglu K, Karli N, Ozcan O (2006) Low-dose botulinum toxin with short-term electrical stimulation in poststroke spastic drop foot: a preliminary study. Am J Phys Med Rehabil 85:75–81

Benjamin EJ, Blaha MJ, Chiuve SE et al (2017) Heart disease and stroke statistics-2017 update: a report from the American Heart Association. Circulation 135:e146–e603

Bensmail D, Robertson JV, Fermanian C, Roby-Brami A (2010) Botulinum toxin to treat upper-limb spasticity in hemiparetic patients: analysis of function and kinematics of reaching movements. Neurorehabil Neural Repair 24:273–281

Bensmail D, Hanschmann A, Wissel J (2014) Satisfaction with botulinum toxin treatment in post-stroke spasticity: results from two cross-sectional surveys (patients and physicians). J Med Econ 17:618–625

Bhadane M, Liu J, Rymer WZ, Zhou P, Li S (2016) Re-evaluation of EMG-torque relation in chronic stroke using linear electrode array EMG recordings. Sci Rep 6:28957

Bhakta BB, Cozens JA, Bamford JM, Chamberlain MA (1996) Use of botulinum toxin in stroke patients with severe upper limb spasticity. J Neurol Neurosurg Psychiatry 61:30–35

Booth BJ, Doyle M, Montgomery J (1983) Serial casting for the management of spasticity in the head-injured adult. Phys Ther 63:1960–1966

Bovend'Eerdt TJ, Newman M, Barker K, Dawes H, Minelli C, Wade DT (2008) The effects of stretching in spasticity: a systematic review. Arch Phys Med Rehabil 89:1395–1406

Brown P (1994) Pathophysiology of spasticity. J Neurol Neurosurg Psychiatry 57:773–777

Burke D, Wissel J, Donnan GA (2013) Pathophysiology of spasticity in stroke. Neurology 80: S20–S26

Burridge JH, Wood DE, Hermens HJ et al (2005) Theoretical and methodological considerations in the measurement of spasticity. Disabil Rehabil 27:69–80

Caleo M, Restani L (2018) Direct central nervous system effects of botulinum neurotoxin. Toxicon 147:68–72

Chang SH, Francisco GE, Li S (2012) Botulinum toxin (BT) injection improves voluntary motor control in selected patients with post-stroke spasticity. Neural Regen Res 7:1436–1439

Childers MK, Brashear A, Jozefczyk P et al (2004) Dose-dependent response to intramuscular botulinum toxin type A for upper-limb spasticity in patients after a stroke. Arch Phys Med Rehabil 85:1063–1069

de Paiva A, Meunier FA, Molgó J, Aoki KR, Dolly JO (1999) Functional repair of motor endplates after botulinum neurotoxin type A poisoning: biphasic switch of synaptic activity between nerve sprouts and their parent terminals. Proc Natl Acad Sci 96:3200–3205

Elovic EP, Brashear A, Kaelin D, Liu J, Millis SR, Barron R, Turkel C (2008) Repeated treatments with botulinum toxin type a produce sustained decreases in the limitations associated with focal upper-limb poststroke spasticity for caregivers and patients. Arch Phys Med Rehabil 89:799–806

Farina S, Migliorini C, Gandolfi M et al (2008) Combined effects of botulinum toxin and casting treatments on lower limb spasticity after stroke. Funct Neurol 23:87–91

Field M, Splevins A, Picaut P, van der Schans M, Langenberg J, Noort D, Foster K (2018) Abobotulinumtoxina (Dysport®), onabotulinumtoxinA (Botox®), and incobotulinumtoxina (Xeomin®) neurotoxin content and potential implications for duration of response in patients. Toxins 10:E535

Fietzek UM, Kossmehl P, Schelosky L, Ebersbach G, Wissel J (2014) Early botulinum toxin treatment for spastic pes equinovarus – a randomized double-blind placebo-controlled study. Eur J Neurol 21:1089–1095

Filippi GM, Errico P, Santarelli R, Bagolini B, Manni E (1993) Botulinum A toxin effects on rat jaw muscle spindles. Acta Otolaryngol 113:400–404

Fonfria E, Elliott M, Beard M, Chaddock JA, Krupp J (2018) Engineering botulinum toxins to improve and expand targeting and SNARE cleavage activity. Toxins (Basel) 10:278

Fortuna R, Vaz MA, Youssef AR, Longino D, Herzog W (2011) Changes in contractile properties of muscles receiving repeat injections of botulinum toxin (Botox). J Biomech 44:39–44

Fortuna R, Horisberger M, Vaz MA, Herzog W (2013) Do skeletal muscle properties recover following repeat onabotulinum toxin A injections? J Biomech 46:2426–2433

Francisco GE (2004) Botulinum toxin: dosing and dilution. Am J Phys Med Rehabil 83:S30–S37

Francisco GE, Li S (2015) Spasticity. In: Physical medicine and rehabilitation, 5th edn. Elsevier, Amsterdam

Fraser JR, Laurent TC, Laurent UB (1997) Hyaluronan: its nature, distribution, functions and turnover. J Intern Med 242:27–33

Fridman EA, Crespo M, Argüello SG et al (2010) Kinematic improvement following botulinum toxin-A injection in upper-limb spasticity due to stroke. J Neurol Neurosurg Psychiatry 81:423–427

Gordon MF, Brashear A, Elovic E, Kassicieh D, Marciniak C, Liu J, Turkel C (2004) Repeated dosing of botulinum toxin type A for upper limb spasticity following stroke. Neurology 63:1971–1973

Gracies JM (2005a) Pathophysiology of spastic paresis. I: paresis and soft tissue changes. Muscle Nerve 31:535–551

Gracies JM (2005b) Pathophysiology of spastic paresis. II: emergence of muscle overactivity. Muscle Nerve 31:552–571

Gracies JM, Lugassy M, Weisz DJ, Vecchio M, Flanagan S, Simpson DM (2009) Botulinum toxin dilution and endplate targeting in spasticity: a double-blind controlled study. Arch Phys Med Rehabil 90:9–16.e12

Gracies JM, Bayle N, Goldberg S, Simpson DM (2014) Botulinum toxin type B in the spastic arm: a randomized, double-blind, placebo-controlled, preliminary study. Arch Phys Med Rehabil 95:1303–1311

Greene P, Fahn S, Diamond B (1994) Development of resistance to botulinum toxin type A in patients with torticollis. Mov Disord 9:213–217

Gunduz A, Kumru H, Pascual-Leone A (2014) Outcomes in spasticity after repetitive transcranial magnetic and transcranial direct current stimulations. Neural Regen Res 9:712–718

Herrmann J, Geth K, Mall V et al (2004) Clinical impact of antibody formation to botulinum toxin A in children. Ann Neurol 55:732–735

Hesse S, Reiter F, Konrad M, Jahnke MT (1998) Botulinum toxin type A and short-term electrical stimulation in the treatment of upper limb flexor spasticity after stroke: a randomized, double-blind, placebo-controlled trial. Clin Rehabil 12:381–388

Hesse S, Mach H, Frohlich S, Behrend S, Werner C, Melzer I (2012) An early botulinum toxin A treatment in subacute stroke patients may prevent a disabling finger flexor stiffness six months later: a randomized controlled trial. Clin Rehabil 26:237–245

Hofstoetter US, McKay WB, Tansey KE, Mayr W, Kern H, Minassian K (2014) Modification of spasticity by transcutaneous spinal cord stimulation in individuals with incomplete spinal cord injury. J Spinal Cord Med 37:202–211

Holman Barden HL, Baguley IJ, Nott MT, Chapparo C (2014) Measuring spasticity and fine motor control (pinch) change in the hand following botulinum toxin-A injection using dynamic computerised hand dynamometry. Arch Phys Med Rehabil 95:2402–2409

Hyman N, Barnes M, Bhakta B et al (2000) Botulinum toxin (Dysport) treatment of hip adductor spasticity in multiple sclerosis: a prospective, randomised, double blind, placebo controlled, dose ranging study. J Neurol Neurosurg Psychiatry 68:707–712

Jankovic J (2017) Botulinum toxin: state of the art. Mov Disord 32:1131–1138

Jankovic J, Vuong KD, Ahsan J (2003) Comparison of efficacy and immunogenicity of original versus current botulinum toxin in cervical dystonia. Neurology 60:1186–1188

Jankovic J, Truong D, Patel AT et al (2018) Injectable daxibotulinumtoxinA in cervical dystonia: a phase 2 dose-escalation multicenter study. Mov Disord Clin Pract 5:273–282

Karri J, Mas MF, Francisco GE, Li S (2017) Practice patterns for spasticity management with phenol neurolysis. J Rehabil Med 49:482–488

Kim HS, Hwang JH, Jeong ST, Lee YT, Lee PK, Suh YL, Shim JS (2003) Effect of muscle activity and botulinum toxin dilution volume on muscle paralysis. Dev Med Child Neurol 45:200–206

Kumru H, Murillo N, Samso JV et al (2010) Reduction of spasticity with repetitive transcranial magnetic stimulation in patients with spinal cord injury. Neurorehabil Neural Repair 24:435–441

Lagalla G, Danni M, Reiter F, Ceravolo MG, Provinciali L (2000) Post-stroke spasticity management with repeated botulinum toxin injections in the upper limb. Am J Phys Med Rehabil 79:377–384

Lampire N, Roche N, Carne P, Cheze L, Pradon D (2013) Effect of botulinum toxin injection on length and lengthening velocity of rectus femoris during gait in hemiparetic patients. Clin Biomech 28:164–170

Lance JW (1980) Symposium synopsis. In: Feldman RG, Young RR, Koella WP (eds) Spasticity: disordered motor control. Year Book Medical Publishers, Chicago, pp 485–494

Lannin NA, Novak I, Cusick A (2007) A systematic review of upper extremity casting for children and adults with central nervous system motor disorders. Clin Rehabil 21:963–976

Lapatki B, van Dijk J, van de Warrenburg B, Zwarts M (2011) Botulinum toxin has an increased effect when targeted toward the muscle's endplate zone: a high-density surface EMG guided study. Clin Neurophysiol 122:1611–1616

Lee LR, Chuang YC, Yang BJ, Hsu MJ, Liu YH (2004) Botulinum toxin for lower limb spasticity in children with cerebral palsy: a single-blinded trial comparing dilution techniques. Am J Phys Med Rehabil 83:766–773

Lee JM, Gracies JM, Park SB, Lee KH, Lee JY, Shin JH (2018) Botulinum toxin injections and electrical stimulation for spastic paresis improve active hand function following stroke. Toxins (Basel) 10:426

Li S (2017) Spasticity, motor recovery, and neural plasticity after stroke. Front Neurol 8:120

Li S, Francisco G (2015) New insights into the pathophysiology of post-stroke spasticity. Front Hum Neurosci 9:192

Malhotra S, Pandyan AD, Day CR, Jones PW, Hermens H (2009) Spasticity, an impairment that is poorly defined and poorly measured. Clin Rehabil 23:651–658

Marchand-Pauvert V, Aymard C, Giboin LS, Dominici F, Rossi A, Mazzocchio R (2013) Beyond muscular effects: depression of spinal recurrent inhibition after botulinum neurotoxin A. J Physiol 591:1017–1029

Mas MF, Li S, Francisco GE (2017) Centrally mediated late motor recovery after botulinum toxin injection: case reports and a review of current evidence. J Rehabil Med 49:609–619

Mathevon L, Declemy A, Laffont I, Perennou D (2019) Immunogenicity induced by botulinum toxin injections for limb spasticity: a systematic review. Ann Phys Rehabil Med 62:241–251

Mayer NH, Esquenazi A (2003) Muscle overactivity and movement dysfunction in the upper motoneuron syndrome. Phys Med Rehabil Clin N Am 14:855–883

Mayer NH, Whyte J, Wannstedt G, Ellis CA (2008) Comparative impact of 2 botulinum toxin injection techniques for elbow flexor hypertonia. Arch Phys Med Rehabil 89(5):982–987

Maynard FM, Karunas RS, Waring WP 3rd (1990) Epidemiology of spasticity following traumatic spinal cord injury. Arch Phys Med Rehabil 71:566–569

Miller DM, Klein CS, Suresh NL, Rymer WZ (2014) Asymmetries in vestibular evoked myogenic potentials in chronic stroke survivors with spastic hypertonia: evidence for a vestibulospinal role. Clin Neurophysiol 125:2070–2078

Mills PB, Finlayson H, Sudol M, O'Connor R (2016) Systematic review of adjunct therapies to improve outcomes following botulinum toxin injection for treatment of limb spasticity. Clin Rehabil 30:537–548
Moeini-Naghani I, Hashemi-Zonouz T, Jabbari B (2016) Botulinum toxin treatment of spasticity in adults and children. Semin Neurol 36:64–72
Mortenson PA, Eng JJ (2003) The use of casts in the management of joint mobility and hypertonia following brain injury in adults: a systematic review. Phys Ther 83:648–658
Mukherjee A, Chakravarty A (2010) Spasticity mechanisms – for the clinician. Front Neurol 1:149
Murphy MP, Carmine H (2012) Long-term health implications of individuals with TBI: a rehabilitation perspective. NeuroRehabilitation 31:85–94
Naumann M, Carruthers A, Carruthers J et al (2010) Meta-analysis of neutralizing antibody conversion with onabotulinumtoxinA (BOTOX(R)) across multiple indications. Mov Disord 25:2211–2218
Nielsen JB, Crone C, Hultborn H (2007) The spinal pathophysiology of spasticity – from a basic science point of view. Acta Physiol 189:171–180
Oo WM (2014) Efficacy of addition of transcutaneous electrical nerve stimulation to standardized physical therapy in subacute spinal spasticity: a randomized controlled trial. Arch Phys Med Rehabil 19:00432–00438
Owen M, Ingo C, Dewald JPA (2017) Upper extremity motor impairments and microstructural changes in bulbospinal pathways in chronic hemiparetic stroke. Front Neurol 8:257
Patejdl R, Zettl UK (2017) Spasticity in multiple sclerosis: contribution of inflammation, autoimmune mediated neuronal damage and therapeutic interventions. Autoimmun Rev 16:925–936
Phadke CP, On AY, Kirazli Y, Ismail F, Boulias C (2013) Intrafusal effects of botulinum toxin injections for spasticity: revisiting a previous paper. Neurosci Lett 541:20–23
Picelli A, Lobba D, Midiri A, Prandi P, Melotti C, Baldessarelli S, Smania N (2014a) Botulinum toxin injection into the forearm muscles for wrist and fingers spastic overactivity in adults with chronic stroke: a randomized controlled trial comparing three injection techniques. Clin Rehabil 28:232–242
Picelli A, Roncari L, Baldessarelli S, Berto G, Lobba D, Santamato A, Smania N (2014b) Accuracy of botulinum toxin type A injection into the forearm muscles of chronic stroke patients with s spastic flexed wrist and clenched fist: manual needle placement evaluated using ultrasonography. J Rehabil Med 46:1042
Picelli A, Tamburin S, Gajofatto F et al (2014c) Association between severe upper limb spasticity and brain lesion location in stroke patients. Biomed Res Int 162754:25
Pinter MM, Gerstenbrand F, Dimitrijevic MR (2000) Epidural electrical stimulation of posterior structures of the human lumbosacral cord: 3. Control of spasticity. Spinal Cord 38:524–531
Pirazzini M, Rossetto O, Eleopra R, Montecucco C (2017) Botulinum neurotoxins: biology, pharmacology, and toxicology. Pharmacol Rev 69:200–235
Ploumis A, Varvarousis D, Konitsiotis S, Beris A (2014) Effectiveness of botulinum toxin injection with and without needle electromyographic guidance for the treatment of spasticity in hemiplegic patients: a randomized controlled trial. Disabil Rehabil 36:313–318
Pohl M, Mehrholz J, Ruckriem S (2003) The influence of illness duration and level of consciousness on the treatment effect and complication rate of serial casting in patients with severe cerebral spasticity. Clin Rehabil 17:373–379
Preissner KS (2002) The effects of serial casting on spasticity: a literature review. Occup Ther Health Care 14:99–106
Pundik S, McCabe J, Skelly M, Tatsuoka C, Daly JJ (2018) Association of spasticity and motor dysfunction in chronic stroke. Ann Phys Rehabil Med. https://doi.org/10.1016/j.rehab.2018.07.006
Raghavan P, Lu Y, Mirchandani M, Stecco A (2016) Human recombinant hyaluronidase injections for upper limb muscle stiffness in individuals with cerebral injury: a case series. EBioMedicine 9:306–313

Rosales RL, Chua-Yap AS (2008) Evidence-based systematic review on the efficacy and safety of botulinum toxin-A therapy in post-stroke spasticity. J Neural Transm 115:617–623
Rosales RL, Arimura K, Takenaga S, Osame M (1996) Extrafusal and intrafusal muscle effects in experimental botulinum toxin-A injection. Muscle Nerve 19:488–496
Rosales RL, Kong KH, Goh KJ et al (2012) Botulinum toxin injection for hypertonicity of the upper extremity within 12 weeks after stroke a randomized controlled trial. Neurorehabil Neural Repair 26:812–821
Rosales RL, Balcaitiene J, Berard H et al (2018) Early abobotulinumtoxinA (Dysport((R))) in post-stroke adult upper limb spasticity: ONTIME pilot study. Toxins (Basel) 10:E253
Santamato A, Notarnicola A, Panza F et al (2013) SBOTE study: extracorporeal shock wave therapy versus electrical stimulation after botulinum toxin type A injection for post-stroke spasticity-a prospective randomized trial. Ultrasound Med Biol 39:283–291
Santamato A, Micello MF, Ranieri M et al (2015) Employment of higher doses of botulinum toxin type A to reduce spasticity after stroke. J Neurol Sci 350:1–6
Santamato A, Panza F, Intiso D et al (2017) Long-term safety of repeated high doses of incobotulinumtoxinA injections for the treatment of upper and lower limb spasticity after stroke. J Neurol Sci 378:182–186
Schnitzler A, Roche N, Denormandie P, Lautridou C, Parratte B, Genet F (2012) Manual needle placement: accuracy of botulinum toxin A injections. Muscle Nerve 46:531–534
Shaari CM, Sanders I (1993) Quantifying how location and dose of botulinum toxin injections affect muscle paralysis. Muscle Nerve 16:964–969
Shaw LC, Price CI, van Wijck FM et al (2011) Botulinum toxin for the upper limb after stroke (BoTULS) trial: effect on impairment, activity limitation, and pain. Stroke 42:1371–1379
Sheean G, Lannin NA, Turner-Stokes L, Rawicki B, Snow BJ (2010) Botulinum toxin assessment, intervention and after-care for upper limb hypertonicity in adults: international consensus statement. Eur J Neurol 17:74–93
Simpson DM, Alexander DN, O'Brien CF et al (1996) Botulinum toxin type A in the treatment of upper extremity spasticity: a randomized, double-blind, placebo-controlled trial. Neurology 46:1306–1310
Simpson DM, Gracies JM, Graham HK, Miyasaki JM, Naumann M, Russman B, So Y (2008) Assessment: botulinum neurotoxin for the treatment of spasticity (an evidence-based review) report of the therapeutics and technology assessment Subcommittee of the American Academy of Neurology. Neurology 70:1691–1698
Simpson DM, Hallett M, Ashman EJ et al (2016) Practice guideline update summary: botulinum neurotoxin for the treatment of blepharospasm, cervical dystonia, adult spasticity, and headache report of the guideline development Subcommittee of the American Academy of Neurology. Neurology 86:1818–1826
Stecco A, Gesi M, Stecco C, Stern R (2013) Fascial components of the myofascial pain syndrome. Curr Pain Headache Rep 17(352):352
Stecco A, stecco C, Raghavan P (2014) Peripheral mechanisms contributing to spasticity and implications for treatment. Curr Phys Med Rehabil Rep 2:121–127
Tao L, Peng L, Berntsson RP et al (2017) Engineered botulinum neurotoxin B with improved efficacy for targeting human receptors. Nat Commun 8(53):53
Tenniglo MJ, Nederhand MJ, Prinsen EC, Nene AV, Rietman JS, Buurke JH (2014) Effect of chemodenervation of the rectus femoris muscle in adults with a stiff knee gait due to spastic paresis: a systematic review with a meta-analysis in patients with stroke. Arch Phys Med Rehabil 95:576–587
Thompson AJ, Jarrett L, Lockley L, Marsden J, Stevenson VL (2005) Clinical management of spasticity. J Neurol Neurosurg Psychiatry 76:459–463
Trompetto C, Currà A, Buccolieri A, Suppa A, Abbruzzese G, Berardelli A (2006) Botulinum toxin changes intrafusal feedback in dystonia: a study with the tonic vibration reflex. Mov Disord 21:777–782

Turkoglu AN, Huijing PA, Yucesoy CA (2014) Mechanical principles of effects of botulinum toxin on muscle length-force characteristics: an assessment by finite element modeling. J Biomech 47:1565–1571

Vattanasilp W, Ada L, Crosbie J (2000) Contribution of thixotropy, spasticity, and contracture to ankle stiffness after stroke. J Neurol Neurosurg Psychiatry 69:34–39

Watkins CL, Leathley MJ, Gregson JM, Moore AP, Smith TL, Sharma AK (2002) Prevalence of spasticity post stroke. Clin Rehabil 16:515–522

Webb RP (2018) Engineering of botulinum neurotoxins for biomedical applications. Toxins (Basel) 10:231

Wedekind C, Lippert-Grüner M (2005) Long-term outcome in severe traumatic brain injury is significantly influenced by brainstem involvement. Brain Inj 19:681–684

Weise D, Weise CM, Naumann M (2019) Central effects of botulinum neurotoxin – evidence from human studies. Toxins 11:21

Wheeler A, Smith HS (2013) Botulinum toxins: mechanisms of action, antinociception and clinical applications. Toxicology 306:124–146

Wilkenfeld AJ (2013) Review of electrical stimulation, botulinum toxin, and their combination for spastic drop foot. J Rehabil Res Dev 50:315–326

Wissel J, Ward AB, Erztgaard P, Bensmail D, Hecht MJ, Lejeune TM, Schnider P (2009) European consensus table on the use of botulinum toxin type A in adult spasticity. J Rehabil Med 41:13–25

Wissel J, Manack A, Brainin M (2013) Toward an epidemiology of poststroke spasticity. Neurology 80:S13

Wissel J, Bensmail D, Ferreira JJ et al (2017) Safety and efficacy of incobotulinumtoxinA doses up to 800 U in limb spasticity: the TOWER study. Neurology 88:1321–1328

Wu D, Qian L, Zorowitz RD, Zhang L, Qu Y, Yuan Y (2013) Effects on decreasing upper-limb poststroke muscle tone using transcranial direct current stimulation: a randomized sham-controlled study. Arch Phys Med Rehabil 94:1–8

Yablon SA, Brashear A, Gordon MF et al (2007) Formation of neutralizing antibodies in patients receiving botulinum toxin type A for treatment of poststroke spasticity: a pooled-data analysis of three clinical trials. Clin Ther 29:683–690

Zhang C, Peng Y, Liu Y, Li S, Zhou P, Rymer WZ, Zhang Y (2017) Imaging three-dimensional innervation zone distribution in muscles from M-wave recordings. J Neural Eng 14:036011

Zhang C, Chen YT, Liu Y, Zhou P, Li S, Zhang Y (2019) Three dimensional innervation zone imaging in spastic muscles of stroke survivors. J Neural Eng 16:034001

Zorowitz RD, Gillard PJ, Brainin M (2013) Poststroke spasticity: sequelae and burden on stroke survivors and caregivers. Neurology 80:S45–S52

Zuber M, Sebald M, Bathien N, De Recondo J, Rondot P (1993) Botulinum antibodies in dystonic patients treated with type A botulinum toxin: frequency and significance. Neurology 43:1715–1718

Use of Botulinum Toxin in Ophthalmology

Michael J. Wan, Sara AlShaker, and David G. Hunter

Contents

Abstract

Botulinum toxin is an important treatment for many conditions in ophthalmology, including strabismus, nystagmus, blepharospasm, hemifacial spasm, spastic and congenital entropion, corneal exposure, and persistent epithelial defects. The mechanism of action of botulinum toxin for both strabismus and nystagmus is the neuromuscular blockade and transient paralysis of extraocular muscles, but

The original version of this chapter was revised. A correction to this chapter can be found at https://doi.org/10.1007/164_2020_413

M. J. Wan
Department of Ophthalmology and Vision Sciences, University of Toronto, Toronto, ON, Canada

Department of Ophthalmology and Vision Sciences, The Hospital for Sick Children, Toronto, ON, Canada

S. AlShaker
Department of Ophthalmology and Vision Sciences, University of Toronto, Toronto, ON, Canada

D. G. Hunter (✉)
Department of Ophthalmology, Harvard Medical School, Boston, MA, USA

Department of Ophthalmology, Boston Children's Hospital, Boston, MA, USA
e-mail: david.hunter@childrens.harvard.edu

S. M. Whitcup, M. Hallett (eds.), *Botulinum Toxin Therapy*,
Handbook of Experimental Pharmacology 263, https://doi.org/10.1007/164_2019_325

when botulinum toxin is used for some forms of strabismus, a single injection can convey indefinite benefits. There are two unique mechanisms of action that account for the long-term effect on ocular alignment: (1) the disruption of a balanced system of agonist-antagonist extraocular muscles and (2) the reestablishment of central control of alignment by the binocular visual system. For other ocular conditions, botulinum toxin acts through transient paralysis of periocular muscles. Botulinum toxin is a powerful tool in ophthalmology, achieving its therapeutic effects by direct neuromuscular blockade of extraocular and periocular muscles and by unique mechanisms related to the underlying structure and function of the visual system.

Keywords
Binocular vision · Blepharospasm · Botulinum toxin · Depth perception · Oculomotor muscles · Strabismus

1 Introduction

The first ever use of botulinum toxin as a therapeutic agent was in the field of ophthalmology for the treatment of strabismus (misalignment of the eyes). In the late 1960s and early 1970s, Alan B. Scott injected the neurotoxin from the *Clostridium botulinum* bacterium into the extraocular muscles of monkeys to alter binocular alignment (Scott et al. 1973). Scott published the first human trial of botulinum toxin type A in 1980, injecting botulinum toxin into the extraocular muscles of adults with strabismus (Scott 1980). Based on evidence from subsequent human trials, botulinum toxin type A was approved for clinical use by the US Food and Drug Administration (FDA) in 1989 for the treatment of strabismus (eye misalignment) and blepharospasm (involuntary spasm of the periocular muscles).

It was during the early trials on botulinum toxin for strabismus that the potential cosmetic applications of botulinum toxin were discovered. While the cosmetic use of botulinum toxin is probably the most well-known application, botulinum toxin remains an important treatment for several conditions in ophthalmology (Table 1, Fig. 1). The mechanism of action of botulinum toxin for strabismus involves not only a short-term change in eye alignment due to transient paralysis of extraocular muscles but also in some cases by permanently altering the balance of muscle tension which controls binocular alignment. The binocular visual system – that is, the central coordination of alignment of the two eyes, mediated by pathways in the cerebral cortex, white matter, and brainstem – likely also contributes to the long-term therapeutic effect of botulinum toxin for strabismus.

Other non-cosmetic, therapeutic applications of botulinum toxin in ophthalmology include nystagmus, benign essential blepharospasm, hemifacial spasm, spastic and congenital entropion, and corneal exposure or persistent epithelial defects. For select cases of nystagmus, botulinum toxin can be injected into the orbit to paralyze all of the extraocular muscles in an attempt to improve vision or decrease oscillopsia. For benign essential blepharospasm and hemifacial spasm, botulinum toxin is

Table 1 Key ophthalmic uses of botulinum toxin

Indication	Reference
Strabismus	
Retreatment after surgery for infantile esotropia	Tejedor and Rodriguez (1999)
Retreatment after surgery for acquired esotropia	Tejedor and Rodriguez (1998)
Acute-onset, acquired comitant esotropia (Fig. 1a)	Wan et al. (2017) and Dawson et al. (1999a)
Augmentation of surgery for large-angle strabismus	Minguini et al. (2012), Wan et al. (2018), Lueder et al. (2012), and Khan (2005)
Prevention of muscle contraction in sixth nerve palsy	Kao and Chao (2003)
Postoperative, consecutive strabismus	Dawson et al. (1999b)
Nystagmus	
Congenital or acquired nystagmus	Lennerstrand et al. (1998) and Tomsak et al. (1995)
Eyelid	
Blepharospasm and hemifacial spasm	Ross et al. (2011), Bilyk et al. (2018), Ababneh et al. (2014), and Cillino et al. (2010)
Corneal exposure	Kasaee et al. (2010), Adams et al. (1987), Ellis and Daniell (2001), Naik et al. (2008), and Sarkies (1999)
Spastic/involutional/congenital entropion	Cillino et al. (2010), Clarke and Spalton (1988), Steel et al. (1997), Deka and Saikia (2011), and Lee et al. (2005)

injected into periocular protractor muscles (orbicularis oculi, procerus, and corrugator and depressor supercilii) to prevent involuntary muscle spasm. For spastic entropion, botulinum toxin is injected into the lower eyelid to prevent spastic infolding. For corneal exposure or persistent epithelial defects, botulinum toxin is injected into the levator muscle to create ptosis (drooping of the upper eyelid). For these indications, the mechanism of action is the transient paralysis of the facial muscles around the eye.

This chapter will describe the mechanisms of action and clinical uses of botulinum toxin in ophthalmology.

2 Pharmacology and Mechanism

The direct mechanism of action of botulinum toxin is to block the release of acetylcholine from the presynaptic terminal of the neuromuscular junction leading to muscle paralysis (Huang et al. 2000). The paralysis is transient, as muscle function is restored by the generation of new nerve terminals and reestablishment of synaptic transmission, which usually takes about 3 months. This transient neuromuscular blockade is critical in all ophthalmological applications of botulinum toxin. In strabismus, the transient paralysis of an extraocular muscle causes a change in eye

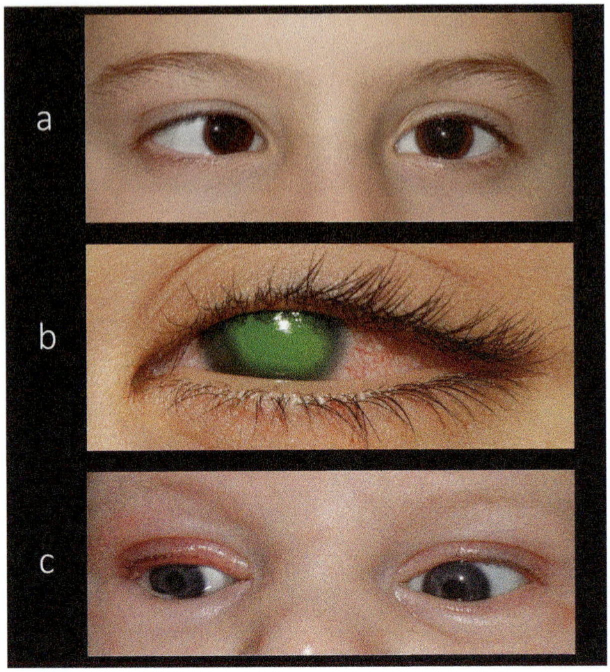

Fig. 1 Common indications for botulinum toxin in ophthalmology. These include strabismus ((**a**) acute-onset comitant esotropia), corneal disease ((**b**) left eye persistent corneal epithelial defect), and entropion ((**c**) right eye congenital lower lid entropion)

position. In nystagmus, the transient paralysis of all extraocular muscles decreases the severity of the abnormal eye movements. In blepharospasm and hemifacial spasm, the transient paralysis of the periocular protractor muscles inhibits involuntary muscle contractions. In corneal exposure and epithelial defects, the transient paralysis of the levator muscle leads to upper eyelid ptosis to cover and protect the surface of the eye. Finally, in spastic and congenital entropion, the transient paralysis of the orbicularis muscle prevents the muscle contraction that causes the lower eyelid to fold inward.

For many of the applications of botulinum toxin in ophthalmology, transient muscle paralysis is sufficient to achieve treatment goals. For instance, for a non-healing corneal epithelial defect, botulinum toxin can create a transient ptosis that protects the surface of the eye and aids in corneal healing (Kasaee et al. 2010). Once the epithelial defect has resolved, the ptosis is no longer needed and becomes a potential cosmetic and functional issue. Therefore, the fact that the ptosis completely resolves a few months after botulinum toxin injection is an advantage for corneal conditions that are not expected to be chronic. A major exception to this is the treatment of strabismus, as a long-lasting change in ocular alignment is the ultimate goal of any effective treatment of strabismus.

The mechanism of action of botulinum toxin for strabismus deserves special consideration. In the early trials on botulinum toxin for strabismus, Alan Scott noted that although muscle paresis was invariably transient, permanent changes in ocular alignment were common (Scott et al. 1973). This indicated that it might be possible to achieve a long-term change in binocular alignment despite the transient nature of the direct neuromuscular blockade. There are two unique mechanisms of action that help to explain the lasting effect of botulinum toxin on eye alignment: (1) disruption of the balanced system of agonist-antagonist extraocular muscles that control eye movement and position and (2) reestablishment of central control of alignment by the binocular visual system.

2.1 Disruption of the Balanced System of Extraocular Muscles

Within each orbit, there are six extraocular muscles that move the eye – four rectus muscles (inferior, superior, medial, and lateral) and two oblique muscles (inferior and superior). Eye alignment and eye movements are based on an agonist-antagonist system (Shumway and Wade 2019). For horizontal eye movements, when an extraocular muscle activates to move an eye in a certain direction, its antagonist muscle relaxes, while in the fellow eye, the opposite muscles activate and relax. For example, when the right eye looks toward the left, the right medial rectus muscle contracts to pull the eye inward toward the nose, while its antagonist right lateral rectus muscle automatically relaxes, reducing resistance to inward movement. This phenomenon is referred to as Sherrington's law of reciprocal innervation, and it potentiates smooth and rapid movement of the eye in one direction. At the same time, the contralateral left lateral rectus muscle contracts, while the left medial rectus muscle relaxes, allowing both eyes to track simultaneously to the left and maintain binocular alignment during eye movements. At rest, the lateral and medial rectus muscles of both eyes are under baseline tension ("tonus"), and this tension is balanced on both sides to keep the eyes looking straight ahead. It is this balanced system of extraocular muscles that controls eye movements and determines (1) the alignment and (2) the position of the eyes.

Regardless of whether the eyes are moving or stationary, the fine-tuning of binocular alignment to maintain fixation of both eyes on the same target at the same time requires additional high-level control by the binocular visual system. Binocular alignment and eye movements are believed to be controlled by separate innervation pathways within the brain, one primarily served by multiply innervated fiber (MIF) motoneurons and the other served by singly innervated fiber (SIF or "twitch") motoneurons (Staube and Büttner 2007). MIF motoneurons receive inputs from areas related to gaze holding, while SIF motoneurons receive inputs from brain areas involved in eye movement generation (Leigh and Zee 2015). As a result, the MIF system is believed to control muscle tension (binocular alignment, gaze holding), while the SIF system controls fast eye movements (saccades). Thus, binocular alignment is not generally under voluntary control, and eye exercises utilizing the SIF system (voluntary eye movements) would not be expected to have impact on the MIF system (binocular alignment).

When botulinum toxin is injected into an extraocular muscle, the neuromuscular blockade causes paralysis of all affected muscle fibers – those that are part of the MIF system and those that are part of the SIF system. Since the extraocular muscle system is based on the balanced tension and force of agonist and antagonist muscles, the paralysis of an agonist muscle causes relative overaction of its antagonist muscle. This disruption in muscle balance results in eye position shifting in the direction of action of the antagonist muscle. For instance, injection of botulinum toxin into the medial rectus muscle leads to relative lack of opponency to the lateral rectus muscle, causing the eye to deviate outward toward the ear (Wan et al. 2017). Depending on the amount injected and the sensitivity of the muscle to botulinum toxin, the result may be a change in alignment with little or no impact on eye movements (the more common outcome) or complete paralysis of the eye muscle with loss of movement in the direction of gaze toward the injected muscle. It is not known, however, whether MIF motoneurons are preferentially more sensitive to botulinum toxin than SIF motoneurons.

During the transient period of direct neuromuscular blockade by botulinum toxin, the eye alignment remains altered. Although the position can be quite variable, the injected agonist muscle is always in a relatively stretched position, while the antagonist muscle is in a relatively contracted position. Although the direct neuromuscular blockade is transient, the change in eye position can lead to long-term structural changes within the extraocular muscles. This structural change was demonstrated in monkey model (Scott 1994), in which researchers sutured one eye to the orbital wall to change the position of the eye from straight to an outwardly deviated (exotropic) position of 30–45° in three animals. Immediately following suturing, histological examination of two monkeys showed lengthened sarcomeres in the stretched medial rectus muscle and shortened sarcomeres in the contracted lateral rectus muscle, as anticipated. After 2 months of maintaining the eye in this exotropic position in the third monkey, histological examination showed similar changes to the lengths of the muscles, but the sarcomeres were no longer stretched in the medial rectus muscle nor contracted in the lateral rectus muscle. The authors interpreted this to mean that, within a 2-month period, sarcomeres had been added to the stretched muscles and removed from the contracted muscle. In effect, the injected muscle had become anatomically longer, and the antagonist muscle had become shorter.

One presumed mechanism of action of botulinum toxin injections, then, is that the transient imbalance of tension causes lengthening of the injected muscle (through addition of sarcomeres) and a complementary shortening of its antagonist (through removal of sarcomeres). These change in anatomic muscle length would maintain the shift in eye position long after the direct effects of botulinum toxin had dissipated. Such changes, which presumably alter both MIF myofibers and SIF myofibers, could not be achieved to the same degree with eye exercises, no matter how vigorously pursued, since these voluntary eye movements affect tension of SIF myofibers only.

2.2 Reestablishment of Control by the Binocular Visual System

In a normally functioning visual system, the brain integrates signals from both eyes in order to achieve binocular vision. Since the eyes are horizontally separated, each sends a similar but slightly disparate image from the retina to the visual cortex.

Binocular vision and stereopsis are achieved when the brain fuses the two slightly disparate images into a single coherent image. Effective binocular fusion requires two separate events: motor fusion, in which muscle tonus is adjusted to align the two images, and sensory fusion, in which the two aligned images are joined into a single percept. Once fusion is achieved, the slight differences in the images from the eyes are processed by the higher-order centers of the brain to achieve stereopsis (enhanced depth perception). Achieving fusion generally requires that the baseline position of the two eyes is relatively aligned, allowing motor fusion to reconcile the smaller disparities and align the images, presumably via the MIF motoneuron system. For an individual with manifest strabismus, the eyes are misaligned beyond the range where motor fusion is possible. The result is either diplopia (double vision) or suppression.

It is this binocular visual system that underlies another mechanism of action of botulinum toxin that is specific to strabismus treatment. When botulinum toxin is injected into an extraocular muscle, the neuromuscular blockade leads to muscle paresis and a change in the alignment of the eyes. If this change in alignment brings the eyes to within an acceptable anatomic range, then a functioning binocular visual system will be able to reestablish motor (and sensory) fusion. Even as the direct effects of botulinum toxin slowly dissipate, the binocular system receives feedback of disparities in motor alignment and is still capable of revising input to the muscles via the slow MIF system, thus maintaining alignment. As such, a transient change in eye alignment by botulinum toxin injection can lead to a long-term effect on strabismus by "rebooting" the binocular visual system, allowing it to regain control of motor fusion during the period of improved alignment and then to sustain that control as the direct effects of the injection wear off. Clinical experience suggests that patients with an intact binocular visual system (such as those who develop acute, comitant esotropia) are most likely to respond indefinitely to a single injection (Fig. 1a), whereas those with no binocular potential (such as adults with sensory strabismus) are most likely to require repeated injections over time (Dawson et al. 1999a; Gardner et al. 2008; Wan et al. 2017).

3 Clinical Uses in Ophthalmology

3.1 Strabismus

While many applications for botulinum toxin have emerged since it was first used in humans, the original therapeutic target – injection into the extraocular muscles to treat strabismus – remains an important clinical indication (Rowe and Noonan 2017). In the years since FDA approval, the utility of botulinum toxin for strabismus has been expanded and refined (Table 1). A recent Cochrane review summarized the findings from all published randomized controlled trials on the use of botulinum toxin for strabismus (Rowe and Noonan 2017). The randomized trials have had varying results, and the strongest evidence for botulinum toxin is as an alternative to incisional strabismus surgery for children requiring retreatment for infantile or acquired esotropia (Tejedor and Rodriguez 1998, 1999). Other trials have suggested

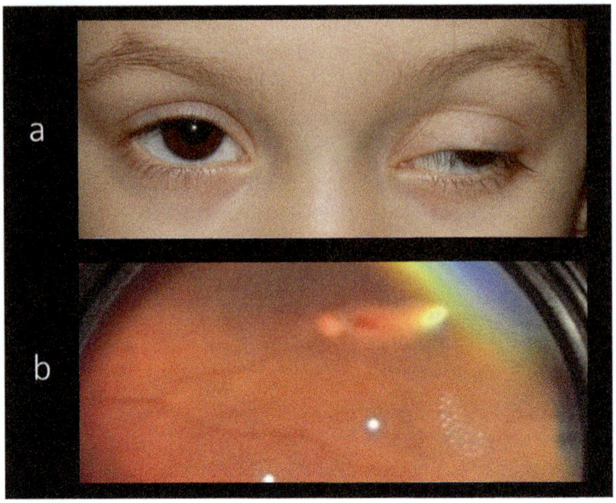

Fig. 2 Select complications of botulinum toxin treatment for strabismus. These include (**a**) commonly occurring, transient overcorrection and ptosis (botulinum toxin treatment of esotropia leading to exotropia and left upper eyelid ptosis) and (**b**) rarely occurring, retinal injury or intraocular injection (linear chorioretinal scar from intraocular penetration during botulinum toxin injection)

that botulinum toxin may be slightly better than observation for acute-onset sixth nerve palsy (Lee et al. 1994) and that botulinum toxin may help to augment the effect of surgery in large-angle strabismus (Minguini et al. 2012). Despite the lack of randomized, controlled clinical trials, there is still a significant body of evidence supporting the effectiveness of botulinum toxin in the treatment of various types of strabismus (Table 1). Most of the evidence for the use of botulinum toxin in strabismus comes from retrospective studies, such as cohort studies and comparative case series.

There are many techniques for injecting botulinum toxin into the extraocular muscles to treat strabismus. All involve injecting botulinum toxin into the belly of the target muscle while trying to minimize the spread of toxin elsewhere in the orbit. If the botulinum toxin injection is performed at the time of strabismus surgery, it can be injected into the extraocular muscle under direct visualization, though some experts prefer to inject before the surgical incision out of concern that injection after incision and dissection may allow for greater spread to other sites. If the botulinum toxin is being used in isolation, the injection is generally administered transconjunctivally (typically with local anesthesia in adults and under general anesthesia in children). Electromyographic (EMG) guidance can be used in awake patients to confirm that the needle is intramuscular prior to injection, but comparable results have been achieved (for medial rectus muscle injections) without using EMG guidance (Sanjari et al. 2008).

The main complications of botulinum toxin treatment of strabismus are transient overcorrection and ptosis (Fig. 2a). Although overcorrection can be quite drastic, it may enhance the beneficial long-term changes in sarcomere length described above (Wan et al. 2017). Lasting overcorrection is rare, but has been described (Tejedor and Rodriguez 2007). In contrast, ptosis is undesirable and, when severe enough to block the pupil, has the potential to cause or exacerbate amblyopia in young children. There is no definite way to avoid ptosis, although some authors recommend minimizing the injection volume (see formula below) and elevating the head of the bed immediately after injection to reduce the likelihood of spread to the levator muscle. Severe complications from botulinum toxin treatment for strabismus are rare but have been reported. These include intraocular injection of botulinum toxin (Pehere et al. 2011), retinal injury (Fig. 2b) (Liu et al. 2004), and permanent pupillary dilation (Pediatric Eye Disease Investigator Group et al. 2016).

Typical dosages for treatment of strabismus vary. At least one study suggests that there is a dose-response curve, with larger amounts leading to greater degrees of long-term correction (Wan et al. 2018). The amount of onabotulinumtoxinA injected into the extraocular muscle for treatment of strabismus ranges in most studies from a minimum of 2–3 units to a maximum of 10 units. Larger amounts are injected for larger deviations and in the presence of inflammation or muscle contracture. To minimize the volume of medication injected, a sliding scale of diluent can be used; for example, if a 100 unit vial is reconstituted in 1 cc of normal saline, then an injection of 0.05 mL will administer (0.05 cc $*$ (100 units / 1 cc)) = 5 units. The formula for administration of varying dosages in a particular volume for any formulation of botulinum toxin is thus

$$D = T * V/U$$

where:

D = diluent injected into the vial for reconstitution (in cc)
T = units of toxin in the vial (in IU)
V = volume of medication to be injected (in cc)
U = units of botulinum toxin to be injected (in IU)

For example, if 7.5 units is to be injected in 0.05 cc after reconstituting a 100 unit vial, the botulinum toxin should be reconstituted with (100 $*$ 0.05 / 7.5) = 0.66 cc of normal saline.

3.2 Nystagmus

Botulinum toxin has shown some effectiveness in the treatment of oscillopsia secondary to acquired nystagmus. The concept is that injection of botulinum toxin into the retrobulbar space will reduce the motility of all of the extraocular muscles and thus damp the nystagmus. All studies of botulinum treatment for nystagmus are small retrospective case series, with a typical dosage of 20–30 units per eye. To date, the results have been mixed. While one study showed some benefit in improving vision or decreasing oscillopsia (Lennerstrand et al. 1998), another reported unsatisfactory results (Tomsak et al. 1995). Even when improvement in the nystagmus is achieved, the side effects are typically severe, including complete ophthalmoplegia, ptosis, and induced strabismus (Lennerstrand et al. 1998). In patients who benefit from treatment, repeat injections are required – there are no reports of indefinite improvement following retrobulbar injections for nystagmus.

3.3 Eyelid

3.3.1 Benign Essential Blepharospasm

Benign essential blepharospasm (BEB) is a focal dystonia that results in involuntary spasms of the periocular protractor muscles. Contractions can be quite debilitating; the forceful contractions generally progress over time, and can ultimately result in functional blindness. Botulinum toxin gained FDA approval for blepharospasm in 1989 and has since become the preferred first-line treatment for BEB (Ababneh et al. 2014; Bilyk et al. 2018; Ross et al. 2011).

For treatment, onabotulinumtoxinA (Botox) is reconstituted to a dose of 2.5–5 units per 0.1 mL and injected at multiple periorbital sites with the aim of achieving symptomatic relief using the smallest dose possible (Bilyk et al. 2018; Ross et al. 2011). Other formulations, abobotulinumtoxinA (Dysport) and incobotulinumtoxinA (Xeomin), have also been used. A standardized conversion ratio is lacking; however, onabotulinumtoxinA-to-abobotulinumtoxinA ratios of 1:2 to 1:4 and onabotulinumtoxinA-to-incobotulinumtoxinA ratios of 1:1 to 1:2 have been reported (Hallett et al. 2013; Marion et al. 1995; Sampaio et al. 1997; Scaglione 2016).

Repeat injections are carried out at 3-month intervals or longer, as required. Side effects include ptosis, lagophthalmos, entropion, ectropion, bruising, and excessive tearing (Ababneh et al. 2014; Bilyk et al. 2018; Ross et al. 2011). Ptosis can be minimized by targeting the pretarsal orbicularis rather than the orbital portion and avoiding injection into the center of the lid, though the pretarsal orbicularis injections are more painful (Albanese et al. 1996; Cakmur et al. 2002; Dutton and Fowler 2007; Jankovic 1996).

3.3.2 Hemifacial Spasm

Hemifacial spasm (HFS) is characterized by involuntary contractions of the muscles innervated by the ipsilateral facial nerve. In contrast with BEB, contractions in HFS persist during sleep. In some cases, the orbicularis oculi muscles are believed to serve as a trigger point for spasm in the lower facial muscles. Compression of the facial nerve at its root must be ruled out by neuroimaging. A starting dose of 12.5 U onabotulinumtoxinA to the orbicularis muscle is usually effective, and the first treatment usually targets the orbicularis oculi muscles alone (Ababneh et al. 2014; Bilyk et al. 2018; Ross et al. 2011).

3.3.3 Exposure Keratopathy

A temporary suture tarsorrhaphy is often used to promote corneal healing in conditions with persistent epithelial defects or to protect the cornea in patients susceptible to developing ulceration secondary to exposure, neuropathy, or infectious causes (Fig. 1b). A "chemical tarsorrhaphy" with botulinum toxin has been described as an effective alternative to a suture tarsorrhaphy (Adams et al. 1987; Ellis and Daniell 2001; Kasaee et al. 2010; Naik et al. 2008). In addition to avoiding the risk of scarring associated with suture tarsorrhaphy, chemical tarsorrhaphy does not impede a thorough examination of the eye. It has also been described in children, and it can be performed at the bedside or in the office.

A chemical tarsorrhaphy should be performed using sterile technique but without need for either local or general anesthesia except in extremely anxious patients. Botulinum toxin is injected into the levator palpebrae superioris using a subcutaneous approach (Ellis and Daniell 2001; Kasaee et al. 2010; Naik et al. 2008). A transconjunctival approach has also been described, with the hypothesis that direct intramuscular inoculation with limited diffusion to other muscles is less likely to cause diplopia (Sarkies 1999). Reports on dosing vary between 5 and 15 units of onabotulinumtoxinA or 30 units of abobotulinumtoxinA to achieve sufficient ptosis, with further titration depending on the clinical response. Desired ptosis can occur on the first day; however, onset can range between 1 and 9 days with loss of effect by 6–12 weeks. Side effects include transient diplopia secondary to superior rectus muscle underaction and preseptal hemorrhage (Adams et al. 1987; Ellis and Daniell 2001; Kasaee et al. 2010; Naik et al. 2008). Permanent diplopia has been noted in patients with preexisting vertical deviations, presumably due to loss of fusion secondary to occlusion of the pupil (Heyworth and Lee 1994).

3.3.4 Entropion

Spastic entropion results from a sustained contraction in the orbicularis oculi muscle in the setting of ocular irritation. This commonly occurs after intraocular surgery, and there are usually underlying mild or unrecognized involutional eyelid changes. The acute entropion resolves with treatment of the underlying cause; however, injection of botulinum toxin to temporarily paralyze the overriding preseptal orbicularis muscle can be a useful temporizing measure (Cillino et al. 2010; Foster et al. 2017).

Involutional entropion is the result of a complex interplay of eyelid laxity, disinsertion of eyelid retractors, and superiorly overriding preseptal orbicularis muscle. Given that eyelid laxity is the primary culprit, it might seem unlikely that botulinum toxin would be helpful; however, it has been used in select cases of involutional entropion, owing to the multifactorial etiology of this condition. The ideal candidates are patients awaiting surgery or those unable to undergo definitive surgical repair, and results have been variable. Injection of 12.5–20 units of onabotulinumtoxinA temporarily paralyzes the preseptal orbicularis muscle. Improvement can be immediate secondary to the volume effect of the injection, with true paralysis occurring in 1–4 days. The duration of effect lasts 12–15 weeks on average (Clarke and Spalton 1988; Deka and Saikia 2011; Lee et al. 2005; Steel et al. 1997).

In congenital entropion, keratopathy from persistent friction of lashes against the cornea can lead to corneal ulceration (Fig. 1c). Injection of 5–7.5 units of botulinum toxin into the pretarsal orbicularis has been described with good results, with a single injection being effective indefinitely, likely related to the changes in anatomy that occur with rapid facial growth in newborns (Christiansen et al. 2004; Deka and Saikia 2011). Reports of side effects are uncommon, but can include epiphora secondary to punctal eversion, strabismus secondary to inferior oblique underaction

(Steel et al. 1997), and facial droop secondary to dissipation of the botulinum toxin inferiorly (Clarke and Spalton 1988).

Acknowledgments/Disclosures *Funding/Support*: Children's Hospital Ophthalmology Foundation, Inc., Boston, MA (DGH).

Financial Disclosures: Dr. Hunter is founder and board member of Rebion, Inc. and advisor to Luminopia, Inc. Neither is of relevance to the content of this chapter.

References

Ababneh OH, Cetinkaya A, Kulwin DR (2014) Long-term efficacy and safety of botulinum toxin A injections to treat blepharospasm and hemifacial spasm. Clin Exp Ophthalmol 42(3):254–261

Adams GG, Kirkness CM, Lee JP (1987) Botulinum toxin A induced protective ptosis. Eye 1 (Pt 5):603–608

Albanese A, Bentivoglio AR, Colosimo C, Galardi G, Maderna L, Tonali P (1996) Pretarsal injections of botulinum toxin improve blepharospasm in previously unresponsive patients. J Neurol Neurosurg Psychiatry 60(6):693–694

Bilyk JR, Yen MT, Bradley EA, Wladis EJ, Mawn LA (2018) Chemodenervation for the treatment of facial dystonia: a report by the American Academy of Ophthalmology. Ophthalmology 125 (9):1459–1467

Cakmur R, Ozturk V, Uzunel F, Donmez B, Idiman F (2002) Comparison of preseptal and pretarsal injections of botulinum toxin in the treatment of blepharospasm and hemifacial spasm. J Neurol 249(1):64–68

Christiansen G, Mohney BG, Baratz KH, Bradley EA (2004) Botulinum toxin for the treatment of congenital entropion. Am J Ophthalmol 138(1):153–155

Cillino S, Raimondi G, Guepratte N et al (2010) Long-term efficacy of botulinum toxin A for treatment of blepharospasm, hemifacial spasm, and spastic entropion: a multicentre study using two drug-dose escalation indexes. Eye 24(4):600–607

Clarke JR, Spalton DJ (1988) Treatment of senile entropion with botulinum toxin. Br J Ophthalmol 72(5):361–362

Dawson EL, Marshman WE, Adams GG (1999a) The role of botulinum toxin A in acute-onset esotropia. Ophthalmology 106(9):1727–1730

Dawson EL, Marshman WE, Lee JP (1999b) Role of botulinum toxin A in surgically overcorrected exotropia. J AAPOS 3(5):269–271

Deka A, Saikia SP (2011) Botulinum toxin for lower lid entropion correction. Orbit 30(1):40–42

Dutton JJ, Fowler AM (2007) Botulinum toxin in ophthalmology. Surv Ophthalmol 52(1):13–31

Ellis MF, Daniell M (2001) An evaluation of the safety and efficacy of botulinum toxin type A (BOTOX) when used to produce a protective ptosis. Clin Exp Ophthalmol 29(6):394–399

Foster JA, Carter KD, Durairaj VD, Kavanagh MC, Korn BS, Nelson CC, Hartstein ME (2017) Basic and clinical science course, section 7: orbit, eyelids, and lacrimal system. American Academy of Ophthalmology, San Francisco

Gardner R, Dawson EL, Adams GG, Lee JP (2008) Long-term management of strabismus with multiple repeated injections of botulinum toxin. J AAPOS 12(6):569–575

Hallett M, Albanese A, Dressler D et al (2013) Evidence-based review and assessment of botulinum neurotoxin for the treatment of movement disorders. Toxicon 67:94–114

Heyworth PL, Lee JP (1994) Persisting hypotropias following protective ptosis induced by botulinum neurotoxin. Eye 8(Pt 5):511–515

Huang W, Foster JA, Rogachefsky AS (2000) Pharmacology of botulinum toxin. J Am Acad Dermatol 43(2 Pt 1):249–259

Jankovic J (1996) Pretarsal injection of botulinum toxin for blepharospasm and apraxia of eyelid opening. J Neurol Neurosurg Psychiatry 60(6):704

Kao LY, Chao AN (2003) Subtenon injection of botulinum toxin for treatment of traumatic sixth nerve palsy. J Pediatr Ophthalmol Strabismus 40(1):27–30

Kasaee A, Musavi MR, Tabatabaie SZ et al (2010) Evaluation of efficacy and safety of botulinum toxin type A injection in patients requiring temporary tarsorrhaphy to improve corneal epithelial defects. Int J Ophthalmol 3(3):237–240

Khan AO (2005) Two horizontal rectus eye muscle surgery combined with botulinum toxin for the treatment of very large angle esotropia. A pilot study. Binocul Vis Strabismus Q 20(1):15–20

Lee J, Harris S, Cohen J, Cooper K, MacEwen C, Jones S (1994) Results of a prospective randomized trial of botulinum toxin therapy in acute unilateral sixth nerve palsy. J Pediatr Ophthalmol Strabismus 31(5):283–286

Lee C, Kikkawa DO, Pasco NY, Granet DB (2005) Advanced functional oculofacial indications of botulinum toxin. Int Ophthalmol Clin 45(3):77–91

Leigh RJ, Zee DS (2015) The neurology of eye movements, 5th edn. Oxford University Press, New York

Lennerstrand G, Nordbo OA, Tian S, Eriksson-Derouet B, Ali T (1998) Treatment of strabismus and nystagmus with botulinum toxin type A. An evaluation of effects and complications. Acta Ophthalmol Scand 76(1):27–27

Liu M, Lee HC, Hertle RW, Ho AC (2004) Retinal detachment from inadvertent intraocular injection of botulinum toxin A. Am J Ophthalmol 137(1):201–202

Lueder GT, Galli M, Tychsen L, Yildirim C, Pegado V (2012) Long-term results of botulinum toxin-augmented medial rectus recessions for large-angle infantile esotropia. Am J Ophthalmol 153(3):560–563

Marion MH, Sheehy M, Sangla S, Soulayrol S (1995) Dose standardisation of botulinum toxin. J Neurol Neurosurg Psychiatry 59(1):102–103

Minguini N, de Carvalho KM, Bosso FL, Hirata FE, Kara-Jose N (2012) Surgery with intraoperative botulinum toxin-A injection for the treatment of large-angle horizontal strabismus: a pilot study. Clinics 67(3):279–282

Naik MN, Gangopadhyay N, Fernandes M, Murthy R, Honavar SG (2008) Anterior chemodenervation of levator palpebrae superioris with botulinum toxin type-A (Botox) to induce temporary ptosis for corneal protection. Eye 22(9):1132–1136

Pediatric Eye Disease Investigator Group, Christiansen SP, Chandler DL et al (2016) Tonic pupil after botulinum toxin-A injection for treatment of esotropia in children. J AAPOS 20(1):78–81

Pehere N, Jalali S, Mathai A, Naik M, Ramesh K (2011) Inadvertent intraocular injection of botulinum toxin A. J Pediatr Ophthalmol Strabismus 48:e1–e3

Ross AH, Elston JS, Marion MH, Malhotra R (2011) Review and update of involuntary facial movement disorders presenting in the ophthalmological setting. Surv Ophthalmol 56(1):54–67

Rowe FJ, Noonan CP (2017) Botulinum toxin for the treatment of strabismus. Cochrane Database Syst Rev 3:CD006499

Sampaio C, Ferreira JJ, Simoes F et al (1997) DYSBOT: a single-blind, randomized parallel study to determine whether any differences can be detected in the efficacy and tolerability of two formulations of botulinum toxin type A – Dysport and Botox – assuming a ratio of 4:1. Mov Disord 12(6):1013–1018

Sanjari MS, Falavarjani KG, Kashkouli MB, Aghai GH, Nojomi M, Rostami H (2008) Botulinum toxin injection with and without electromyographic assistance for treatment of abducens nerve palsy: a pilot study. J AAPOS 12(3):259–262

Sarkies NJ (1999) A clinical algorithm for the management of facial nerve palsy from an oculoplastic perspective. Eye 13(Pt 2):273

Scaglione F (2016) Conversion ratio between Botox(R), Dysport(R), and Xeomin(R) in clinical practice. Toxins 8(3):65

Scott AB (1980) Botulinum toxin injection into extraocular muscles as an alternative to strabismus surgery. Ophthalmology 87(10):1044–1049

Scott AB (1994) Change of eye muscle sarcomeres according to eye position. J Pediatr Ophthalmol Strabismus 31(2):85–88

Scott AB, Rosenbaum A, Collins CC (1973) Pharmacologic weakening of extraocular muscles. Investig Ophthalmol 12(12):924–927

Shumway CL, Wade M (2019) Anatomy, head and neck, eye extraocular muscles. StatPearls, Treasure Island

Staube A, Büttner U (2007) Neuro-ophthalmology: neuronal control of eye movements, vol 40. Karger, Basel

Steel DH, Hoh HB, Harrad RA, Collins CR (1997) Botulinum toxin for the temporary treatment of involutional lower lid entropion: a clinical and morphological study. Eye 11(Pt 4):472–475

Tejedor J, Rodriguez JM (1998) Retreatment of children after surgery for acquired esotropia: reoperation versus botulinum injection. Br J Ophthalmol 82(2):110–114

Tejedor J, Rodriguez JM (1999) Early retreatment of infantile esotropia: comparison of reoperation and botulinum toxin. Br J Ophthalmol 83(7):783–787

Tejedor J, Rodriguez JM (2007) Management of nonresolving consecutive exotropia following botulinum toxin treatment of childhood esotropia. Arch Ophthalmol 125(9):1210–1213

Tomsak RL, Remler BF, Averbuch-Heller L, Chandran M, Leigh RJ (1995) Unsatisfactory treatment of acquired nystagmus with retrobulbar injection of botulinum toxin. Am J Ophthalmol 119(4):489–496

Wan MJ, Mantagos IS, Shah AS, Kazlas M, Hunter DG (2017) Comparison of botulinum toxin with surgery for the treatment of acute-onset comitant esotropia in children. Am J Ophthalmol 176:33–39

Wan MJ, Gilbert A, Kazlas M et al (2018) The effect of botulinum toxin augmentation on strabismus surgery for large-angle infantile esotropia. Am J Ophthalmol 189:160–165

Hyperhidrosis and Aesthetics

Jordan V. Wang, Nazanin Saedi, and Christopher B. Zachary

Contents

Abstract

When one considers the avalanche of new indications and uses for botulinum toxins, it is truly surprising that this has all happened in such a short time. And the safety and dependability of these products are profound, when used appropriately. There is still much to be discovered about the potential of this agent when you contemplate the profound non-cosmetic benefits reported by clinicians and scientists from around the world. The mechanism of action has been studied in depth, and yet the benefits appreciated by people with chronic migraine or major depressive disorder, for instance, are unlikely to be explained by our current

J. V. Wang · N. Saedi
Department of Dermatology and Cutaneous Biology, Thomas Jefferson University Hospital, Philadelphia, PA, USA
e-mail: Jordan.Wang@jefferson.edu; Nazanin.Saedi@jefferson.edu

C. B. Zachary (✉)
University of California-Irvine, Irvine, CA, USA
e-mail: czachary@uci.edu

S. M. Whitcup, M. Hallett (eds.), *Botulinum Toxin Therapy*,
Handbook of Experimental Pharmacology 263, https://doi.org/10.1007/164_2019_347

mechanistic understanding. Given that these toxins control acetylcholine at the motor end plates, and given that acetylcholine is central to practically every cell in the body, it will not be surprising to find that botulinum toxin researchers will be enjoying many decades of fruitful studies. The advent of the non-surgical aesthetic physician has helped push the clinical utilization of botulinum toxins well beyond its original adoption by oculoplastic surgeons in their patients with blepharospasm. We can expect that the next edition of this book to have a dozen or more new indications which will surprise us all.

Keywords
Brow ptosis · Crows feet · Facial expression · Glabellar lines · Hyperhidrosis · Platysmal bands · Prejuvenation · Rejuvenation · Sweating

1 Introduction

Botulinum toxin (BoNT) treatment is one of the most commonly performed noninvasive aesthetic procedures today. It has been shown to be safe, effective, and predictable. The origins date back to the 1980s, when patients were receiving BoNT-A for the treatment of strabismus, hemifacial spasm, and benign essential blepharospasm. However, it was Dr. Jean Carruthers, an ophthalmologist and oculoplastic surgeon, who discovered that one of her blepharospasm patients noted a marked reduction in the appearance of glabellar furrows. In partnership with her husband, Dr. Alastair Carruthers, a dermatologist and aesthetic medicine specialist, they began assessing the potential benefits on their staff and patients. In 1992, their landmark report on the efficacy of BoNT-A for the treatment of glabellar furrows was published (Carruthers and Carruthers 1992). Since then, the aesthetic benefits of BoNT have been the subject of intense investigation resulting in worldwide usage of this agent and in many new indications.

2 Indications

BoNT was granted its first approval by the US Food and Drug Administration (FDA) for cosmetic use in 2002 for the temporary improvement in the appearance of moderate to severe glabellar lines associated with corrugator and procerus muscle activity. Its effects on wrinkles are mediated by directly acting on motor neurons to reduce muscle activity. In 2004, it was then approved for the treatment of primary axillary hyperhidrosis that is inadequately managed by topical agents. For this indication, it disrupts neurotransmitter release at autonomic endings and reduces the responsiveness of sweat glands to acetylcholine, which differs from known neuromuscular mechanisms (Shibasaki et al. 2009). The effective duration of impaired sweat secretion is also longer relative to impaired muscle contraction. BoNT was later approved in 2013 to temporarily improve the appearance of moderate to severe lateral canthal lines ("crow's feet") associated with orbicularis oculi activity. Further, BoNT has been used off-label for many other indications, both medical and cosmetic.

Contraindications to BoNT include prior allergic reaction, injection into areas of infection, injection into areas of inflammation, breastfeeding, pregnancy (category C), and in those patients with neuromuscular diseases, such as myasthenia gravis and Lou Gehrig's disease. Caution should be given if patients are on certain medications, such as cholinesterase inhibitors and calcium channel antagonists, which can alter its metabolism.

3 Hyperhidrosis

In response to increased body temperature or stress, sweating is a normal physiologic response. It is controlled by the sympathetic nervous system. When initiated, muscarinic receptors on eccrine glands are activated by acetylcholine from postganglionic neurons to release sweat. Sites of high eccrine gland density include the palms, soles, forehead, and axillae. In hyperhidrosis, patients experience excess sweating beyond what is considered to be physiologically normal in response to stimuli. This can vary between individuals. The most common sites of hyperhidrosis include the axillae, palms, and soles in decreasing order (Lear et al. 2007). Hyperhidrosis can be significantly debilitating to routine activities of daily living for some patients, and it can greatly affect patient quality of life (Hamm et al. 2006). This is particularly the case when other treatments, such as topical aluminum chloride, have failed, since BoNT can offer a very effective treatment strategy, if only temporary.

When patients present with a primary focal hyperhidrosis, a thorough history should be elicited by the practitioner. To make the diagnosis, the patient must have excessive focal sweating for at least 6 months without apparent secondary cause for this. Secondary causes can include medications or systemic health problems, and they are usually associated with generalized hyperhidrosis. Relevant workup is indicated if there is any concern about potential causative conditions, such as endocrine or metabolic conditions, neurologic disorders, and neoplastic disease. There should also be at least two of the following characteristics: age of onset less than 25 years old, positive family history, cessation during sleep, frequency at least once a week, bilateral and relatively symmetric, and all sufficient to impair daily activities (Hornberger et al. 2004). It is important for the clinician to obtain information pertaining to the extent that activities are affected, since this condition can be quite debilitating for patients. Oftentimes, patients will have already failed topical medications, such as aluminum chloride, and may want to avoid systemic medications, such as glycopyrrolate.

Focal sites of hyperhidrosis are typically injected superficially with BoNT. Each injection is placed about 1–2 cm apart to allow for diffusion into surrounding tissue. Deeper injections should be avoided in order to prevent denervation to deeper structures, such as local nerves and muscles, which will cause temporary weakness. Pain is minimal when injecting the axillae, but quite an ordeal on the hands and feet.

However, the procedure is usually well tolerated. Different techniques can be utilized to minimize pain, such as icing the area to be treated or the application of a pre-procedural topical anesthetic. A variety of dilutions have been used, but dilute concentrations to allow for diffusion are generally preferred. In one study with 320 patients, 94% of those who received 50 units of BoNT-A into each axilla experienced a 50% or greater reduction in sweat at 4 weeks compared to 36% in the placebo group (Naumann and Lowe 2001). By 16 weeks, these rates became 82% for the treatment arm and 21% for the placebo arm. For palmar hyperhidrosis, doses of either 50 or 100 units into each hand decreased sweating for at least 2 months and for 6 months in most patients (Saadia et al. 2001). Another study demonstrated 80–90% improvement with effects lasting for up to 12 months (Grunfeld et al. 2009). Temporary hand weakness lasting several days or weeks can be experienced. Hyperhidrosis of the soles is more difficult to manage even with higher doses of BoNT. Although some studies have shown similar results between varying dosages of BoNT, 50 units of BoNT-A per single site is considered to be the standard starting dose. If necessary, this dose can be increased to 100 units for each side. While responses have been demonstrated to be durable for several months, this can vary between patients.

4 Glabella

One of the most popular uses of BoNT has been for the treatment of glabellar lines. It has long been used clinically as a safe and effective option even years before it was officially approved by the FDA for cosmetic use. Multiple studies have demonstrated the efficacy and predictable outcomes for patients treated with BoNT for glabellar lines. In 2002, Carruthers et al. demonstrated a significant reduction in glabellar line severity at maximum frown and rest in a multicenter, double-blind, randomized, placebo-controlled study, and effects were maintained by many patients through 120 days (Carruthers et al. 2002). More recently, Carruthers et al. also showed that repeated and regular treatments over time were associated with progressive improvement in glabellar lines at rest (Carruthers et al. 2016).

The glabellar complex consists of two corrugator supercilii muscles laterally and the vertical procerus muscle medially, which serve to pull the brows inferiorly and medially. The corrugator supercilii muscles lie somewhat parallel to the medial eyebrows and insert deeply into the bone medially and laterally extend to about the mid-pupillary line where they insert into the deep dermis. However, no two patients are absolutely identical, and it is best to visualize these muscles during maximum contraction with frowning to optimize the injection points. For treatment of the glabellar complex, a standard approach involves five injection points consisting of one site centrally for the procerus, two sites for the medial portion of the corrugators, and two points for the lateral portion of the corrugators. The medial head is injected more deeply, and the lateral component should be injected intradermally. Traditionally, it has been taught that injections for the corrugators should be performed at least 1 cm above the orbital rim in order to prevent unwanted downward diffusion. However, the intradermal technique for the lateral corrugator component seems to avoid the eyelid ptosis problem. Dosing depends on the strength of the glabellar complex, size of the muscles, and desired outcomes. For most patients, 18–24 units of BoNT-A is sufficient. An additional 10–30 units may often be necessary for patients with stronger glabellar complexes, especially in males. Carruthers and Carruthers demonstrated that males with glabellar rhytides may benefit from starting doses of at least 40 units of BoNT-A (Carruthers and Carruthers 2005).

For patients who desire lateral brow lifting, 4–6 units of BoNT-A can typically be injected into the lateral tail of the brows. Depending on the dosing and technique, brow elevations of 1–4 mm have been achieved (Huang et al. 2000; Ahn et al. 2000). However, even without this lateral brow injection, lifting of the brows can be seen with treatment of the glabellar complex due to outward diffusion of BoNT (Carruthers and Carruthers 2007). It is important to note that this technique may be more appropriate for female patients as opposed to males, who often desire straighter and less arched brows.

5 Frontalis

Treatment of the frontalis, which induces the horizontal forehead lines, represents another popular use of BoNT in facial aesthetics. The frontalis is generally a thin and broad muscle that covers 60–80% of the forehead. It serves to raise the eyebrows and upper eyelids. Horizontal forehead lines are a normal part of aging, which is due to repeated contraction of the frontalis over time. BoNT therapy has long been used safely to help reduce the appearance of these lines and, if used early enough in life, can prevent these from forming in the first place.

Over time, several different injection techniques have been described, consisting of varying patterns and dosing strategies. It is important to note that any one pattern may not be perfect for all patients and that each patient requires personalization in the treatment approach. Remember that some patients have two somewhat obliquely placed frontalis muscles with a central divarication over the mid forehead, while others might have a continuous sheet of muscle across much of the forehead. Thus, each patient needs to be assessed and treated accordingly. The injection points should generally be at least 1–2 cm above the orbital rim in order to decrease the risk of subsequent brow ptosis (Carruthers et al. 2004). Significant relaxation of part of the forehead might appear to induce some "new" horizontal lines by the hairline. The typical number of injection points can vary between 6 and 12 sites, and the total dose might range from 10 to 20 units of BoNT-A. It may be best to start with lower doses and slowly increase as needed. Men may also require higher starting doses due to increased musculature strength. The total dose is divided by the number of injection points, while each point is typically injected with 1–2 units each. A multicenter, randomized trial demonstrated that BoNT-A treatment of forehead lines was tolerable, effective, and sustained (Solish et al. 2016). Treatment of forehead lines has also been associated with increased patient satisfaction and significant improvements in appearance-related emotional and psychological issues (Ogilvie et al. 2019).

6 Crow's Feet

Patients often desire periocular rejuvenation for prominent lateral canthal lines, or so-called crow's feet. These represent a prominent and easily identifiable sign of aging and result from degenerative changes in the bone and soft tissues, increased skin laxity, photodamage, and smoking. Crow's feet originate from the lateral canthus and often fan outward. Four common lateral canthal rhytid patterns have been described and include full fan pattern in 47% of patients, lower lid and upper cheek area alone in 25% of patients, upper eyelid skin down to the lateral canthus in 18% of patients, and only the skin immediately surrounding the lateral canthus in 10% of patients (Kane 2003). Lateral canthal lines are clinically at their maximum when patients are instructed to squint or smile. Hyperkinetic movements by the lateral orbital portion of the orbicularis oculi muscle encircling the orbital rim play a large role. Therefore, it's to be expected that BoNT therapy is useful in helping to ameliorate the appearance of crow's feet.

Due to different patterns of crow's feet and varying levels of severity, individualized treatment approaches utilizing BoNT are always recommended. Standard injection techniques can only offer a helpful starting point for practitioners. Any baseline asymmetry should be noted, photographed, and discussed with the patient prior to procedure. If this is not pointed out, the patient is going to think that the asymmetry was caused by the procedure when they scrutinize their face following the procedure. Injections are placed superficially into the subcutaneous space immediately under the dermis in order to minimize the risk of diffusion to deeper muscles and to avoid bruising. Various injection patterns have been described. Standard technique involves injections that are placed about 1–1.5 cm from the lateral canthus using three or four injection points spaced about 1 cm apart in an arcuate pattern. Typical dosing regimens range from using 4–8 units of BoNT-A per side. A multicenter, double-blind, placebo-controlled study by Carruthers et al. demonstrated effective and well-tolerated treatment with 12 units of BoNT-A to each side for moderate to severe crow's feet lines (Carruthers et al. 2014). Response duration for the treatment of crow's feet was proven to be greater than 4 months using data from trials that included 833 patients receiving BoNT-A (Baumann et al. 2016). Compared to placebo, BoNT-A treatment of crow's feet lines was associated with patients experiencing significant improvements in perceived appearance, attractiveness, age, tiredness, and satisfaction (Dayan et al. 2015).

7 Orbicularis Oris and Depressor Anguli Oris

While BoNT was originally promoted for the upper face, it has since grown in popularity for treatment of the lower face, especially around the mouth. Perioral lip lines, downward turning of the angles of the mouth, and lengthening, thinning, and inversion of the upper lip are all common concerns. Perioral lines can be exaggerated not only from photodamage but also from repetitive pursing of the lips, which can be associated with habitual smoking and drinking through straws. The perioral region comprises an interdigitating complex of muscles capable of creating profoundly subtle yet recognizable facial expressions, such as fear, anger, disgust, sadness, and literally hundreds of others. Two of the most significant muscles include orbicularis oris and depressor anguli oris (DAO). The orbicularis oris surrounds the opening of the mouth and serves to aid with speech, mastication, expression, sucking, and puckering. The DAO is a triangular-shaped muscle that extends from the inferior mandible to the angle of the mouth, and its main function is to depress the angle of the mouth.

For patients with clinically significant perioral lines and exaggerated downward turn of the oral commissures, BoNT therapy can prove to be an effective treatment option. While full relaxation of some muscles of the upper face may often be desired and implemented, such a request for the lower face would be incompatible with normal social activities where use of these muscles is important for vital activities, such as mastication, drinking, and phonation. Practitioners should be conservative with placement and dosage. For the orbicularis oris, typical treatment consists of four symmetrical superficial injections of 1–2 units of BoNT-A to the upper lip and two injections to the lower lip along the cutaneous aspect of the vermillion border. The midline and corners of the lips should be avoided to prevent flattening of Cupid's bow and weakness in muscles used for elevating the lips (Kaplan et al. 2007). Treatment with BoNT not only can smooth vertical lines but can also enhance lip fullness and lip eversion (Semchyshyn and Sengelmann 2003). For treatment of the DAO, injections should be in the mid to lower third of the muscle. The oral commissure can be traced inferiorly to just above the mandible, where the injection should be located about 1 cm posteriorly from this point in order to avoid treatment of the depressor labii inferioris (DLI), which sits more medially. Effects on the DLI can cause lower lip dysfunction and flattening of its contour. A dose of 4–6 units of BoNT-A per side is typical. Best results are observed in young patients who have greater muscle strength without significant laxity or adjacent lipodystrophy (Goldman and Wollina 2010). Whenever treating the perioral region, be aware that this may impact the patient's livelihood for those who depend on precise movement of their lips while singing, whistling, or playing a wind instrument.

8 Platysma

During the aging process, platysmal hypertrophy and separation can be pronounced, especially due to the dermal and subcutaneous atrophy. This can be accentuated with speech and movement, which becomes clinically evident as prominent platysmal bands and perpendicular horizontal lines. The presence of platysmal bands at rest is the result of increased muscular resting tone. This is a cosmetic rather than a medical problem and can be ameliorated with BoNT. Total dosage can be based upon how many bands are present in addition to their length and severity. Modest dosing levels are considered to be 30–50 units of BoNT-A. Higher doses from 60 to 250 units were previously associated with adverse effects (Chen and Cohen 2015). Modest dosing can typically offer noticeable cosmetic improvement without increased risk for complications. During the injection procedure, the patient should be asked to contract their platysmal bands by grimacing. The non-dominant hand can be used to pinch the platysmal band, which not only assures precise injection into the muscle but also helps to distance it from deeper underlying structures. Intramuscular deposition of 2–5 units of BoNT-A can be spaced about 1–2 cm apart along the length of the platysmal band. Relaxation of the vertical platysmal bands, improvement of the horizontal neck lines, and enhancement of the mandibular line can all be expected after 1–2 weeks (Trévidic et al. 2015). Duration of BoNT can vary depending on injection technique, dosage, and patient, but typically ranges between 3 and 4 months. A recent systematic review has demonstrated BoNT to be relatively safe and effective for the treatment of mild to moderate platysmal bands (Sugrue et al. 2019). However, practitioners must be cautious not to inject too deeply or to use high dosages in order to avoid diffusion to deeper neck muscles, which may affect neck movement and create problems with phonation and glutition.

9 Masseter

One out of five people have bruxism, which is a nighttime grinding of the teeth that is often associated with sleepless nights for them (and their partners). This can leave them with an aching pain in their jaws and temporomandibular joints, headaches, and significant masseter hypertrophy. BoNT therapy has been found to be of great benefit by relaxing these muscles of mastication after which the patients sleep better, wake up more refreshed, and have fewer bruxism symptoms. More recently, BoNT has been used to improve the aesthetic appearance of those with a heavy-set lower face related to masseter hypertrophy unassociated with bruxism. This has become an especially popular treatment for patients of Asian descent, where many would like a more oval-shaped face.

Prior to injection, the patient is instructed to clench the jaw so that the borders of the masseter can be best palpated. Various techniques have been described to help map the rough outline of where the masseter resides. However, more pronounced masseters can be typically palpated with relative ease. About two to three injection points are typically performed in the area of maximal bulge. Injections should be placed intramuscularly. Caution should be advised to avoid injecting too cephalically as the facial nerves might be negatively impacted. Dosing depends on the muscle bulk, but it typically ranges from 10 to 30 units of BoNT-A per side. Repeat injections are often necessary at intervals of 3–4 months in order to reach the desired outcome and facial shape. In a study of 50 patients treated with repeat injections, patients experienced durable responses that were maintained at 4-year follow-up (Shome et al. 2019). When performed appropriately, procedures have been shown to offer adequate results and safety profiles (Yeh et al. 2018). Most complications appeared within 2–4 weeks and disappeared within 12 weeks. The so-called Popeye sign, where the relaxed masseter muscle following BoNT treatment is projected outwards with contraction, might last for about 1 week until the rest of the masseter is impacted equally. Practitioners should familiarize themselves with regional anatomy for safe injections.

10 Conclusion

BoNT has surprised many by how effective this agent has been, both on- and off-label, in medical and aesthetic conditions while providing efficacious and safe outcomes. And now that millions of people have received BoNT, a second wave of interest exists in conditions, such as chronic depression, which are quite separate from the aesthetic area. There seems no end to the discovery of benefit. It is so rewarding to be able to treat people, some of whom feel past their best, with such a safe and relatively simple treatment. For those who are averse to surgery, and even in conjunction with surgery, fillers and neurotoxins used in combination seem to have no end in popularity. Safety is of prime importance with any treatment, and so it is true also in aesthetic medicine. Practitioners must know their injection anatomy and understand the pharmacology of BoNT in order to appreciate how best to help their patients and how to both prevent and treat complications. In the end, the field might well look at the youth of this country to determine when best to start treatment with these agents. Prejuvenation, which is "treatment to prevent the appearance of aging," is taking hold. The earlier that patients start, the more effective these treatments will be (Spanogle et al. 2014). The shifting of goals to prevent the appearance of fine lines and wrinkles may offer improved cosmetic outcomes in the long term than simply working to ameliorate deep wrinkles after the fact.

References

Ahn MS, Catten M, Maas CS (2000) Temporal brow lift using botulinum toxin A. Plast Reconstr Surg 105(3):1129–1135

Baumann L, Dayan S, Connolly S et al (2016) Duration of clinical efficacy of onabotulinumtoxinA in crow's feet lines: results from two multicenter, randomized, controlled trials. Dermatol Surg 42(5):598–607

Carruthers JD, Carruthers JA (1992) Treatment of glabellar frown lines with C. botulinum-A exotoxin. J Dermatol Surg Oncol 18(1):17–21

Carruthers A, Carruthers J (2005) Prospective, double-blind, randomized, parallel-group, dose-ranging study of botulinum toxin type A in men with glabellar rhytids. Dermatol Surg 31 (10):1297–1303

Carruthers A, Carruthers J (2007) Eyebrow height after botulinum toxin type A to the glabella. Dermatol Surg 33(1 Spec No.):S26–S31

Carruthers JA, Lowe NJ, Menter MA et al (2002) A multicenter, double-blind, randomized, placebo-controlled study of the efficacy and safety of botulinum toxin type A in the treatment of glabellar lines. J Am Acad Dermatol 46(6):840–849

Carruthers J, Fagien S, Matarasso SL et al (2004) Consensus recommendations on the use of botulinum toxin type A in facial aesthetics. Plast Reconstr Surg 114(6 Suppl):1S–22S

Carruthers A, Bruce S, de Coninck A et al (2014) Efficacy and safety of onabotulinumtoxinA for the treatment of crows feet lines: a multicenter, randomized, controlled trial. Dermatol Surg 40 (11):1181–1190

Carruthers A, Carruthers J, Fagien S et al (2016) Repeated onabotulinumtoxinA treatment of glabellar lines at rest over three treatment cycles. Dermatol Surg 42(9):1094–1101

Chen DL, Cohen JL (2015) Botulinum toxin-A chemical denervation for platysmal bands: maximal dosing considerations. J Drugs Dermatol 14(9):931

Dayan S, Coleman WP 3rd, Dover JS et al (2015) Effects of onabotulinumtoxinA treatment for crow's feet lines on patient-reported outcomes. Dermatol Surg 41(Suppl 1):S67–S74

Goldman A, Wollina U (2010) Elevation of the corner of the mouth using botulinum toxin type A. J Cutan Aesthet Surg 3(3):145–150

Grunfeld A, Murray CA, Solish N (2009) Botulinum toxin for hyperhidrosis: a review. Am J Clin Dermatol 10(2):87–102

Hamm H, Naumann MK, Kowalski JW et al (2006) Primary focal hyperhidrosis: disease characteristics and functional impairment. Dermatology 212(4):343–353

Hornberger J, Grimes K, Naumann M et al (2004) Recognition, diagnosis, and treatment of primary focal hyperhidrosis. J Am Acad Dermatol 51(2):274–286

Huang W, Rogachefsky AS, Foster JA (2000) Browlift with botulinum toxin. Dermatol Surg 26(1):55–60

Kane MA (2003) Classification of crow's feet patterns among Caucasian women: the key to individualizing treatment. Plast Reconstr Surg 112(5 Suppl):33S–39S

Kaplan SE, Sherris DA, Gassner HG et al (2007) The use of botulinum toxin A in perioral rejuvenation. Facial Plast Surg Clin North Am 15(4):415–421

Lear W, Kessler E, Solish N et al (2007) An epidemiological study of hyperhidrosis. Dermatol Surg 33(1 Spec No.):S69–S75

Naumann M, Lowe NJ (2001) Botulinum toxin type A in treatment of bilateral primary axillary hyperhidrosis: randomised, parallel group, double blind, placebo controlled trial. BMJ 323 (7313):596–599

Ogilvie P, Rivkin AZ, Dayan S et al (2019) OnabotulinumtoxinA for treatment of forehead and glabellar lines: subject-reported satisfaction and impact from a phase 3 double-blind study. Dermatol Surg 45(5):689–699

Saadia D, Voustianiouk A, Wang AK et al (2001) Botulinum toxin type A in primary palmar hyperhidrosis: randomized, single-blind, two-dose study. Neurology 57(11):2095–2099

Semchyshyn N, Sengelmann RD (2003) Botulinum toxin A treatment of perioral rhytides. Dermatol Surg 29(5):490–495

Shibasaki M, Davis SL, Cui J et al (2009) Botulinum toxin abolishes sweating via impaired sweat gland responsiveness to exogenous acetylcholine. Br J Dermatol 161(4):757–761

Shome D, Khare S, Kapoor R (2019) Efficacy of botulinum toxin in treating Asian Indian patients with masseter hypertrophy: a 4-year follow-up study. Plast Reconstr Surg 144(3):390e–396e

Solish N, Rivers JK, Humphrey S et al (2016) Efficacy and safety of onabotulinumtoxinA treatment of forehead lines: a multicenter, randomized, dose-ranging controlled trial. Dermatol Surg 42 (3):410–419

Spanogle J, Glaser DA, Zachary CB (2014) Cosmetic uses of botulinum neurotoxins. In: Truong D, Dressler D, Hallett M, Zachary CB (eds) Manual of botulinum toxin therapy. Cambridge University Press, Cambridge, pp 181–193

Sugrue CM, Kelly JL, McInerney N (2019) Botulinum toxin treatment for mild to moderate platysma bands: a systematic review of efficacy, safety, and injection technique. Aesthet Surg J 39(2):201–206

Trévidic P, Sykes J, Criollo-Lamilla G (2015) Anatomy of the lower face and botulinum toxin injections. Plast Reconstr Surg 136(5 Suppl):84S–91S

Yeh YT, Peng JH, Peng HP (2018) Literature review of the adverse events associated with botulinum toxin injection for the masseter muscle hypertrophy. J Cosmet Dermatol 17 (5):675–687

Use of Botulinum Toxin in the Genitourinary System

Michael B. Chancellor and Christopher P. Smith

Contents

Abstract

Botulinum toxin injection has been widely accepted by the urology and urogynecology medical communities as a safe and effective treatment for refractory urinary incontinence. There are two approved genitourinary indications for botulinum toxin. OnabotulinumtoxinA (onaBoNTA) 200 U for the treatment of urinary incontinence due to detrusor overactivity associated with a neurologic condition (e.g., spinal cord injury, multiple sclerosis) in adults who have an inadequate response to or are intolerant of an anticholinergic medication. In addition, onaBoNTA 100 U is used for the treatment of overactive bladder with symptoms of urinary incontinence, urgency, and frequency, in adult patients who

M. B. Chancellor (✉)
Aikens Center for Neurourology Research, Department of Urology, Beaumont Health System and Oakland University William Beaumont School of Medicine, Rochester, MI, USA
e-mail: Michael.chancellor@beaumont.org

C. P. Smith
Scott Department of Urology, Baylor College of Medicine, Houston, TX, USA

S. M. Whitcup, M. Hallett (eds.), *Botulinum Toxin Therapy*,
Handbook of Experimental Pharmacology 263, https://doi.org/10.1007/164_2019_308

have an inadequate response to or are intolerant of an anticholinergic medication. We will discuss the application of botulinum toxin for genitourinary indications with a focus on bladder injection and on potential use of BoNT use in the prostate and pelvic floor.

Keywords
Benign prostatic hyperplasia · Interstitial cystitis · Neurogenic detrusor overactivity · Overactive bladder

1 Introduction

Botulinum toxin's mechanism of action and rationale for clinical utility are described by the world experts in this book, so we will focus on the rationale and use in urology (Billante et al. 2002; Chancellor et al. 2008; Apostolidis et al. 2006). The use of botulinum toxin (BoNT) in the genitourinary system was initially described by Dykstra et al. (1988) as a treatment for detrusor sphincter dyssynergia (DSD) in spinal cord-injured (SCI) patients. Bladder injection of BoNT followed as Schurch et al. (2000) first reported bladder intradetrusor injections of BoNT in 21 SCI patients with neurogenic detrusor overactivity (NDO) refractory to high-dose anticholinergic medications. The exciting and promising initial off-label use of BoNT led to registry trial and two Phase 3 multicenter, double-blind, placebo-controlled trials that led to the 2011 regulatory approval of onabotulinumtoxinA (onaBoNTA), at 100 and 200 U, for the treatment of urge incontinence due to NDO (Cruz et al. 2011; Ginsberg et al. 2012). Subsequently Phase 3 multicenter trials led to the 2013 regulatory approval of onaBoNTA for the treatment of idiopathic overactive bladder (OAB) without neurological diseases and refractory to anticholinergics (Nitti et al. 2013; Chapple et al. 2013). Compared to placebo, a 100 U dose of onaBoNTA injected into the detrusor decreased mean urge incontinence in both studies.

Other applications for BoNT include benign prostatic hyperplasia (BPH) and interstitial cystitis/bladder pain syndrome (IC/BPS). McVary et al. (2014) reported on a Phase 2 randomized clinical trial comparing onaBoNTA 200 U to placebo for the treatment of BPH, but no differences were seen in the primary and majority of secondary outcome parameters. For the treatment of IC/BPS, Kuo and Chancellor (2009) reported a signal of efficacy in off-label use of BoNT in bladder pain score in IC/BPS patients. BoNT is currently listed as a fourth-line treatment in the American Urological Association guideline for the treatment of IC/BPS (Hanno et al. 2015).

2 Personal Story with Botulinum Toxin

I (MC) started using botulinum toxin in 1998. I was frustrated with three women suffering from multiple sclerosis with high post-residual urine volume, recurrent urinary tract infections, and dyssynergia of the urethral sphincter. My nurse and I tried but were just unable to teach them to perform self-catheterization because of poor hand dexterity. After consultation with their neurologists, the patients asked to try BoNT instead of indwelling catheters or reconstructive surgery. The neurology staff taught me how to prepare the toxin for injection. The procedure of sphincter BoNT injections was easy, and all three patients voided with significantly lower residual urine volume and avoided needing self-catheterization.

I discussed these remarkable results with my mentor, Professor William (Chet) de Groat, at the Department of Pharmacology of the University of Pittsburgh, and he encouraged me to translate the bedside observations to benchtop research. Dr. Christopher Smith (CS), who was training with me during his NIH Physician Scientist fellowship, was interested, and we started experiments on BoNT in the urinary tract. We found that BoNT not only effectively relaxed urethral strips but also detrusor muscle strips. We further found that toxin blocked acetylcholine and norepinephrine release from bladder and urethral strips (Chancellor and Smith 2011).

We found significant decreases in the release of labelled acetylcholine in normal rat bladders injected with onaBoNTA, suggesting that BoNT could reduce cholinergic nerve-induced bladder activity (Smith et al. 2003). In addition, the release of other transmitters, including impair ATP release in addition to acetylcholine release from isolated bladder tissue (Chuang et al. 2004). BoNT might also inhibit afferent neurotransmission and have analgesic properties. In a model of somatic pain associated with an acetic acid-induced bladder pain model, rats pre-treated with onaBoNTA had significantly reduced pain behavior and glutamate release approximately 1 week after the injection (Chuang et al. 2004). Bladder urothelium is important in the sensory transduction mechanisms modulating micturition, particularly in conditions of increased sensory nerve transmission after chronic inflammation and spinal cord injury. We have also shown onaBoNTA to inhibit ATP release from the urothelium in spinal cord-injured rat bladders (Smith et al. 2008).

3 Neurogenic Detrusor Overactivity Indication

Neurogenic detrusor overactivity, most common in MS and SCI, but also seen in other neurological diseases including stroke and Parkinson's disease, is characterized by the presence of involuntary detrusor contractions (IDCs) during filling cystometry (Chancellor and Smith 2011). Because NDO causes reduced bladder capacity and incontinence, quality of life is often greatly impaired. In addition, long-term anticholinergic treatment for NDO has only modest efficacy and often causes intolerable side effects such as dry mouth and constipation (Chancellor et al. 2006).

Cruz et al. (2011) and Ginsberg et al. (2013) reported the Phase 3 studies that resulted in NDO regulatory approval in the USA and EU in 2011. A total of 691 MS or SCI patients who had inadequate response to or were intolerant to ≥ 1 anticholinergic medication were enrolled. These patients were randomized to receive onaBoNTA 200 U ($n = 227$), onaBoNTA 300 U ($n = 223$), or placebo ($n = 241$). Significant improvement in the primary outcome, change from baseline in weekly frequency of UI episodes, was achieved with onaBoNTA 200 U compared with placebo. 300 U onaBoNTA was about equally effective but had more side effects than 200 U onaBoNTA. The improvement was seen after 2 weeks, and the average duration of response was 9–10 months. OnaBoNTA treatment was associated with significant improvements in maximal cystometric capacity of approximately 150 mL. Among patients who were not catheterizing at baseline prior to treatment, catheterization for urinary retention was initiated in 31% patients following treatment with 200 U onaBoNTA compared with 7% of those on placebo. The most frequently reported adverse reactions included urinary tract infection (24%), urinary retention (17%), hematuria (4%), fatigue (4%), and insomnia (2%).

Denys et al. (2016) reported the efficacy and safety of two administration modes of bladder injection of Dysport (abobotulinumtoxinA) 750 U in patients suffering from refractory NDO in a randomized placebo-controlled Phase 2 study. 47 MS or SCI patients were treated with 15 or 30 bladder injections of aboBoNTA 750 U or placebo. Primary endpoint was the change from baseline in mean number of daily incontinence episode frequency after about 3 months. The authors concluded that both 15 and 30 injection administration modes of aboBoNTA decreased daily incontinence frequency and resulted in significant improvements in urodynamic parameters in NDO patients. Reduction to 15 injection sites did not appear to be associated with any impact on efficacy.

Patients from both MS and SCI groups found significant improvement in quality of life index after either 200 U or 300 U of onaBoNTA. In addition, the change in quality of life index scores from baseline was analyzed in patients who didn't perform clean intermittent catheterization at baseline to determine if self-catheterization initiation after onaBoNTA affected life quality. The results showed that the improvement in quality of life index was similar regardless of whether self-catheterization was begun or not after onaBoNTA.

Urinary tract infection (UTI) was the most common adverse event (AE) observed after onaBoNTA injection, both in SCI and MS patients. Clean intermittent catheterization (CIC) was performed at baseline in >80% of the SCI patients in all three groups. The incidence of UTIs within the MS population was 54% in the onaBoNTA 200 U group compared to 29% in the placebo group ($p < 0.001$), and this increase correlated directly with the need to start self-catheterization because of de novo urinary retention. At baseline 29% of the MS patients were on CIC. For those MS patients who did not perform catheterization at baseline, the rate of de novo urinary retention that required CIC was 31% and 47% in the onaBoNTA 200 U and 300 U groups, respectively. There were no reports of increased risk of MS exacerbation or respiratory compromise following onaBoNTA injection. No patient developed serum neutralizing antibodies to onaBoNTA (Ginsberg et al. 2012; Cruz et al. 2011).

4 Refractory Idiopathic Detrusor Overactivity Indication

OAB is defined as urinary urgency, with or without urge urinary incontinence (UUI), usually accompanied with urinary frequency and nocturia (Abrams et al. 2002). The prevalence of OAB in the general population is 12–17%, and about half of OAB patients have incontinence (Stewart et al. 2003). The current guidelines for the management of OAB list first- and second-line therapies as behavioral therapies and pharmacotherapy, respectively (Gormley et al. 2015). A meta-analysis of several RCTs of different anticholinergic drugs used for the treatment of OAB demonstrated improvements in both symptoms and QOL (Chapple et al. 2008). Unfortunately, most individuals discontinue anticholinergic therapy because of either inadequate long-term efficacy or intolerable side effects.

In two clinical studies, the safety and efficacy of onaBoNTA was evaluated in patients with overactive bladder whose symptoms were not adequately managed with anticholinergic medications (Nitti et al. 2013). OnaBoNTA reduced daily frequency of urinary leakage episodes from baseline by approximately 50% or more by week 12 compared to placebo.

The efficacy of onaBoNTA at reducing urinary leakage and other symptoms of overactive bladder was up to 6 months duration. Improvements in other symptoms of overactive bladder, daily frequency of urination, and the amount of urine voided also occurred with onaBoNTA treatment compared to placebo at week 12.

The most common side effects reported with onaBoNTA treatment in the clinical studies included urinary tract infection (18%, vs. 6% with placebo); dysuria (9%, vs. 7% with placebo), which means painful or difficult urination; and urinary retention (6%, vs. 0% with placebo), which is a temporary inability to fully empty the bladder requiring clean intermittent catheterization. Patients with diabetes mellitus treated with onaBoNTA were more likely to develop urinary retention.

The US National Institutes of Health sponsored a comparative study between onaBoNTA and neuromodulation (Amundsen et al. 2016). For this study, conducted at nine centers, only women with refractory urgency urinary incontinence were randomized to an injection of onaBoNTA ($n = 192$) or sacral neuromodulation ($n = 189$). Of the 364 women, mean age 63 years, the onaBoNTA group had a statistically significant greater reduction in 6-month average number of episodes of urgency incontinence per day than did the sacral neuromodulation group (−3.9 vs. −3.3 episodes per day). There were no cases of urinary retention with sacral neuromodulation while onaBoNTA increased the risk of UTI, retention, and need for self-catheterization. Although subjects treated with onaBoNTA noted greater improvement for symptom bother and treatment satisfaction than neuromodulation, there was no significant difference for quality of life or for measures of treatment preference, convenience, or adverse effects.

5 Bladder Injection Technique

Preparation Before injection, the urine should be analyzed before toxin is reconstituted to rule out infection. Anticoagulation medicine should be stopped at least 5 days before injection. Empty the bladder and apply local anesthesia (i.e., 1–2% lidocaine solution) for at least 15 min. Partially fill the bladder to approximately 150–200 mL for visualization but avoid overdistention. Take latex allergy precaution in at-risk population of neurogenic detrusor overactivity. In spinal cord-injured or even possibly multiple sclerosis patients with injury and/or lesions above the T6 spinal cord level, precautions to deal with and procedures to minimize the risk of autonomic dysreflexia should be in place. Patient's blood pressure should be monitored, and bladder overfilling should be avoided. The use of urethral and bladder local anesthesia is helpful, and general anesthesia may need to be considered in selective patients.

Injection Paradigm For the neurogenic detrusor overactivity indication, the recommended dose is 200 U onaBoNTA; the recommended reconstitution volume is 30 mL of sterile injectable saline with 1 mL volume per injection at a depth of injection of approximately 2 mm intradetrusor. It is recommended to inject at 30 sites spaced 1 cm apart starting 1 cm above the trigone. For the idiopathic OAB indication, the recommended dose is 100 U onaBoNTA, reconstituted in 10 mL of saline and injected into 20 sites of 0.5 mL per site.

Cystoscope Both rigid and flexible cystoscopic techniques work well for BoNT injection. Surgeon preference and institutional practice usually decide what technique is used.

Rigid Cystoscope The rigid scope is more painful for the patients but allows for easier orientation within the bladder compared to a flexible cystoscope. The larger working port of a rigid cystoscope facilitates quicker injection, and the 25-gauge needle minimizes bleeding and potential backflow from the injection sites. The bladder volume is typically kept at 150–200 mL, and blood vessels are avoided during injection.

Flexible Cystoscope We use flexible cystoscopy in the office for the majority of our cases in men and women. The flexible scope accommodates a 27-gauge 4-mm-long flexible injection needle. Office procedures with only local anesthesia are adequate for most of our patients. Patients appreciate the convenience of an in-office procedure.

Depth, Location, and Amount of Injection Abdel-Meguid (2010) reported a prospective, randomized, controlled trial of trigonal injection in 36 patients. The patients were evenly randomized to either only detrusor injection excluding the trigone (300 U onaBoNTA) or intradetrusor (200 U onaBoNTA) plus intratrigonal injection (100 U onaBoNTA). Abdel-Meguid reported that all parameters improved significantly in each group with greater improvement in decreased incontinence episodes and complete dry rate in the trigonal injected group. There were no de novo or worsening of vesicoureteral reflux in either group. Although vesicoureteral reflux might be a potential complication after BoNT trigone injection, there is no evidence of it so far. It should be emphasized that no standardized injection technique exists for botulinum toxin injection in lower urinary tract tissues. Different bladder injection paradigms have been described (i.e., trigone vs. trigone-sparing) although none has been proven to be superior.

Follow-Up Patients are advised that they may notice some pain and blood-tinged urine, as well as possible difficulty urinating following treatment. These symptoms should resolve within 24–48 h, and they should call and contact the office if they have any questions or concerns. We discuss the appropriate antibiotic coverage and risk of infection in these patients who often have more bladder infections. It may take several days to notice a gradual improvement in overactive bladder symptoms. Similarly, it generally takes several days for a patient to notice impaired voiding, and I would instruct that patient to start self-catheterization if clinically necessary. Office follow-up in about 2 weeks with urine analysis and post-void residual urine check is recommended.

Subsequent Dose Selection For the majority of patients who notice a benefit with bladder BoNT therapy, the same dose can be used with repeat injections. Most neurologically impaired patients have had consistent improvement using the same dose, and we have observed this for almost 20 years. If the patient finds benefit but incontinence did not completely resolve with 200 U onaBoNTA, we may consider going up to 250 or 300 U onaBoNTA with the next injection. Alternatively, in MS patients who do not perform self-catheterization but have noticed retention or incomplete bladder emptying, we generally start at 100 U onaBoNTA. If a patient who received 200 U onaBoNTA noticed more difficulty with bladder emptying, we will decrease the dose from 200 to 100 or 150 U onaBoNTA on the next injection. We have even used 50 or 75 U on the first injection in a few special cases. In summary, dose titration is possible and helpful, but the percent of patients who will need dose adjustment up or down is small.

6 Other Genitourinary Experience with BoNT: Urinary Sphincter and Pelvic Floor, Prostate, and Bladder Pain

Urinary Sphincter There is one Class I and two Class II studies of BoNT in detrusor sphincter dyssynergia (DSD). In the Class I study, the effects of BoNT vs. placebo were studied on DSD in 86 patients with multiple sclerosis (MS) (Gallien et al. 2005). The study employed a single transperineal injection of onaBoNTA, 100 U in 4 mL normal saline, or placebo into the striated sphincter with EMG guidance. A single injection of BoNT did not decrease residual urine volume in this group of MS patients. These findings differ from those in patients with spinal cord injury and may be due to lower detrusor pressures observed in patients with MS. The American Academy of Neurology recommends BoNT to be considered for DSD but recognizes the limited head-to-head comparisons of treatment options in DSD.

Technique in Men (Fig. 1) In men we usually use 200 U of onaBoNTA diluted in 4 mL of preservative free saline under local or general anesthesia using a rigid cystoscope loaded with a 25-gauge endoscopic injection needle. OnaBoNTA is injected in equal aliquots at 12, 3, 6, and 9 o'clock positions. The injection is directed deeper than urethral bulking agent injections to target the nerve terminals

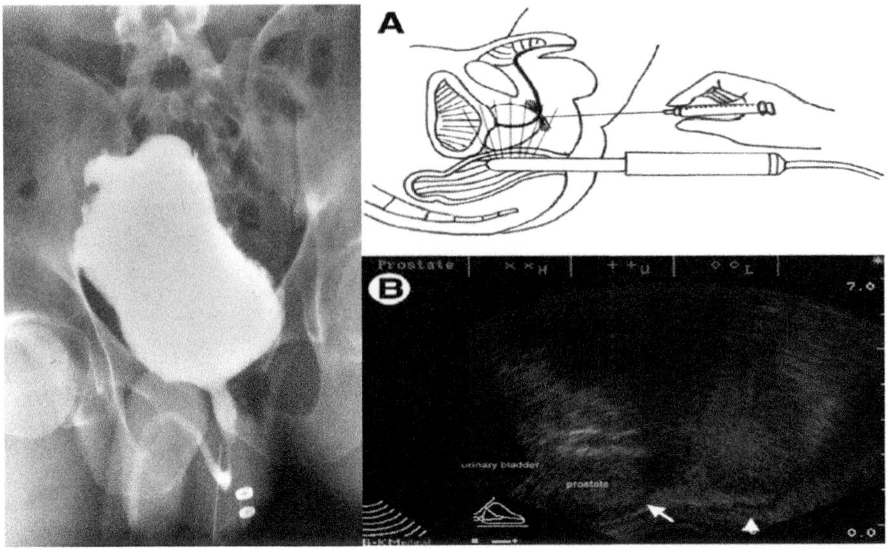

Fig. 1 Left: Voiding cystourethrogram image demonstrating detrusor sphincter dyssynergia in a male patient. Note the open bladder neck and prostatic urethral but abrupt cutoff of the contrast at the level of the external urethral sphincter. Diagram (**a**) and (**b**) image depicting transrectal ultrasound-guided perineal injection for the external urethral sphincter in the male (Chancellor and Smith 2011)

innervating the external (skeletal muscle) sphincter. Other methods described in the literature include perineal and/or transrectal ultrasound-guided external urethral sphincter injections.

Technique in Women (Fig. 2) In women we usually use 100 U of onaBoNTA diluted in 2 mL of saline. We use a 25-gauge spinal needle that is inserted for 1.5 cm at 3 o'clock and 9 o'clock positions 1 cm lateral to the urethral meatus in the periurethral folds. The female urethra is short, and it is easier to inject with a spinal needle rather than using a needle through the sheath of a cystoscope. 1 mL of onaBoNTA (i.e., 50 U) is injected at each site. We typically use a rigid cystoscope during the injection to help with positioning.

Pelvic Floor Pain For the treatment of levator spasm, direct transvaginal injection into the levator muscles can be done using 100–200 U onaBoNTA. For levator injections, a disposable pudendal nerve block kit allows for easy transvaginal finger-guided injections of a standardized depth into the pelvic floor muscles. The trocar is guided to the appropriate landmark with the fingertip, the needle is engaged, and the vaginal wall is pierced with the tip of the needle targeting the underlying levator muscles. 0.75–1.5 mL is injected into each of four sites, typically posterolateral, following aspiration to ensure avoidance of intravascular injection. Proximal injections target the pubococcygeus muscles, at 5 o'clock and 7 o'clock positions,

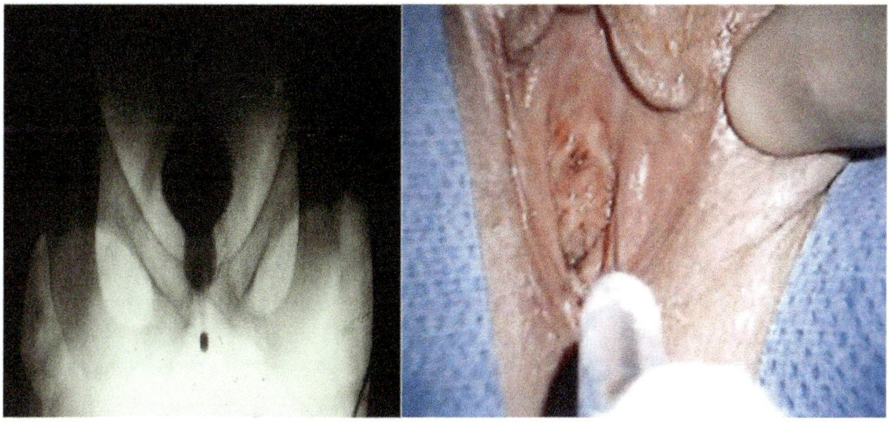

Fig. 2 Left: Voiding cystourethrogram image of the detrusor sphincter dyssynergia in a female patient. Note the open bladder neck and proximal urethra but closure at the level of the external sphincter. Right: Needle periurethral injection of the external sphincter (Chancellor and Smith 2011)

just proximal medial and distal to the ischial spine. Distal injections target the puborectalis muscle, at 5 o'clock and 7 o'clock positions just inside the hymenal ring. For women with vulvodynia, an injection (i.e., 25–50 U onaBoNTA) can be performed under direct vision targeting superficial perineal muscles using a small-gauge spinal needle. Injection sites are placed posterolaterally at 5 o'clock and 7 o'clock positions within the posterior fourchette and vulva (Chancellor and Smith 2011).

7 Pelvic Floor Injection Results

Kuo (2007) evaluated the effects of onaBoNTA urethral injection in 27 patients with idiopathic low detrusor contractility. Detrusor contractility recovered in 48% of those treated. Patients with normal bladder sensation combined with poor relaxation or hyperactive urethral sphincter activity were most likely to respond to urethral injections with onaBoNTA. Complications of BoNT injection into the external sphincter are rare except for transient stress urinary incontinence. In 38% of patients, the therapeutic effect of restoring detrusor contractility lasted over 1 year. Ghazizadeh and Nikzad (2004) injected 150–400 U of abobotulinumtoxinA into the levator ani of 24 women with refractory vaginismus. Symptoms significantly improved such that 75% of patients could have satisfactory intercourse. In contrast, a double-blind randomized clinical trial of onaBoNTA vs. saline in 60 patients with 2 years or more of chronic pelvic pain that received either onaBoNTA 80 U (20 U/mL) or normal saline injections into the puborectalis and pubococcygeus muscles (Abbott et al. 2006) showed mixed results. After 26 weeks of follow-up, quality of life measures were improved in both the BoNT and placebo groups, but the difference between BoNT and placebo groups did not reach statistical significance.

However, the authors found a reduction in resting pelvic muscle tone in the women injected with onaBoNTA compared to placebo ($p < 0.001$), and this translated into significant improvements in both dyspareunia ($p < 0.001$) and nonmenstrual pelvic pain ($p = 0.009$). Adelowo et al. (2013) reported on their experience using onaBoNTA (100–300 U) in 29 women with chronic myofascial pelvic pain. In this retrospective study, the authors placed several onaBoNTA 10 U injections (total 300 U) into the pelvic floor muscles. Pain improvement was seen in 79% of patients at <6 weeks postinjection. After a median of 4 months from the first injection, 52% requested repeat onaBoNTA. Urinary retention (defined by PVR > 100 mL) and fecal incontinence resulted in three patients and two patients, respectively, and these AEs completely resolved. Larger placebo-controlled RCTs and patient-reported outcomes are needed to support the use of onaBoNTA for women with myofascial pelvic pain refractory to standard PFPT. The use of sphincter and pelvic floor BoNT injections is currently off-label.

8 Benign Prostatic Hyperplasia (BPH)

Application of BoNT to treat BPH was reported by Maria et al. (2003). Thirty men with symptomatic BPH were randomized to receive either 200 U of onaBoNTA (n-15) or placebo saline injection ($n = 5$). OnaBoNTA 100 U in 2 mL of saline or saline alone in the placebo arm was injected into each lobe of the prostate through the perineum via a 22-gauge spinal needle with transrectal ultrasound guidance. Clinical improvement was evident after 1 month. The investigators noted that the American Urological Association symptom score, a common index for the assessment of BPH, decreased by 65% compared to baseline in the onaBoNTA patients ($p = 0.00001$). Also, maximum flow rate increased from 8.1 to 14.9 mL/s with onaBoNTA ($p = 0.00001$). There was no significant improvement in the saline alone injected patients. No urinary incontinence or systemic side effects were reported over 18 months follow-up.

Chuang et al. (2006) stratified drug treatment refractory BPH with either prostate size <30 g or >30 g and injected them with either 100 U onaBoNTA or 200 U onaBoNTA, respectively, via ultrasound-guided perineal injection. At 12 months, the percent improvements in the International Prostate Symptom Score (IPSS), maximum flow rate, and post-void residual urine volume were similar to those of Maria et al. (2003), except that the percent shrinkage of prostate size was substantially smaller (13–19% versus 61%). In 29% of men, there was no change in prostate volume, yet 58% of these men still had a >30% improvement in IPSS, maximum flow rate, and post-void residual urine volume, suggesting that onaBoNTA may relieve BPH symptoms by an effect on sensory nerve pathways rather than reducing the prostate size alone.

McVary et al. (2014) performed a Phase 2 multicenter, placebo-controlled, randomized clinical trial using a onaBoNTA 200 U to treat men with BPH and moderate lower urinary tract symptoms. The men had an IPSS of 14 or >, a maximum flow rate of 4–15 mL/s, and a post-void residual urine volume ≤200 mL. 315 men were randomized to either onaBoNTA 200 U ($n = 158$) or placebo ($n = 157$). The primary endpoint was the change from baseline in IPSS at week 12. Although a significant decrease from baseline in IPSS was seen with both onaBoNTA (−6.3 points) and placebo (−5.6 points), there was no difference between the groups; however, onaBoNTA showed efficacy over placebo in improving maximum flow rate at week 6 postinjection ($p \leq 0.01$). The most common adverse events in both groups were hematuria and hematospermia. The authors concluded that intraprostatic injection of onaBoNTA was not more efficacious compared to placebo in improving lower urinary tract symptoms and the commercial development of onaBoNTA for BPH indication was subsequently stopped at this time.

Adverse Events There are little adverse events reported in the literature secondary to prostate BoNT injection (Chancellor and Smith 2011). Dysuria and occasional minor hematuria and epididymitis have been reported. No negative effect on sperm function has been reported.

Prostate Injection Approach 200 U of onaBoNTA has typically been used for prostate injection and diluted in a volume of 4 mL of preservative free saline. Prostatic injections of BoNT can be carried out transperineally, transrectally, or transurethrally with preference often dictated by regional practice habits. Transperineal injection minimizes the risk of infection, but ultrasound-guided transrectal prostatic injection is the procedure that urologists in Europe and North America are most familiar with. The preparation and positioning of the patient are identical to that used for transrectal or transperineal ultrasound-guided prostate biopsy. Some urologists may prefer transurethral prostate injection using a familiar cystoscopy and injecting needle to approach the enlarged prostate glands. This method may be more effective for managing trilobar prostate enlargement (Fig. 3).

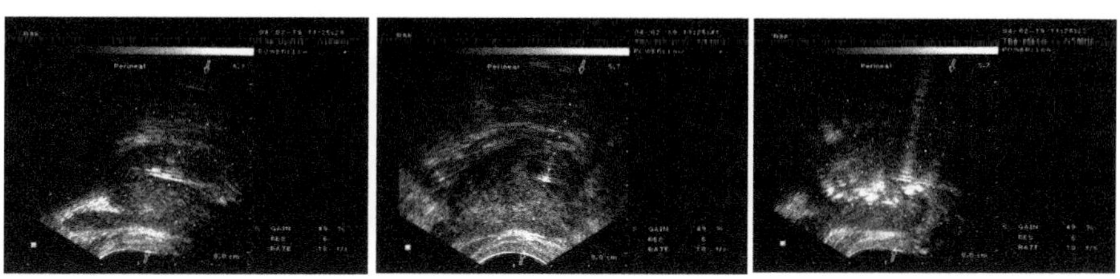

Fig. 3 Left: Confirmation of needle (black arrow) within the prostate, longitudinal view. Middle: Transverse view. Right: Diffusion of hyperechoic BoNT (black arrow) over the prostate immediate post injection (Chancellor and Smith 2011)

9 Bladder Pain

Interstitial cystitis/bladder pain syndrome (IC/BPS) is defined as pain perceived to be related to the urinary bladder, associated with lower urinary tract symptoms greater than 6-month duration, in the absence of infection or other identifiable causes (Kuo and Chancellor 2009). The first report was a case series of 13 women with NIDDK-defined IC (Smith et al. 2004). The patients underwent submucosal transurethral injections of 100–200 U of abobotulinumtoxinA (7 patients) or onaBoNTA 100 U (6 patients) into 20–30 sites in the trigone and bladder base. Validated questionnaire (Interstitial Cystitis Symptom Index, Interstitial Cystitis Problem Index) or voiding charts and a visual analog pain scale were evaluated at baseline, 1-month and subsequently at 3-month intervals. Statistically significant improvements in frequency, nocturia, and pain were observed 1 month following treatment, with improvements in first desire to void and cystometric capacity in those patients so evaluated. Onset of symptom relief was 5–7 days following treatment, and mean duration of symptom relief was 3.7 months.

Kuo and Chancellor (2009) performed a randomized trial in IC/BPS patients comparing bladder hydrodistention (HD) with either 100 U or 200 U doses of onaBoNTA versus hydrodistention alone. At 3 months, the bladder pain visual analog scale, functional bladder capacity, cystometric bladder capacity, and global response assessment significantly improved only in the onaBoNTA groups vs. the control group. The 200 U dose didn't provide better efficacy compared to 100 U, and there were more side effects, including urinary retention, with using 200 U onaBoNTA. The study and other non-randomized studies suggested the promise of using botulinum toxin for treating bladder pain.

10 Conclusion

The use of botulinum toxin for the treatment of neurogenic and refractory idiopathic overactive bladder has resulted in improved continence and quality of life. The intraprostatic injection of botulinum toxin for benign prostatic hypertrophy to date has not shown efficacy in improving lower urinary tract symptoms. Treating detrusor sphincter dyssynergia, myofascial pain, and interstitial cystitis/bladder pain syndrome with botulinum toxin has showed some promising results in controlled trials, but they are currently an off-label use of the product. Application of botulinum toxin for lower urinary tract dysfunction is exciting, expanding, and evolving. We believe there will be further exciting advances in the application of botulinum toxin in the genitourinary system.

References

Abbott JA, Jarvis SK, Lyons SD et al (2006) Botulinum toxin type A for chronic pain and pelvic floor spasm in women. Obstet Gynecol 108:915–923

Abdel-Meguid TA (2010) Botulinum toxin-A injections into neurogenic overactive bladder--to include or exclude the trigone? A prospective, randomized, controlled trial. J Urol 184 (6):2423–2428

Abrams P, Cardozo L, Fall M et al (2002) The standardization of terminology of lower urinary tract function: report from the standardisation sub-committee of the international continence society. Neurourol Urodyn 21:167–178

Adelowo A, Hacker MR, Shapiro A et al (2013) Botulinum toxin type A (BOTOX) for refractory myofascial pelvic pain. Female Pelvic Med Reconstr Surg 19:288–292

Amundsen CL, Richter HE, Menefee SA et al (2016) OnabotulinumtoxinA vs sacral neuromodulation on refractory urgency urinary incontinence in women: a randomized clinical trial. JAMA 316:1366–1374

Apostolidis A, Dasgupta P, Fowler CJ (2006) Proposed mechanism for the efficacy of injected botulinum toxin in the treatment of human detrusor overactivity. Eur Urol 49:644–650

Billante CR, Zealear DL, Billante M et al (2002) Comparison of neuromuscular blockade and recovery with botulinum toxins A and F. Muscle Nerve 26:395–403

Chancellor MB, Smith CP (2011) Botulinum toxin in urology. Springer, Berlin. http://www.springer.com/medicine/urology/book/978-3-642-03579-1?changeHeader

Chancellor MB, Anderson RU, Boone TB (2006) Pharmacotherapy for neurogenic detrusor overactivity. Am J Phys Med Rehabil 85:536–545

Chancellor MB, Fowler CJ, Apostolidis A et al (2008) Drug insight: biological effects of botulinum toxin A in the lower urinary tract. Nat Clin Pract Urol 5:319–328

Chapple CR, Khullar V, Gabriel Z et al (2008) The effects of antimuscarinic treatments in overactive bladder: an update of a systematic review and meta-analysis. Eur Urol 54:543–562

Chapple C, Sievert K-D, MacDiarmid S et al (2013) OnabotulinumtoxinA 100 u significantly improves all idiopathic overactive bladder symptoms and quality of life in patients with overactive bladder and urinary incontinence: a randomized, double-blind, placebo-controlled trial. Eur Urol 64:249–256

Chuang YC, Yoshimura N, Huang CC, Chiang PH, Chancellor MB (2004) Intravesical botulinum toxin A administration produces analgesia against acetic acid induced bladder pain responses in rats. J Urol 172:1529–1532

Chuang YC, Chiang PH, Yoshimura N, De Miguel F, Chancellor MB (2006) Sustained beneficial effects of intraprostatic botulinum toxin type A on lower urinary tract symptoms and quality of life in men with benign prostatic hyperplasia. BJU Int 98(5):1033–1037

Cruz F, Herschorn S, Aliotta P et al (2011) Efficacy and safety of onabotulinumtoxinA in patients with urinary incontinence due to neurogenic detrusor overactivity: a randomized, double-blind, placebo-controlled trial. Eur Urol 60:742–750

Denys P, Del Popolo G, Amarenco G, Dysport Study Group et al (2016) Efficacy and safety of two administration modes of an intra-detrusor injection of 750 units dysport® (abobotulinumtoxinA) in patients suffering from refractory neurogenic detrusor overactivity (NDO): a randomised placebo-controlled phase IIa study. Neurourol Urodyn. https://doi.org/10.1002/nau.22954

Dykstra D, Sidi A, Scott A et al (1988) Effects of botulinum A toxin on detrusor-sphincter dyssynergia in spinal cord injury patients. J Urol 139:919–922

Gallien P, Reymann J-M, Amarenco G et al (2005) Placebo controlled, randomized, double blind study of the effects of botulinum A toxin on detrusor sphincter dyssynergia in multiple sclerosis patients. J Neurol Neurosurg Psychiatry 76:1670–1676

Ghazizadeh S, Nikzad M (2004) Botulinum toxin in the treatment of refractory vaginismus. Obstet Gynecol 104:922–925

Ginsberg D, Gousse A, Keppenne V et al (2012) Phase 3 efficacy and tolerability study of onabotulinumtoxinA for urinary incontinence from neurogenic detrusor overactivity. J Urol 187:2131–2139

Ginsberg D, Cruz F, Herschorn S et al (2013) OnabotulinumtoxinA is effective in patients with urinary incontinence due to neurogenic detrusor activity regardless of concomitant anticholinergic use or neurologic etiology. Adv Ther 30:819–833

Gormley EA, Lightner DJ, Faraday M et al (2015) Diagnosis and treatment of overactive bladder (non-neurogenic) in adults: AUA/SUFU guideline amendment. J Urol 193:1572–1580

Hanno PM, Erickson D, Moldwin R et al (2015) Diagnosis and treatment of interstitial cystitis/bladder pain syndrome: AUA guideline amendment. J Urol 193:1545–1553

Kuo HC (2007) Recovery of detrusor function after urethral botulinum A toxin injection in patients with idiopathic low detrusor contractility and voiding dysfunction. Urology 69:57–61; discussion 61–52

Kuo HC, Chancellor MB (2009) Comparison of intravesical botulinum toxin type A injections plus hydrodistention with hydrodistention alone for the treatment of refractory interstitial cystitis/painful bladder syndrome. BJU Int 104:657–661

Maria G, Brisinda G, Civello IM et al (2003) Relief by botulinum toxin of voiding dysfunction due to benign prostatic hyperplasia; results of a randomized, placebo-controlled study. Urology 62:259–264

McVary KT, Roehrborn CG, Chartier-Kastler E et al (2014) A multicenter, randomized, double-blind, placebo controlled study of onabotulinumtoxinA 200 U to treat lower urinary tract symptoms in men with benign prostatic hyperplasia. J Urol 192:150–156

Nitti VW, Dmochowski R, Herschorn S et al (2013) OnabotulinumtoxinA for the treatment of patients with overactive bladder and urinary incontinence: results of a phase 3, randomized, placebo controlled trial. J Urol 189:2186–2193

Schurch B, Schmid D, Stohrer M et al (2000) Treatment of neurogenic incontinence with botulinum toxin. N Engl J Med 342:665

Smith CP, Boone TB, de Groat WC, Chancellor MB, Somogyi GT (2003) Effect of stimulation intensity and botulinum toxin isoform on rat bladder strip contractions. Brain Res Bull 61:165–171

Smith CP, Radziszewski P, Borkowski A et al (2004) Botulinum toxin A has antinociceptive effects in treating interstitial cystitis. Urology 64(5):871–875

Smith CP, Gangitano DA, Munoz A et al (2008) Botulinum toxin type A normalizes alterations in urothelial ATP and NO release induced by chronic spinal cord injury. Neurochem Int 52:1068–1075

Stewart WF, Van Rooyen JB, Cundiff GW et al (2003) Prevalence and burden of overactive bladder in the United States. World J Urol 20:327–336

Gastrointestinal Uses of Botulinum Toxin

Maria Cariati, Maria Michela Chiarello, Marco Cannistra',
Maria Antonietta Lerose, and Giuseppe Brisinda

Contents

Abstract

Botulinum toxin (BT), one of the most powerful inhibitors that prevents the release of acetylcholine from nerve endings, represents an alternative therapeutic approach for "spastic" disorders of the gastrointestinal tract such as achalasia, gastroparesis, sphincter of Oddi dysfunction, chronic anal fissures, and pelvic floor dyssynergia.

M. Cariati · M. M. Chiarello · M. Cannistra' · M. A. Lerose
Department of Surgery, "San Giovanni di Dio" Hospital, Crotone, Italy

G. Brisinda (✉)
Department of Surgery, "San Giovanni di Dio" Hospital, Crotone, Italy

Department of Surgery, "Agostino Gemelli" Hospital, Catholic School of Medicine, Rome, Italy

S. M. Whitcup, M. Hallett (eds.), *Botulinum Toxin Therapy*,
Handbook of Experimental Pharmacology 263, https://doi.org/10.1007/164_2019_326

BT has proven to be safe and this allows it to be a valid alternative in patients at high risk of invasive procedures but long-term efficacy in many disorders has not been observed, primarily due to its relatively short duration of action. Administration of BT has a low rate of adverse reactions and complications. However, not all patients respond to BT therapy, and large randomized controlled trials are lacking for many conditions commonly treated with BT.

The local injection of BT in some conditions becomes a useful tool to decide to switch to more invasive therapies. Since 1980, the toxin has rapidly transformed from lethal poison to a safe therapeutic agent, with a significant impact on the quality of life.

Keywords
Achalasia · Autonomic nervous system diseases · Biliary diseases · Botulinum toxin · Cholinergic nerve ending · Enteric nervous system · Esophageal diseases · Fissures · Gastric emptying · Hirschsprung · Motility · Neuromuscular agents · Obesity · Spasm · Therapeutic agents

Local injection of botulinum toxin A (BT) is an effective treatment for many different diseases of the gastrointestinal tract because it inhibits contraction of smooth muscles and sphincters by blocking cholinergic nerve endings in the autonomic nervous system (ANS). Normal gastrointestinal (GIT) motility depends on intrinsic neurons contained in the enteric nervous system (ENS), with significant modulatory input being provided by the central nervous system (CNS) via autonomic sympathetic and parasympathetic nerves (Civelek et al. 1985; Albanese et al. 2000). Immediate control of muscle tone in the gut reflects a balance between both excitatory (predominantly cholinergic) and inhibitory (predominantly nitrinergic). In some disease states, this balance is disrupted, usually due to a relatively selective loss of inhibitory neurons (Poulain et al. 1988; Grumelli et al. 2010; Akaike et al. 2013). In this setting, BT, by blocking excitatory neurotransmitter release, can restore the balance and cause a decrease in the resting tone of the muscle involved.

The ENS provides the intrinsic innervation. It is a highly complex system, responsible for the coordination of motility in the GIT. A deficiency of enteric neurons causes obstruction and lack of intestinal propulsion (Miftakhov and Wingate 1993). The ENS is composed of two main ganglionated plexuses (Auerbach's myenteric plexus and Meissner's submucous plexus) and non-ganglionated plexuses (the longitudinal muscle plexus, the circular muscle plexus, the plexus of the muscularis mucosae, and the mucosal plexus) (Kuhn and Belafsky 2013). Intraparietal neurons encompass motor excitatory and inhibitory neurons, interneurons, and intrinsic sensory neurons. Sympathetic and parasympathetic neurons also innervate the GIT. ACh is the primary excitatory transmitter. Inhibitory motor neurons relax smooth muscles; these neurons release a combination of at least three transmitters: NO, adenosine triphosphate (ATP), and VIP (Albanese et al. 2000). At cellular level, smooth muscle contraction and relaxation are regulated by changes in cytosol calcium levels (Hansen 2003). These functions depend on the intrinsic electrical and mechanical properties of GIT smooth muscles and are regulated by the ENS and by sympathetic and parasympathetic influences (Albanese et al. 2000). Hormones also influence GIT motility (Lourenssen et al. 2009). Interstitial cells of Cajal act as local pacemakers to generate the rhythmic activity of the circular muscle layer throughout the GIT. Motor neurons control the musculature indirectly, through their action on the Cajal cells. Substances, such as histamine, serotonin, adenosine, and eicosanoids, produced by nonneural cells, can influence smooth muscle activity (Walzer and Hirano 2008).

At esophageal level, muscle tone of the lower esophageal sphincter (LES) results from the interaction of neurogenic and myogenic conditions. Neurogenic tone in humans is partly due to cholinergic innervation. The modulation of LES tone is largely mediated through the vagus nerve. Acetylcholine (ACh) is the presynaptic neurotransmitter; postsynaptic transmission is mediated by NO, but vasoactive intestinal polypeptide (VIP) is also thought to contribute (Walzer and Hirano 2008).

At anal level, the sphincter complex consists of two overlapping sphincters (Brisinda et al. 2004b). The external anal sphincter (EAS) that forms the outer layer is composed of voluntary, striated, skeletal muscle. The internal anal sphincter (IAS) is the inner, involuntary, smooth muscle component. It is in a state of continuous maximal contraction, due to a combination of intrinsic myogenic and autonomic neurogenic properties. Being of visceral origin, IAS is supplied both by sympathetic and parasympathetic nerves; in addition, the ENS modulates its tonic activity (Albanese et al. 2000). Noradrenergic sympathetic nerves are considered excitatory and the parasympathetic inhibitory to the IAS. Vagal neurons do not act directly but rather form synaptic connections with neurons whose cell bodies are in the intrinsic GIT ganglia. This transmission is principally mediated by ACh acting on nicotinic receptors (Brisinda et al. 2007b). Recently, it has been shown that the longitudinal layer and the circular smooth muscle in the human rectum receive an intrinsic NO-mediated inhibitory innervation.

Although BT can clearly inhibit the release of acetylcholine, little else is known about its effects in GIT muscle. Thus, while nitric oxide (NO) release is not affected – which is to be expected, since this is not a vesicular process – the specific effects on other potentially important neurotransmitters have not been well documented (Mariotti and Bentivoglio 1996; Lepiarczyk et al. 2015). Further, there is some suggestion that it may also inhibit the responsiveness of smooth muscle to exogenous stimuli, an effect that is quite unique to the GIT.

1 Esophageal Applications

1.1 Cricopharyngeal Dysphagia

Dysphagia associated with failed relaxation of the upper esophageal sphincter (UES) has been observed in patients suffering from different types of neurological disease. The absent relaxation of the cricopharyngeal (CP) muscle during bolus swallowing prevents the UES from opening; consequently, the bolus cannot progress into the esophagus. This may result in penetration or aspiration of ingested food into the airways. Many reports in the literature demonstrate that neurogenic dysphagia associated with UES spasms or dyskinesia can be effectively treated by injecting BT into the CP muscle (Alberty et al. 2000; Haapaniemi et al. 2001; Moerman 2006; Krause et al. 2008; Alfonsi et al. 2010; Regan et al. 2014) (Table 1). Most of these reports are case series formed by a low number of patients, and randomized control trials are lacking. Moreover, because of different methodological approaches, study designs, and outcome measures, the results obtained by different authors are not absolutely comparable. Indeed patient selection criteria vary greatly from study to study, and the same is also true for follow-up times: some authors have focused only on short-term safety and efficacy of BT treatment, while others have investigated long-term effects (Haapaniemi et al. 2001; Shaw and Searl 2001; Zaninotto et al. 2004a). A number of injection techniques have been employed including rigid endoscopy with electromyographic control, flexible endoscopy, and an open technique with various doses (10–50 units Onabotulinumtoxin A, Ona-A). Endoscopically, 3–4 injections of BT can be delivered to the dorsomedial and bilateral ventromedial compartments of CP muscle. CP injection of BT has distinct appeal in patients who are not ideal candidates for longer general anesthesia or in whom the temporary nature of BT injection is warranted. It may be advantageous to pursue CP injection of BT in patients in whom multilevel dysphagia is suspected and in whom the clinician suspects that there may be some detriment to treatment directed at the UES. Additionally, CP injection of BT is a diagnostic tool used to identify patients who may potentially benefit from CP myotomy (Kelly et al. 2013; Regan et al. 2014; Kuhn and Belafsky 2013; Blitzer and Brin 1997).

Only two series included more than 20 patients; the largest study included 34 patients. The causes of CP dysfunction in these published series encompassed several diagnoses, including neurological diseases, diabetic neuropathy, external-beam radiation treatment, cerebrovascular accident, and others. The dosage and administration techniques of BT were also quite variable (Kelly et al. 2013). There were also different types of BT administered (Kelly et al. 2013; Moerman 2006).

In general, the majority of patients reported improved swallowing function, approximately 75% in combined analysis. Complications were infrequent and included transient vocal fold paresis, temporary worsening of dysphagia, neck cellulitis, and aspiration pneumonia. There were no reported deaths in the literature that were directly related to CP injection of BT. Kelly and coworkers demonstrated that CP injection of BT is a well-tolerated treatment for dysphagia related to CP dysfunction, with good efficacy in the majority of their 49 patients (Kelly et al. 2013).

Alfonsi et al. enrolled 67 patients with neurogenic dysphagia associated with incomplete or absent opening of the UES (24 with brain stem or hemispheric stroke, 21 with parkinsonian syndromes, 12 with multiple sclerosis, and 10 with spastic-dystonic syndromes secondary to post-traumatic encephalopathy), and they were treated with the injection of incobotulinumtoxin A (Inco-A, Xeomin) (dose 15–20 U) into the CP muscle under electromyographic guidance. The patients were assessed at baseline and after the first and second treatment through clinical

Table 1 Review of the literature on the treatment of cricopharyngeal dysphagia with BT injection

Authors	Pts	Ona-A (Unit)	Abo-A (Unit)	Improvement	Method of delivery	Causes	Complications
Schneider et al. (1994)	7	80–120		5/7 (71%)	GA, EGD	Stroke, CN palsies, supraglottic or oropharyngeal cancer, reflux disease	None
Atkinson and Rees (1997)	5	5–20		4/5 (80%)	CT-guided injection	Stroke, CN palsies, bulbar palsy	Left vocal fold paresis, aspiration pneumonia
Blitzer and Brin (1997)	6	10		6/6 (100%)	Percutaneous injection	CVA, partial pharyngectomy, small Zenker's diverticulum	None
Brant et al. (1999a)	1	100		1 (100%)	Flexible EGD	CVA	None
Alberty et al. (2000)	10	30		10/10 (100%)	GA, EGD	CVA, idiopathic polymyositis	None
Shaw and Searl (2001)	12	25–50		10/12 (83%)	GA, EGD, open technique	Progressive neuropathy, oculopharyngeal dysphagia, skull base tumor resection, total laryngectomy, CVA, partial pharyngectomy, CNS neuropathy	Pharyngeal tear, worsening dysphagia
Haapaniemi et al. (2001)	4	14–50		3/4 (75%)	GA, EGD	Brain stem stroke, inclusion body myositis, peripheral motor neuropathy, CVA	None
Moerman et al. (2002)	4	100		4/4 (100%)	GA	Head and neck cancer resection including total laryngectomy, radiation	None
Parameswaran and Soliman (2002)	12	10–30		11/12 (92%)	EGD with mask ventilation and apneic technique	Idiopathic, radiation, CVA, total laryngectomy, ALS, Parkinson's disease	Neck cellulitis (concurrent thyroglossal duct excision)

(continued)

Table 1 (continued)

Authors	Pts	Ona-A (Unit)	Abo-A (Unit)	Improvement	Method of delivery	Causes	Complications
Zaninotto et al. (2004a)	21	4–10		9/21 (43%)	Percutaneous with EMG	CNS disease, peripheral neuropathies, idiopathic	Death of aspiration (attributed to underlying disease)
Liu et al. (2004)	2	100		2/2 (100%)	Flexible EGD under sedation	Inclusion body myositis	None
Chiu et al. (2004)	1		120	1/1 (100%)	GA and direct laryngoscopy	Brain stem stroke	None
Murry et al. (2005)	13	100		11/13 (85%) 2/13 improvement after second injection	EMG-guided transcutaneous approach	Stroke, head and neck surgery, cranial neuropathies, MVC, chemical inhalation, radiation therapy or lymphoma	None
Kim et al. (2006)	8	100		5/8 (62.5%)	Flexible endoscopy	CVA	None
Masiero et al. (2006)	2	25, 30		2/2 (100%)	Percutaneous injection	CVA	None
Restivo et al. (2006)	12		60	12/12 (100%)	EMG-guided transcutaneous approach	Diabetic neuropathy	None
Suzukia et al. (2007)	1	5		1/1 (100%)	Percutaneous injection	Spinal muscular atrophy type 2	Transient worsening of dysphagia
Krause et al. (2008)	1		180, 150	1/1 (100%) 0/1 (0%)	Endoscopic injection with propofol sedation	Spasticity secondary to SAH	None
Alfonsi et al. (2010)	34	15		17/34 (50%)	EMG-guided transcutaneous approach	MS, multiple system atrophy, Parkinson's disease, progressive supranuclear palsy, ataxia-telangiectasia	None

Restivo et al. (2011)	14	20		14/14 (100%)	Percutaneous injection with EMG guidance	MS	None

ALS amyotrophic lateral sclerosis, *CN* cranial nerve, *CNS* central nervous system, *CT* computed tomography, *CVA* cerebrovascular accident or stroke, *EGD* esophagogastroduodenoscopy, *EMG* electromyography, *GA* general anesthesia, *MVC* motor vehicle collision, *SAH* subarachnoid hemorrhage, *MS* multiple sclerosis, *Ona-A* onabotulinumtoxin A (Botox), *Abo-A* abobotulinumtoxin A (Dysport) evaluation and fiber-optic endoscopy of swallowing, while their dysphagia was quantified using the Dysphagia Outcome and Severity Scale (DOSS). An electrokinesiographic/electromyographic study of swallowing was performed at baseline. Most patients responded to the first BT treatment: 35 patients (52.2%) were classified as high responders (DOSS score increase >2 levels), while other 19 patients (28.4%) were low responders (DOSS score increase of ≤2 levels). The effect of the first treatment usually lasted longer than 4 months (67%) and in some cases up to a year. The treatment efficacy remained high also after the second injection: 31 patients (46.3%) qualified as high responders, and other 22 patients (32.8%) showed a low response. Only in the parkinsonian syndrome group, they observed a reduction in the percentage of high responders as compared with the first treatment. Side effects were mostly mild and reported in non-responders following the first injection. A severe side effect, consisting of ingestion pneumonia, was observed following the second BT injection in two patients who had both been non-responders to the first (Alfonsi et al. 2017).

On the basis of these results, CP injection of BT appears to be effective in patients with UES dysfunction. Response to BT injection may select out a group of patients with higher likelihood of a more durable response to surgical myotomy (Allen et al. 2010). Further work, however, is needed to define the population of patients who might have a poor response to BT treatment. Furthermore, non-response may indicate another etiology of dysphagia, i.e., stricture.

1.1.1 Cricopharyngeal Achalasia (CPA) in Children

CPA is a condition characterized by an incomplete relaxation of the UES or by a lack of coordination of the UES opening with pharyngeal contractions (Drendel et al. 2013; Hussain et al. 2002). Both etiologies can lead to choking, cough, and aspiration. CPA is a different entity than the CP dysphagia that was seen in adults. Although an exact cause of CPA is unknown, it is considered to be associated with an immature neuromuscular system. Immaturity of the interstitial intraparietal cells of Cajal may explain why there have been reports of spontaneous resolution of CPA seen in infants (Scholes et al. 2014). CPA has also been associated with gastroesophageal reflux disease and CNS abnormalities (Kuhn and Belafsky 2013; Drendel et al. 2013; Hussain et al. 2002; Scholes et al. 2014; Shogan et al. 2014; Huoh and Messner 2013).

Recently, six children were identified with CPA (Drendel et al. 2013). The decision to proceed with BT therapy was based on ongoing severe symptoms, the necessity of altered feeds, and parent preference over a surgical myotomy. The number of injections ranged from 1 to 3 per patient. The mean dose was 5.6 units/kg of Ona-A, with a range of 1.6–7.9 units/kg and a median of 6.0 units/kg. In those patients with multiple injections, the mean time between injections was approximately 13 months. The mean time to return to normal radiographic swallow study was 8.2 weeks. Two of the children benefited from BT injections and went on to have CP myotomy, while four of the children did not require myotomy, and their symptoms resolved after one or two injections. The authors concluded that BT injection of CP muscle is a useful tool to help diagnose and treat CPA (Drendel et al. 2013). It is a feasible alternative to more invasive surgical procedures. However, more research is needed to elucidate the optimal dosing, frequency of injections, and when to move on to surgical intervention.

1.2 Achalasia

The major pathophysiological lesion in achalasia results from a relatively specific loss of nitrergic inhibitory neurons of the LES, resulting in an inability of the sphincter to relax after swallowing (Woltman et al. 2005). This results in a functional obstruction and dysphagia. Although no cure exists for achalasia, there are a number of palliative treatments available including surgical myotomy, pneumatic dilation (PD), and BT injections into the LES (Tack and Zaninotto 2015; Maradey-Romero et al. 2014; Marjoux et al. 2014; Vela 2014; Mabvuure et al. 2014; Patti and Fisichella 2014). Surgical myotomy has proven durable but is associated with increased morbidity and mortality in high-risk surgical patients. Pneumatic dilation of the sphincter results in an initial symptomatic improvement in 60–90% of patients, but repeated dilations are often necessary. Furthermore, the procedure carries a small but significant risk of esophageal perforation (Leyden et al. 2014; Jung et al. 2014; Kim do and Jung 2014). Thus, BT provides a potentially attractive alternative to the above treatment methods (Vela 2014).

Endoscopic injection of 25 units of Ona-A BT in four LES quadrants is generally the standard of care. The efficacy of BT in achalasia has been proven by the results of several randomized trials comparing it to either placebo or pneumatic dilation. Table 2 summarizes the response rates to BT in patients with achalasia.

Most patients (75–100%) show an initial response, but more sustained improvement (beyond 6 months) is seen in about two-thirds. For unclear reasons, it appears that patients older than 50 years of age respond at a higher rate (82% vs 43% in younger patients). Similarly, patients with so-called vigorous achalasia (with the esophagus retaining some contractile ability) respond at a higher rate (100% vs 52% with classic achalasia).

Several studies have compared BT to pneumatic dilation with most reporting similar initial clinical or manometric responses. However, the 1-year remission rate after a single injection is markedly inferior for BT, which is to be expected given its pharmacological properties. In the only study comparing the two modalities in a head-to-head comparison, 80 patients were randomized to receive 100 BT Ona-A units or laparoscopic surgical myotomy with fundoplication. After 6 months, symptom scores improved more in surgical patients (82% vs 66%, $P < 0.05$). The drop in LES pressure was similar in the two groups; the reduction in esophageal diameter was greater after surgery (19% vs 5%, $P < 0.05$). The only complication in the surgical group was one patient bled at the trocar site. The probability of being symptom-free at 2 years was 87.5% after surgery and 34% after BT ($P < 0.05$). The same group investigated the cost-effectiveness of the two modalities and concluded that BT was more cost-effective in the short term, but at 2 years, cost between the two groups was similar. The results of a recent meta-analysis suggest

Table 2 Review of experience using BT for the treatment of esophageal achalasia

Authors	Description	Patients	Results/conclusions
Pasricha et al. (1995)	BT vs placebo	21	67% were improved at 6 weeks
Annese et al. (1996)	BT vs placebo vs PBD	16	100% were improved at 1 month, 88% required repeated injections. BT is as effective as pneumatic dilatation
Fiorini et al. (1996)	BT vs placebo	13	72% were improved at 3 months
Pasricha et al. (1996)	BT	31	60% (82% of those aged >50) were improved at 3 months
Fishman et al. (1996)	BT	65	60 idiopathic cases: BT treatment improved symptoms of dysphagia, chest pain, and regurgitation in the majority of patients. Five secondary cases: there was no response to BT in four patients. Patients, who respond to a first BT injection but relapse, may respond to a second treatment
Cuilliere et al. (1997)	BT	55	60% were improved at 6 months
Brant et al. (1999b)	BT in Chagas' disease	3	Clinical improvement occurred in all patients. Mean LES pressure drop by 29%
Kolbasnik et al. (1999)	BT	30	Symptomatic improvement for >3 months was seen in 77% of patients. 7 patients had a sustained response after a single injection; 16 relapsed and required re-treatment
Annese et al. (1999)	Ona-A vs Abo-A	78	Comparable efficacy in esophageal achalasia after up to 6 months after treatment
Muehldorfer et al. (1999)	BT vs PBD	24	The two treatments had equal initial success rate (dilatation 83%, BT 75%). In the long term, the efficacy of BT injection was statistically significant and shorter than that of balloon dilatation
Panaccione et al. (1999)	BT vs PBD	NR	Intrasphincteric BT injection was more costly than pneumatic dilatation (USD 5,033 compared to USD 3,608). BT treatment may be less costly if life expectancy is less than 2 years
Greaves et al. (1999)	BT	11	The relapse rate was 73% within 2 years from treatment. There were a beneficial effect on dysphagia and no improvement in chest pain or regurgitation scores, and no reduction of mean LES pressure was improved at 6 weeks
Wehrmann et al. (1999)	BT in high-risk patients	20	80% were improved at 6 weeks. Mean cardiac diameter was increased from 2.1 to 3.2 mm. The patients who initially had a symptomatic relapse after an average of 5 months. BT reinjections were efficacious
Hurwitz et al. (2000)	BT in children	23	The mean duration of effect in 19 responders was 4.2 months. 50% of the patients required

(continued)

Table 2 (continued)

Authors	Description	Patients	Results/conclusions
			an additional procedure (PD, surgery) on average 7 months after the first treatment
Annese et al. (2000)	BT dose raging study	118	82% of the patients were responders at 1 month. No dose-related effect was observed. Vigorous achalasia was the main determinant of BT response
Ip et al. (2000)	BT in children	7	100% were improved at 4 months. Sustained response beyond 6 months occurred in 43% of patients
Hep et al. (2000)	BT plus PBD	3	Propulsive peristalsis of the esophagus was restored in all patients
Mikaeli et al. (2001)	BT vs PBD	40	Cumulative 12-month remission rate was significantly higher after a single PD (53%) compared to a single BT injection (15%, $P < 0.01$). The 12-month estimated adjusted hazard for relapse and need for re-treatment for BT group was 2.69 times that of the PD group
Allescher et al. (2001)	BT vs PBD	37	After 24 months a single PD was superior to a single BT injection, and after 48 months, all patients treated for BT injection had experienced a symptomatic relapse
Ghoshal et al. (2001)	BT vs PBD	17	Both therapies resulted in a significant reduction in LES pressure
Zarate et al. (2002)	BT	17	The effect of BT injection wanes with time in elderly patients, necessitating repeated injections to keep the patients symptom-free
D'Onofrio et al. (2002)	BT	37	Of the 35 patients followed, 12 had a relapse and were treated; 4 out of 12 did not respond after treatment. One or two BT injections result in a clinical and objective improvement in about 84% of achalasia patients and are not associated with serious side effects; patients over 50 years showed better benefit than younger patients
Neubrand et al. (2002)	BT	25	Good results after 2.5 years of median follow-up in 9 of 25 patients that were significantly older than 14 patients for whom BT treatment was unsuccessful
Brant et al. (2003)	BT in Chagas' disease	24	Over a period of 6 months, clinical improvement of dysphagia was statistically significant ($P < 0.001$) in patients receiving BT when compared with the placebo. Esophageal emptying time in BT group was significantly lower than in the placebo ($P = 0.04$) after 90 days
Bansal et al. (2003)	BT vs PBD	32	After 12-month follow-up, 16 of 18 patients of PBD were in clinical remission despite 6 of 16 of BT group

(continued)

Table 2 (continued)

Authors	Description	Patients	Results/conclusions
Martinek et al. (2003)	BT vs PBD	41	16 patients had BT injection from the antegrade angle only (group A), 15 both from antegrade than retrograde (group B) and 10 had subsequent PD (group C). 93% had an immediate clinical response after 1 month, and 49% were in remission after 22 months. Better responders were older and with lower LES pressure. Patients in group C had better results at 1 and 2 years
Martinek and Spicak (2003)	Modified BT	16	After a single BT injection, 11 responders reported a relapse with a median symptom-free interval of 17 months. After reinjection the median symptom-free interval was 16 months
Vela et al. (2004)	PBD vs HM vs BT PBD vs HM in patient with prior surgery	232	111 patients underwent PBD, 72 HM, and 39 elderly patients BT injection. 48 patients had already surgical treatment and underwent to PBD or redo-HM. PBD and HM are the best treatments for untreated achalasia and are less successful after surgery. BT group needed repeated injections, and their symptoms improving lasted for a mean period of 6.2 months
Zaninotto et al. (2004b)	BT vs HM	80	After 6 months similar results were reported in the 2 groups of 40 patients, but after 2 years, 87.5% of patients of surgical groups were symptom-free vs 34% of BT group ($P < 0.05$)
Mikaeli et al. (2004)	BT + PBD vs PBD	24	BT + PBD (case group) had a significant higher cumulative remission rate compared to control (PBD) group (24.6 vs 12.6 months $P < 0.01$) and a significant reduction in symptom score (76% vs 53% $P < 0.001$). Control group needed a 35 mm PBD vs 30 mm of case group
Dughera et al. (2005)	BT elderly	12	After 12 months of follow-up, up to 70% of patients were considered responders. They underwent 2 BT injection (time 0 and after 1 month). Average age 86 y.o. ASA 3 or 4
Bassotti et al. (2006)	BT elderly	33	Patients underwent 2 BT injections (time 0 and after 1 month). 78% were considered responders after 1 year and 54% after 2 years. No relationship was found between baseline LES pressure and symptom score
Mikaeli et al. (2006)	BT + PBD vs PBD	54	77% of patients of BT + PBD group were in remission after 1 year vs 62% of PBD group and showed a significant reduction in barium volume at the various time intervals post-treatment

(continued)

Table 2 (continued)

Authors	Description	Patients	Results/conclusions
Zhu et al. (2009)	BT vs PBD vs BT + PBD	90	LES pressure and symptom score in group C (BT + PBD) were significantly lower compared with those in group A (BT) or group B (PBD) ($P < 0.05$). At 2 years after treatment, the response rate in group C remained 56.67% vs 35.71% (group B) and 13.79% (group A) ($P < 0.05$)
Kroupa et al. (2010)	BT + PBD vs PBD	91	The mean duration of follow-up was 48 months (12–96 months). 41 of 51 patients were followed up more than 2 years. Effect of therapy lasted in 75% (31/41) of them. The cumulative 5-year remission rate in combined treated patients was higher than in controls but not statistically significant ($P = 0.07$). Injection of BT followed by PD seems to be effective for long-term results, but the combined therapy is not significantly superior to PD alone
Gutschow et al. (2010)	BT vs PBD vs PBD-HM vs HM	41	Patients of BT group ($n = 7$) had the lower mean LES pressure (18.1 mmHg) and higher recurrence rate (71.4%) compared to patients of PBD group ($n = 16$, 34.8 mmHg – 50%), PBD-HM group ($n = 14$, 22.2 mmHg – 35.7%), and HM group ($n = 6$, 36.4 mmHg – 16.7%)
Bakhshipour et al. (2010)	BT + PBD vs PBD	34	Patients of study group already underwent two initial PBD with a low response. They were randomized to receive another PBD or BT injection and PBD by 4 weeks interval. BT + PBD group had higher remission rate at 1, 6, and 12 months compared to PBD group (87.5% vs 67.1%, 87.5% vs 61.1%, 87.5% vs 55.5%, respectively). Difference was not statistically significant
Porter and Gyawali (2011)	BT	36	Response lasted a mean of 12.8 months, and symptom relief for >6 months was seen in 58.3% of patients. Chest pain, younger age, and contraction amplitudes >180 mmHg independently predicted <6 months relief ($P < 0.05$ for each)
Ciulla et al. (2013)	BT	68	36 patients who underwent echo-guided BT injection had complete relief of obstruction compared to 32 patients who underwent blind treatment
Cai et al. (2013)	BT vs SEMS	110	Improvements in global symptom, in dysphagia scores, and in LES pressure were significantly more marked in the SEMS group ($n = 59$) than in the BT group ($n = 51$). Remission rate in the SEMS group was

(continued)

Table 2 (continued)

Authors	Description	Patients	Results/conclusions
			statistically significantly higher than that in the BT group at 12 and 36 months [81.28 vs 64.58 ($P < 0.05$) and 49.1 vs 4.2 ($P < 0.01$)]. No side effects were reported in BT group vs 26 in SEMS group
Jung et al. (2014)	BT vs PBD	37	A significant difference was observed in the mean remission duration between the BT injection ($n = 25$) and PBD ($n = 12$) (13 months vs 29 months). Independent factors predicting long-term remission included treatment type and the difference in the initial LES pressure
Marjoux et al. (2014)	BT	45	22 patients had achalasia, 8 jackhammer esophagus, 7 distal esophageal spasm, 5 esophagogastric junction outflow obstruction, 1 nutcracker esophagus, and 2 unclassified cases. 71% were significantly improved after 2 months, and 57% remained satisfied for more than 6 months. No clear difference was observed in terms of response according to manometric diagnosis. Type 3 achalasia had the worst outcome with none of these patients responding to the endoscopic BT injection

BT botulinum toxin, *HM* Heller myotomy, *LES* lower esophageal sphincter, *NR* not reported, *PBD* pneumatic balloon dilatation, *PD* pneumatic dilatation, *SEMS* self-expanding metal stent

that PD is the more effective endoscopic treatment in the long term (greater than 6 months) for patients with achalasia (Leyden et al. 2014).

BT injections into the upper GIT appear to be quite safe with very few, if any, reports of serious adverse effects. The incidence of gastroesophageal reflux has not been well characterized in most studies but has been reported to be about 20%, by symptoms at least. There has also been some question in recent years whether BT prior to PD or myotomy complicates the more invasive procedures possible secondary to LES fibrosis. However, although previous BT injection (or PD for that matter) may make myotomy more challenging technically because of obliteration of tissue planes, this does not appear to affect the final outcome after myotomy.

Given its favorable safety profile, BT injection is a reasonable option for the short-term treatment of achalasia; it cannot be recommended as a long-term solution for patients who are candidates for more definitive therapies. Thus, this treatment is currently reserved for patients in whom PD or myotomy is precluded by patient-related risk.

HRM (high-resolution manometry) has enabled identification of achalasia subtypes that have important prognostic implications. Pneumatic dilatation is a commonly used and cost-effective method of treating achalasia but has shown poor longevity of symptom relief compared with other modalities and carries a risk of esophageal perforation. LHM (laparoscopic Heller myotomy) is often the preferred, most effective treatment modality; however new studies may show that outcomes are equivalent or even inferior to POEM (peroral endoscopic myotomy). Botulinum toxin injection of the lower esophageal sphincter has a waning and short duration of efficacy and is used primarily for patients unsuitable for more definitive invasive procedures. POEM is considered the most effective treatment for type III achalasia but carries a high risk of iatrogenic gastroesophageal reflux disease that might predispose to the development of Barrett's esophagus (Zaninotto et al. 2019).

1.3 Other Esophageal Disorders

BT has also been used in a variety of less well-characterized esophageal conditions including diffuse esophageal spasm (DES) and patients with non-cardiac chest pain suspected to be on the basis of a dysfunctional esophagus. DES is a condition that is related to achalasia and may be associated with LES dysfunction as well (Marjoux et al. 2013; Burmeister 2013; Achem and Gerson 2013; Sharata et al. 2013; Vaezi 2013; Vanuytsel et al. 2013; Roman and Kahrilas 2013; Spector et al. 2013). In a clinical trial assessing the effect of BT in DES (Storr et al. 2001a, b), each of the nine patients was given 100 Ona-A BT units diluted in 10 mL of saline solution and injected endoscopically at multiple sites along the esophageal wall beginning in the LES region and moving proximally in 1- to 1.5-cm intervals and into endoscopically visible contraction rings. At week 4, eight patients had a significant reduction in symptom score, and four patients required subsequent injections over a 2-year period. A recent study examined 22 patients with DES or nutcracker esophagus who had primarily dysphagia and gave them blinded saline or BT injections in a crossover study design (Vanuytsel et al. 2013). Results showed that symptom scores and weight loss improved after BT treatment, not the saline injections, and this benefit was sustained for over a year in almost half of the patients.

In addition to dysphagia and regurgitation, chest pain can be associated with achalasia, DES, ineffective esophageal motility (IEM), and isolated LES dysfunction which may respond to BT administration as shown in previous studies. A study, with improvement of chest pain as the primary end-point, evaluated 29 patients with non-cardiac chest pain who received 100 Ona-A BT units injection into the LES, same as the treatment regimen for achalasia. Seventy-two percent of the patients responded with at least 50% reduction in chest pain (Miller et al. 2002a).

The response rates of BT injection therapy vary depending on the esophageal motility disorder. Studies have shown that response is transient in achalasia patients, and given the more effective therapies available, it is only recommended in patients who are not surgical candidates. In nonachalasia patients, studies of BT injections have demonstrated improvement in dysphagia symptoms in patients with spastic disorders, though studies are small and largely retrospective. The available literature showed a variable response to BT in esophagogastric junction outlet obstruction (EGJOO) and non-cardiac chest pain patients. Despite advances in diagnosing esophageal motility disorders, there is a need for further research in patient selection for esophageal BT, dose and injection location, and disease-specific outcomes.

Placebo-controlled trials are crucial to evaluate BT efficacy and duration of response. Esophageal-directed BT injections are beneficial in improving dysphagia in spastic motility disorders and in achalasia patients who are elderly or have multiple comorbidities. There is a lack of evidence to support use in patients with EGJOO and non-cardiac chest pain or for young or healthy achalasia patients (Sterling et al. 2018).

2 Gastric Applications

2.1 Gastroparesis

Gastroparesis or delayed gastric emptying resulting in nausea, vomiting, dyspepsia, and abdominal bloating can occur as a result of poorly controlled diabetes mellitus, postsurgical manifestations, or idiopathic causes (Lacy et al. 2004; Friedenberg et al. 2004; Rayner and Horowitz 2005; Bromer et al. 2005). It has been hypothesized that one of the clinical causes of gastroparesis is pylorospasm partially from impaired relaxation and unopposed cholinergic stimulation, thus decreasing pylorospasm may increase gastric emptying. In recent years, BT injection into the pylorus has been investigated as a treatment option in this otherwise debilitating disorder.

The initial study evaluating the BT efficacy in patients with diabetic gastroparesis assessed six patients with abnormal solid phase gastric emptying studies (Ezzeddine et al. 2002). Each patient received 100 BT Ona-A units into the pyloric sphincter, and symptom scores and gastric emptying were assessed after 6 weeks. There was an improvement of subjective symptom scores of 55%, which was maintained at 6 weeks. In addition, there was a 52% improvement in gastric emptying at 6 weeks. Another study investigated the BT use in cases of idiopathic gastroparesis (Miller et al. 2002b). Ten patients were given 80–100 Ona-A BT units, and a 38% reduction in symptom scores were seen at 4 weeks which correlated with findings of increased gastric emptying. A recent study evaluated the effects of BT on diabetic gastroparesis for 12 weeks (Lacy et al. 2004). Eight patients received 200 Ona-A BT units into the pyloric sphincter, and seven patients completed the 12-week follow-up. Mean symptom scores declined from 27 to 12.1 ($P < 0.01$). Furthermore, six of the seven patients gained weight ($P = 0.05$), and gastric emptying scan time improved in four patients (Lacy et al. 2004). The largest study to address this issue retrospectively evaluated 63 patients who met the study criteria (Bromer et al. 2005). Gastroparesis was secondary to diabetes in 26 patients (41.2%), after surgery in two (3.2%), and idiopathic in 35 (55.6%). Twenty-seven of 63 (43%) patients experienced a symptomatic response to treatment (100–200 units Ona-A) with a mean duration of 5 months. Male gender was associated with response to therapy. However, vomiting as a major symptom was predictive of no response to BT (Bromer et al. 2005).

Based on the current available literature, there is conflicting data regarding the efficacy of intrapyloric botulinum injections (IPBIs) for refractory gastroparesis. There have been many open-label trials showing good clinical response, but the only two randomized controlled trials on the matter showed no objective improvement gastric emptying studies. However, both studies were likely underpowered, and changes in gastric emptying may not correlate with symptom improvement. As such, these discouraging findings should not be used to exclude BT from the armamentarium of therapies for refractory GP. More large-scale, double-blinded, multicenter randomized control trials are needed to further validate the long-term efficacy and safety of IPBI, as well as gastric peroral endoscopic myotomy (G-POEM), as compared to gastric electrical stimulation (GES) or surgical intervention (i.e., laparoscopic pyloromyotomy) for refractory gastroparesis (Thomas et al. 2018).

2.2 Obesity

BT injection into the gastric antrum may be used to transiently decrease gastric emptying as a treatment for obesity (Gui et al. 2000; Rollnik et al. 2003; Garcia-Compean et al. 2005; Coskun et al. 2005; Albani et al. 2005). Preliminary data in rats have shown a significant loss of body weight associated with a reduction of dietary intake in the BT-treated group. In a double-blind controlled study, 24 morbidly obese patients [mean body mass index (BMI) 43.6 $\pm$ 1.09 kg/m^2] were blindly randomized to receive 200 Ona-A BT units or placebo into the antrum and fundus of the stomach by intraparietal endoscopic administration (Foschi et al. 2007). The two groups were homogenous for anthropometric characteristics. Eight weeks after the treatment, BT patients had significantly higher weight loss (11 $\pm$ 1.09 kg vs 5.7 $\pm$ 1.1 kg, $P < 0.001$) and BMI reduction (4 $\pm$ 0.36 kg/m^2 vs 2 $\pm$ 0.58 kg/m^2, $P < 0.001$) than controls. No significant side effects or neurophysiologic changes were found. Similar results have been found in an open-label study of ten obese adults (BMI 31–54 kg/m^2) who received 100 units (four patients) or 300 units (six patients) of Ona-A BT and were followed for 16 weeks (Topazian et al. 2008).

Further results demonstrated that BT makes weight loss easier in obese patients (Foschi et al. 2008). It seems conceivable that BT acts by increasing the solid gastric emptying time and reducing the solid eating capacity of the stomach. However, the results in literature are controversial. In several other clinical experiences, intragastric BT injection did not seem to reduce body weight (Garcia-Compean et al. 2005; Cardoso et al. 2006; Mittermair et al. 2007; Topazian et al. 2013; Wiesel et al. 1997; Saliakellis and Fotoulaki 2013; Martin et al. 2009; Bai et al. 2010; Kent et al. 2007; Bagheri et al. 2013; Ballal and Sanford 2000; Shrestha and Pasricha 2001; Mandal and Robinson 2001; Gorelick et al. 2004; Wehrmann et al. 2000; Hackert et al. 2017; Murray 2011; Maria et al. 1999, 2000a, 2001, 2002, 2006; Brisinda et al. 2003a, 2006; Hallan et al. 1988; Joo et al. 1996; Ron et al. 2001; Madalinski et al. 2002; Albanese et al. 1997, 2003; Keshtgar et al. 2007, 2009; Irani et al. 2008; Farid et al. 2009a, b, c; Ahmadi et al. 2013; Zhang et al. 2014; Shafik and El-Sibai 1998; Cadeddu et al. 2005; Emile et al. 2016; Christiansen et al. 2001; Lund and Scholefield 1996; Madoff and Fleshman 2003; Shawki and Costedio 2013;

Lindsey et al. 2004a; Gui et al. 1994; Jost and Schimrigk 1994, 1995; Mason et al. 1996; Jost 1997; Minguez et al. 1999).

2.3 Other Gastropyloric Disorders

BT has been used to facilitate gastric emptying in patients who underwent pylorus-preserving duodenopancreatectomy (Wiesel et al. 1997). Initial studies suggest that BT injection into the pylorus improves both gastric emptying and symptoms.

Infantile hypertrophic pyloric stenosis is a congenital hereditary disorder characterized by a functional gastric outlet obstruction (Saliakellis and Fotoulaki 2013). Obstruction is the result of a gradual hypertrophy of the circular smooth muscle of the pylorus, and the neurons that innervate the circular muscle layer lack NO synthase. Recently lack of response to BT injection has been observed in two patients with pyloric stenosis. Studies have shown that BT injection helps patients suffering from postsurgical pyloric clogging. BT injection is also used as an alternative method for the treatment of gastric emptying disorders (Rayner and Horowitz 2005; Bromer et al. 2005; Ezzeddine et al. 2002). In a recent study, the authors compared the effect of BT injection and pyloroplasty in preventing delayed gastric emptying after esophagectomy for esophageal cancer (Bagheri et al. 2013). In the study 60 patients were included and were randomly divided into two groups. In group A, 30 patients underwent pyloroplasty, and in group B, injection of 200 BT units into the pyloric sphincter muscle was used in 30 patients. Isotope scan 3 weeks after surgery showed that five patients in group A and three in group B had delayed gastric emptying; there was no significant difference between the two groups, and the success rate of BT injection was 90% (Bagheri et al. 2013). BT injection may be used instead of pyloroplasty as a simple, effective, and complication-free method to prevent gastric emptying delay.

3 Duodenal and Biliary Applications

3.1 Sphincter of ODDI Dysfunction (SOD)

SOD is a poorly understood and controversial condition postulated to result in biliary pain, typically in the setting of a previous cholecystectomy. It has also been hypothesized that pancreatic SOD can result in pancreatic-type pain and/or recurrent pancreatitis. The standard of SOD treatment currently is endoscopic sphincterotomy, which is a relatively high-risk procedure that is not uniformly effective. Hence there is interest in the use of a simpler procedure such as BT to serve as a therapeutic trial; patients who respond to this treatment could then go on for more permanent relief using a sphincterotomy (Ballal and Sanford 2000; Shrestha and Pasricha 2001; Mandal and Robinson 2001). This was first suggested in a short report on two patients. Subsequently a larger study was reported evaluating 22 patients who had undergone cholecystectomy and had manometrically confirmed type III SOD

(Gorelick et al. 2004). Six weeks after 100 Ona-A units injected into the sphincter, 12 patients (55%) were symptom-free, but 10 patients (45%) were not. Of the ten patients who did not experience symptomatic benefit from BT injection, five had normal basal sphincter of Oddi pressures (<40 mmHg), and biliary sphincterotomy did not relieve the symptoms of these patients. Two of the remaining five patients with sustained sphincter hypertension after BT injection benefited from biliary sphincterotomy. Of the 12 patients who initially responded to BT injection, 11 patients remained symptom-free for a median duration of 6 months. These patients had recurrence of biliary hypertension and responded to biliary sphincterotomy. The authors concluded that response to BT injection may select a subset of patients who will respond to biliary sphincterotomy. BT has also been used with similar intent, although in an uncontrolled manner in patients with acute recurrent pancreatitis suspected to be due to pancreatic SOD (Wehrmann et al. 2000). Preoperative sphincter of Oddi botulinum toxin injection is a novel and safe approach to decrease the incidence of clinically relevant postoperative pancreatic fistula after distal pancreatectomy. The results of a recent trial suggest its efficacy in the prevention of clinically relevant postoperative pancreatic fistula and are validated currently in the German Federal Government-sponsored, multicenter, randomized controlled PREBOT trial (Hackert et al. 2017).

3.2 Other Biliary Disorders

BT-induced relaxation of the sphincter of Oddi may help to direct appropriate therapy for patients with acalculous biliary pain (Murray 2011). A protocol-based management of 25 patients with acalculous biliary pain who had 100 Ona-A BT units injected into their sphincter of Oddi musculature to relax the sphincter has been audited. Patients whose pain was temporarily relieved after BT injection were offered endoscopic biliary sphincterotomy, and patients who failed to experience benefit after BT injection were assessed for laparoscopic cholecystectomy. A total of 11 patients had a positive response to BT treatment. Of these patients, ten consented to undergo endoscopic biliary sphincterotomy, with relief of biliary pain in all cases. A total of 14 patients had a negative response to BT injection, with 10 of these patients progressing to laparoscopic cholecystectomy, which resulted in biliary pain relief in 8.

4 Pelvic and Anorectal Applications

4.1 Pelvic Floor Dyssynergia

Pelvic floor dyssynergia, also known as anismus, is a common cause of chronic constipation, hallmarked by inappropriate, paradoxical contraction or a failed relaxation of the puborectalis muscle and EAS during defecation (Maria et al. 2002; Brisinda et al. 2003a, 2006). In normal patients, the puborectalis muscle and the

EAS relax to straighten the anorectal angle and open the anal canal. Usually, this alteration in defecation is from maladaptive learning and responds to biofeedback in 60–70% of patients as demonstrated in mostly single group, uncontrolled trials. Surgery has not been shown to be effective and has been largely discouraged as a treatment option. There are a limited number of studies evaluating the BT use in pelvic floor dyssynergia (Table 3).

An initial trial evaluating seven patients with constipation and anismus received BT of unknown dose into the EAS (Hallan et al. 1988). Symptom scores improved significantly correlating with a reduction in the maximum voluntary and anal canal squeeze pressure and a significant increase in the anorectal angle on straining with subsequent fecal incontinence in two patients. In another study with a sample size of four patients with anismus, the dose of Ona-A BT ranged from 6 to 15 units injected into the EAS or puborectalis muscle under electromyography guidance (Maria et al. 2000a). All four patients, who had numerous failed biofeedback sessions, responded to BT with two patients having sustained responses for up to 1 year. A larger study evaluating 15 patients at a dose of 25 Ona-A BT units injected into the EAS showed improvement in 13 patients (87%) for a mean of 4.8 months (Maria et al. 2006). It is unclear whether BT should be injected into the EAS or the puborectalis muscle. Another study evaluated 25 patients who received 10 Ona-A BT units on each side of the puborectalis muscle or 20 Ona-A units in the posterior aspect of the muscle. Manometric relaxation was achieved after the first injection in 18 patients (75%), which endured throughout a 6-month follow-up. Seven of 16 patients who failed the first injection had an additional one. Symptom improvement of 29.2% in straining index was recorded during follow-up with an overall satisfaction rate of 58.3%. Twenty-four consecutive patients with chronic outlet obstruction constipation resulting from puborectalis syndrome were included in a recent study (Maria et al. 2006). The patients were treated with 60 units of Ona-A, injected into two sites on either side of the puborectalis muscle under ultrasonographic guidance. At 2 months, evaluation inspection revealed a symptomatic improvement in 19 patients. Anorectal manometry demonstrated decreased tone during straining from 98 ± 24 to 56 ± 20 mmHg at a 1-month evaluation ($P < 0.01$) and 56 ± 29 mmHg at a 2-month follow-up ($P < 0.01$). Pressure during straining was lower than resting anal pressure at the same time in all patients. Defecography after the treatment showed improvement in anorectal angle during straining, which increased from 98 ± 9° to 121 ± 15° ($P < 0.01$) (Mason et al. 1996). Similar results have been noted in patients with Parkinson's disease (Cadeddu et al. 2005; Albanese et al. 1997).

Recently, in a review of 7 studies including 189 patients, the median dose of Ona-A injected per procedure was 100 IU (range, 20–100 IU). Lateral injection was done in five trails and combined lateral and posterior injections in two trials. Three studies used endorectal ultrasonography-guided technique, one study used EMG-guided technique, whereas the remaining three studies used manual palpation with the index finger. The median percentage of patients who reported initial improvement of symptoms was 77.4% (range 37.5–86.7%), this percentage declined to a median of 46% (range 25–100%) at 4 months after injection of Ona-A. Rates of improvement evaluated by balloon expulsion test, EMG, and defecography ranged

Table 3 Published results of treatment of pelvic floor dyssynergia with BT

Author	Pts	Name of drug/dose (units)	Results	Complication
Hallan et al. (1988)	7	Abo-A – Nr	Maximum voluntary contraction from 70 to 28 cm H_2O. Anorectal angle from 96 to 124°. Symptomatic improvement in four patients	Incontinence in two patients
Joo et al. (1996)	4	Ona-A – 6–15 U	Symptomatic improvement in all treated patients. Two patients relapsed	0
Shafik and El-Sibai (1998)	15	Ona-A – 25 U	Symptomatic improvement in 13 patients, on average 4.8 months after the first treatment	0
Maria et al. (2000a)	4	Ona-A – 30 U	75% were improved at 8 weeks. Anal tone during straining from 96.2 to 42.5 mmHg at 4 weeks and to 63.2 mmHg at 8 weeks. Anorectal angle from 94 to 114°	0
Maria et al. (2001)	14 AR	Ona-A – 30 U	At 2-month evaluation, a symptomatic improvement was found in nine patients. At defecography, the rectocele depth was reduced from 4.3 ± 0.6 cm to 1.8 ± 0.5 ($P < 0.001$), and the rectocele area was reduced from 9.2 ± 1.2 to 2.8 ± 1.6 cm^2 ($P < 0.001$). The anorectal angle measured during straining increased from a mean of 98 ± 15° before treatment to a mean of 121 ± 19 ° ($P = 0.001$). At one-tear evaluation, there was no report of digitally rectal voiding, and rectocele was not found at physical examination	0
Ron et al. (2001)	25	Ona-A – 20 U	Symptomatic improvement in 75% of the patients	Perianal pain in three patients
Madalinski et al. (2002)	39	Ona-A – 25 U Abo-A – 150 U	Nr	Perianal pain in four patients
Albanese et al. (2003)	10 PD	Ona-A – 100 U	Following treatment, anal tone during straining was reduced from 97.4 ± 19.6 mmHg at baseline to 40.7 ± 11.5 mmHg 1 month after treatment ($P = 0.00001$); no further change was observed at 2-month evaluation (38.2 ± 10.4 mmHg; $P = 0.00001$ vs baseline values). The anorectal angle during straining (as measured with defecography) increased from a mean of 90° ± 7.9 before treatment to 122.2° ± 15	0

(continued)

Table 3 (continued)

Author	Pts	Name of drug/dose (units)	Results	Complication
			($P = 0.0004$); nine patients evacuated the barium paste without the need for laxative or enemas	
Cadeddu et al. (2005)	18 PD	Ona-A – 100 U	At 2-month evaluation, inspection revealed a symptomatic improvement in ten patients. Anorectal manometry demonstrated decreased tone during straining from 96.2 ± 17.1 mmHg to 45.9 ± 16.2 mmHg at 1-month evaluation ($P < 0.00001$) and to 56.1 ± 10.7 mmHg at 2 months ($P < 0.00001$). Pressure during straining was lower than resting anal pressure at the same time in all patients. Defecography after the treatment showed improvement in anorectal angle during straining which increased from 99.1° ± 8.4 to 121.7° ± 12.7 at 2 months ($P < 0.00001$)	0
Maria et al. (2006)	24	Ona-A – 60 U	At 2-month evaluation, inspection revealed a symptomatic improvement in 19 patients. Anorectal manometry demonstrated decreased tone during straining from 98 ± 24 mmHg to 56 ± 20 mmHg at 1-month evaluation ($P < 0.01$) and 56 ± 29 mmHg at 2-month follow-up ($P < 0.01$). Defecography after the treatment showed improvement in anorectal angle during straining	0
Keshtgar et al. (2007)	42	Ona-A – 60 U	BT injection ($n = 21$) is equally effective and less invasive than M of IAS ($n = 21$) for chronic idiopathic constipation. At 3 months the median preoperative SS score improved from 34 to 20 in BT group ($P < 0.001$) and from 31 to 18 in the M group ($P < 0.002$). At 12 months the score was 19 and 14.5 in BT and M group, respectively ($P < 0.0001$)	0
Irani et al. (2008)	24	Ona-A – 20 U	Of 24 patients, 22 experienced significant improvement in their constipation lasting greater than 22 weeks. There was a statistically significant improvement from 2.1 to 6.5 bowel movement per week ($P < 0.001$). The benefit of the BTX-A persisted a variable period of time among the	5 fecal soiling

(continued)

Table 3 (continued)

Author	Pts	Name of drug/dose (units)	Results	Complication
			responders, with 12 patient (55%) demonstrating a response lasting 6 months or more	
Farid et al. (2009a)	48	Abo-A – 100 U	In BFB group ($n = 24$) initial improvement was recorded in 12 patients (50%), while long-term success was recorded in 6 patients (25%). In the BT group ($n = 24$), clinical improvement was recorded in 17 patients (70.8%), but the improvement persisted only in 8 patients (33.3%). There is a significant difference between BT group and BFB group regarding the initial success ($P = 0.008$), but this significant difference disappeared at the end of follow-up ($P = 0.23$)	Nr
Farid et al. (2009b)	30	Abo-A – 100 U	BT injection ($n = 15$) achieved initial success in 13 patients (86.7%). Long-term success persisted only in six patients (40%). PDPR ($n = 15$) achieved initial success in all patients (100%) with a long-term success in ten patients (66.6%). However this difference did not produce any significant value. Recurrence was observed in seven patients (53.8%) and five patients (33.4%) following BT injection and PDPR, respectively	0
Keshtgar et al. (2009)	16	Abo-A – 200 U	There were significant improvements in symptoms of constipation, soiling, painful defecation, general health and behavior, and fecal impaction of rectum ($P < 0.05$). Outcome was measured by a validated SS score questionnaire. At 3-month follow-up, the median SS score improved in all children after BT injection from 32.50 to 7.50 ($P < 0.0001$). At 12-month follow-up, the improvement of SS score in BT injection group was significantly more than the control group ($n = 31$) as follows: 4 vs 15, respectively ($P < 0.002$)	0
Farid et al. (2009c)	60	Abo-A – 100 U	The groups differed significantly regarding clinical improvement at 1 month [50% for BFB ($n = 20$), 75% BT injection ($n = 20$), and 95% for	Nr

(continued)

Table 3 (continued)

Author	Pts	Name of drug/dose (units)	Results	Complication
			PDPR ($n = 20$), $P = 0.006$], and differences persisted at 1 year (30% for BFB, 35% BT injection, and 70% for PDPR, $P = 0.02$). BT injection seems to be successful for temporary treatment, but PDPR is found to be effective with lower morbidity in contrast to its higher success rate	
Ahmadi et al. (2013)	88	Abo-A – 160 U	Defecation of painful stool existed in 88% of patients before BT injection, and it was reduced to 15% after BT injection ($P = 0.0001$). Stool was hard in 80% of patients before and was reduced to 28% after BT injection ($P = 0.0001$). Soiling existed in 62% of patients before and was reduced to 8% after BT injection ($P = 0.0001$). Defecation interval was 9.1 days and after BTX-A injection was reduced to 2.6 days ($P = 0.0001$)	Nr
Zhang et al. (2014)	31	Inco-A – 100 U	After treatment, the pressure of the anal canal during rest and defecation was significantly reduced from (93 ± 16.5) mmHg and (105 ± 28.3) mmHg to (63 ± 8.6.3) mmHg and (42 ± 8.9) mmHg, respectively. BT injection combined with pelvic floor biofeedback training achieved success in 24 patients with 23 maintaining persistent satisfaction during a mean period of 8.4 months	8 fecal incontinence

AR anterior rectocele, *BFB* biofeedback training, *BT* botulinum toxin, *M* myectomy, *Nr* non-reported, *PD* Parkinson's disease, *PDPR* partial division of puborectalis, *SS score* symptom severity score

between 37.5–80%, 54–86.7%, and 25–86.6%, respectively. Fourteen (7.4%) patients developed complications after injection of Ona-A. Complication rates across the studies ranged from 0 to 22.6%. Initial satisfactory improvement of symptoms after Ona-A injection remarkably deteriorated after 3 months of the procedure. However, repeated injection may provide better sustained results with no additional morbidities. Further analysis of more patients is necessary to conclude the safety of Ona-A for the treatment of anismus (Emile et al. 2016).

Rectoceles are commonly associated with outlet obstruction, such as pelvic floor dyssynergia. Therefore, decreasing anal sphincter tone during strain may decrease the size of the rectocele and improve symptoms of constipation. In a study of 14 patients with anterior rectocele, each patient received 30 Ona-A BT units into

3 sites, 2 on either side of the puborectalis muscle and the 1 in the anterior portion of the external anal sphincter, under ultrasonographic guidance (Maria et al. 2001). At 2 months, 9 of 14 patients had symptomatic improvement with a decrease in rectocele depth and area and decrease in tone during straining. At 1 year, no patient experienced incomplete or required digitally assisted rectal voiding.

Many questions still remain such as the dose of BT in the treatment of pelvic floor dyssynergia, location of injection, use of ultrasound or electromyography, number of treatments, and combination with biofeedback. These questions need further study using placebo-controlled trials and larger sample sizes.

4.2 Chronic Idiopathic Anal Pain

Chronic idiopathic anal pain is part of a rather ill-defined group of disorders termed chronic idiopathic perineal pain, which also includes proctalgia fugax and coccygodynia (Christiansen et al. 2001). The main feature of these syndromes is that no objective abnormalities are found on clinical examination, and the distinction between the different groups of perineal pain is based solely on the patient's description of the pain and location of tenderness by palpation. In the majority of patients, the pain is present constantly, usually intense, sometimes burning, often with some irradiation; it was usually aggravated by sitting, whereas defecation had no constant effect and is relieved by lying down. The pathogenesis of the syndromes is unknown. There is no satisfactory treatment for chronic anal pain; nonetheless, anal stretch and lateral internal sphincterotomy (LIS) are still used in some patients on the assumption that the pain might be caused by a hypertonic IAS, because no objective changes can be demonstrated. Eighteen patients who met the criteria for chronic idiopathic anal pain were studied. Treatment consisted of analgesics only in four patients, 0.2% nitroglycerin ointment in four, and ultrasound BT injection into the intersphincteric space in nine. Four patients were managed satisfactorily on analgesic treatment under the guidance of the hospital's pain clinic. Nitroglycerin ointment resulted in temporary pain relief in one of four patients. BT injection resulted in a permanent improvement in four patients, a temporary improvement in one patient, and no effect in four patients. Two patients had a colostomy, resulting in complete pain relief (Christiansen et al. 2001). As in other syndromes based on muscular dystonia, some patients may benefit from BT injection.

4.3 Anal Fissure

Anal fissures are tears in the anoderm that start at the anal verge and can extend to the dentate line (Lund and Scholefield 1996; Madoff and Fleshman 2003; Shawki and Costedio 2013). They can manifest into painful defecation and rectal bleeding. These fissures, which most commonly arise in the mid-posterior position of the anus, are thought to occur secondary to ischemia as a result of increased anal sphincter pressures and decreased blood flow (Lindsey et al. 2004a). Once chronic fissures develop, treatment options are aimed at interrupting this cycle by reducing sphincter tone using topical nitroglycerin, BT injection, oral nifedipine, or LIS performed surgically (Lindsey et al. 2004a). There are many reports on the efficacy of BT for this condition (Table 4).

These studies include several controlled trials comparing the toxin to either placebo or other modalities (Gandomkar et al. 2015; Maria et al. 1998a, b). Clinical benefit is seen in the vast majority of patients, typically accompanied by reduction in resting anal sphincter pressure (Brisinda et al. 1999; Maria et al. 2000b).

The exact site and dose of injection remain somewhat unsettled. Most of the trials to this point have evaluated BT administration at the point of the fissure, primarily, the posterior midline area of the anal verge. However, there is evidence that IAS fibrosis exists at the base of the fissure and is more prominent in this zone than other sites in the smooth muscle. This fibrosis may decrease the effects of BT on sphincter relaxation, thus delaying fissure healing. A study to evaluate this theory was conducted on 50 patients with posterior anal fissures who were either given 20 Ona-A BT units lateral to the posterior fissure or 20 Ona-A BT units on each side of the anterior midline (Brisinda et al. 1999). After 2 months, a healing scar was observed in 15 patients (60%) of the posterior midline group and in 22 patients (88%) of the anterior midline group ($P = 0.025$). Resting anal pressure was significantly different from the baseline values at 1 and 2 months in both groups, but the values were significantly lower in patients of the anterior midline group.

Another study evaluated 150 patients with posterior anal fissures who were treated with BT injected in the IAS on each side of the anterior midline. Patients were randomized to receive either 20 Ona-A BT units and, if the fissure persisted, were retreated with 30 units or 30 units and retreated with 50 units, if the fissure persisted (Maria et al. 2000b). One month after the injection, examinations revealed complete healing in 55 patients (73%) in the group receiving the lower dose and 65 patients (87%) in the group receiving the higher dose ($P = 0.04$). Five patients from the second group reported a mild incontinence of flatus that lasted 2 weeks after the treatment and disappeared spontaneously. The values of the resting anal pressure ($P = 0.3$) and the maximum voluntary pressure ($P = 0.2$) did not differ between the two groups. However, after 2 months, healing rates were similar between the two groups (89% and 96%). The authors concluded that the higher dose was more effective, but the improved effectiveness was not seen at 2 months (Maria et al. 2000b).

The gold standard for treatment for anal fissures is surgery, primarily LIS. However, surgical intervention is associated with a low complication rate resulting in fecal incontinence, hematoma, and wound infection. A study compared BT injection (20–30 Ona-A units) and LIS (Brisinda et al. 2002). Overall healing rates were similar in both groups at 6 months with 10 of 61 patients requiring a second BT injection at 2 months. However, the response rate was higher at 1 and 2 months in the sphincterotomy group, 82% (41/50) at day 28 and 98% (49/50) at the second month ($P = 0.023$ and $P < 0.0001$, respectively, compared with the BT group). The response to BT was not as durable as surgery at 12 months falling to a success rate 75.4% (46/61) with seven recurrences in the BT group, whereas it remained

Table 4 Comparison of published results on the treatment of patients with chronic anal fissure

Author	Cases (n)	Units/ injection's site	Healing rate (%)		Reinjection (%)/ dose	Complete healing rate (%)	Temporary incontinence (%)	Recurrence (%)
			1 m	2 m				
Gui et al. (1994)	10	15 B/IAS	60	70	40/20 B	90	10	10
Jost and Schimrigk (1994)	12	5 B/EAS	Nr	83.3	–	83.3	0	8.3
Jost and Schimrigk (1995)	54	5 B/EAS	Nr	78	–	78	6	6
Mason et al. (1996)	5	NR D/IAS	Nr	60	–	60	0	NR
Jost (1997)	100	2.5–5 B/EAS	Nr	82	–	82	7	8
Maria et al. (1998a)	15 15	20 B/IAS Saline	53.3 13.3	73.3 13.3	26.6/25 B	100	4	6.7
Maria et al. (1998b)	23 34	15 B/IAS 20 B/IAS	21.7 50	43.5 67.6	8.7/20 B 20.6/25 B	100 100	0	0
Minguez et al. (1999)	23 27 19	10 B/IAS 15 B/IAS 21 B/IAS	48 74 100	Nr	52 30 37	83 78 90	0	37–52
Jost and Schrank (1999)	25 25	20 D/EAS 40 D/EAS	Nr	76 80	–	76 80	4 12	4 8
Brisinda et al. (1999)	25 25	20 B/IAS 0.2% GTN	88 40	96 60	–	96 60	0	0
Fernandez Lopez et al. (1999)	76	40 B/IAS	56	67	45.2/40 B	67	3	0
Madalinski et al. (1999)	13	20 B/EAS	84.6	Nr	–	–	NR	15.4
Maria et al. (2000b)	25 25	20 B/IAS PI 20 B/IAS AI	48 88	60 88	24/25 B 12/25 B	80 100	0	0
Lysy et al. (2001)	15 15	20 B + ID/IAS 20 B/IAS	66 20	73 60	–	73 60	0	0

(continued)

Table 4 (continued)

Author	Cases (*n*)	Units/ injection's site	Healing rate (%) 1 m	Healing rate (%) 2 m	Reinjection (%)/ dose	Complete healing rate (%)	Temporary incontinence (%)	Recurrence (%)
Madalinski et al. (2001)	14	25–50 B/EAS	Nr	54	–	54	0	8
Tilney et al. (2001)	10	Nr D/IAS	Nr	Nr			NR	NR
Jost (2001)	10	200 NB/EAS	Nr	70	Nr	0	NR	NR
Brisinda et al. (2002)	75 75	20 B/IAS 30 B/IAS	73 87	89 96	10.7/30 B 4/50 B	100 100	0 3	0 4
Brisinda et al. (2003b)	6	150 D/IAS	100	Nr	–	100	0	0
Mentes et al. (2003)	61 50	20–30 B/IAS LIS	62.3 82	73.8 98	–	86.9 98	0 16	11.4 0
Siproudhis et al. (2003)	22 22	100 D/IAS Saline	50 45	32 32	Nr	NR	NR	NR
Brisinda et al. (2004a)	50 50	50 B/IAS 150 D/IAS	82 84	92 94	– 6/150 D	92 94	22 16	0
Giral et al. (2004)	10 11	20 B/IAS LIS	Nr	70 82	–	70 82	0	0
Simms et al. (2004)	47	30 B/IAS	Nr	Nr	17/Nr	78.7	0	27
Lindsey et al. (2004b)	30	25 B/IAS + FIS	Nr	Nr	–	93	7	0
Arroyo et al. (2005a)	40 40	25 B/IAS LIS	Nr	85 97.5	–	45 92.5	5 7.5	55 7.5
Arroyo et al. (2005b)	100	25 B/IAS	–	88	–	47	6	53
De Nardi et al. (2006)	15 15	20 B/IAS 0.2% GTN	33.3 13.3	53.3 66.7	–	33.3 40	0	33 33

Brisinda et al. (2007a)	50 50	30B/90D/IAS 0.2% GTN	82 58	92 70	–	92 46	0 0	0 34
Scholz et al. (2007)	40	10 B/IAS + FIS	95	Nr	5/Nr	79	2.5	10
Witte and Klaase (2007)	100	40–60 D/IAS	Nr	Nr	22/40–100 D	66	1	14
Festen et al. (2009)	37 36	20B/IAS + Poin 1%ISDN+Pinj	Nr	18.9 44.4	21.6/20 B	37.8 58.3	17.8	13.5 25
Nasr et al. (2010)	40 40	20 B/IAS LIS	55 80	62.5 90	–	62.5 90	0 10	40 12.5
Samim et al. (2012)	60 74	20 B/IAS 2% Dz	25 14	43 43	–	32 26	5.5	11.7 17.6
Valizadeh et al. (2012)	25 25	50 B/IAS LIS	28 40	44 88	Nr	48 92	12 48	50 8
Berkel et al. (2014)	27 33	60 D/IAS 1% ID	Nr	66.6 33.3	3.7/Nr	66.6 33.3	18.5 12	28 50
Halahakoon and Pitt (2014)	30	40 B/IAS + AF	86.7	Nr	–	60	3.3	NR
Farouk (2014)	141	100 B/IAS + FIS	Nr	Nr	14/Nr	76	8	18
Gandomkar et al. (2015)	49 50	150D/ IAS + 2%Dz LIS	46.9 74	67.3 92	–	65.3 94	2 7	10.2 0

AI injection in anterior midline, *AF* advancement flap, *B* Ona-A (Botox), *D* Abo-A (Dysport), *DZ* diltiazem, *EAS* external anal sphincter, *FIS* fissurectomy, *GTN* glyceryl trinitrate, *IAS* internal anal sphincter, *ID* isosorbide dinitrate, *LIS* lateral internal sphincterotomy, *NB* neuroblock (trade name of the type b preparation manufactured by Elan Pharma International Ltd, Ireland), *NR* not reported, *PI* injection in posterior midline, *PINJ* placebo injection, *POIN* placebo ointment stable in the LIS group (94%, $P = 0.008$). Sphincterotomy was associated with a significantly higher complication rate, eight cases of anal incontinence versus none in the BT group ($P < 0.001$) (Brisinda et al. 2002). Thus, it appears that surgery is still the more durable treatment option but associated with more complications. These results have been supported in a more recent study. Some investigators have recommended surgery in younger patients and those with high resting anal pressures, as this is a risk factor for recurrence. Older patients may benefit from BT injection as they may be at higher risk of fecal incontinence.

A recent meta-analysis showed that even though LIS is associated with a better healing rate and recurrence rate, BT treatment is superior to LIS in overall complication rates and incontinence rates (Mentes et al. 2003). Thus, some advantages BT offers to patients with anal fissure include a good tolerance of the procedure, an outpatient setting, and a low risk of incontinence. The results of the meta-analysis are in line with previous research (Chen et al. 2014). Furthermore, in a recent study, BT injection was used not only as a therapeutic tool but also as a diagnostic test to identify patients who would not be suitable for further surgical LIS if they developed temporary incontinence after BT injection (Sajid et al. 2008). Combination therapy such as nitroglycerine and BT has also been evaluated; it appears that this only results in a modest increase in the rate of healing (Brisinda et al. 2008; Asim et al. 2014).

BT injection is efficacious in the treatment of chronic anal fissures. With greater than 60% response rates noted at 2 months with further response to re-treatment, BT can be considered a viable treatment option when more conservative treatment fails. In elderly patients, in whom rates of fecal incontinence after surgery may be increased, BT can be considered first-line treatment. Surgery is still the most durable treatment option, but the risks of fecal incontinence must be weighed carefully against the benefits of the procedure.

Thus, according to many authors, we recommend a safety-first approach and treat all patients medically in the first instance. We believe that specific indications for surgical intervention in patients with anal fissure include persistence/recurrence and noncompliance or intolerance to the medical treatment. Patients at higher incontinence risk can be evaluated by anorectal manometric and endoanal sonography test, or, at best, the patient should be offered a sphincter-sparing procedure. The need for further investigations imposes a cost increase. Furthermore, it is difficult to calculate the increased cost in the event of complications. Some of these patients may wish to avoid LIS and persist with an alternative medical therapy.

The recommendations are that simple and readily available therapy associated with fewer complications and requiring no hospitalization should be offered as first line of care. Rational thinking suggests conservative measures as the first-line therapy given that they are simple and have good safety records. Local application of NO donors is readily available, and many reports support these agents as the starting point in the management of these patients. Nevertheless, drawbacks of these drugs are headaches, orthostatic hypotension, and tachyphylaxis, which usually limit their benefits and call for second-line therapy, such as BT. BT injection has an excellent healing rate, can be repeated if necessary, and obviates the patients' compliance. BT potential side effects should be kept in mind, however, including patient aversion to injection.

Recently, Mishra et al. concluded that both treatments (NO donors and BT) may be considered as first-line treatment even if less effective than surgery (Tranqui et al. 2006). However, this view has been challenged by other observations based on smaller series, providing inferior evidence of efficacy. The results of some studies are so disappointing that it led Nelson and coworkers to conclude a Cochrane review stating that "...medical therapy for chronic anal fissure... may be applied with a chance of cure that is only marginally better than placebo..." (Mishra et al. 2005). We think that such conclusion is too pessimistic and welcome further multicenter trials with appropriate methodology (intention-to-treat based selection of patients, doses, and injection technique) and adequate follow-up, to ascertain the safety and efficacy of the therapy. Moreover, the addition of multiple treatment modalities prolonged time to healing from initial evaluation but allowed up to 75% of patients to avoid the need for permanent sphincter division while maintaining the highest rate of healing.

The introduction of these therapies has made the treatment of anal fissure easier, in the outpatient setting, at a lower cost, and without permanent complications. Any conservative treatment used has lower costs than surgery (Nelson et al. 2012). Considering the three hypothetical scenarios reported in a recent paper, we found that the BT approach is more cost-effective than the ointment approach. In addition to cost reduction (on average 62% lower than the association NO donors plus surgery and on average 50% lower than the association CCA plus surgery), BT reduces the number of patients who need further surgery.

4.4 Other Anorectal Conditions

BT into the IAS has been applied both diagnostically and therapeutically after pull-through surgery for Hirschsprung's disease in which it is postulated that IAS spasm can result in persistent obstructive symptoms. Minkes and Langer prospectively evaluated 18 such children who underwent BT injection (total dose 15–60 Ona-A units) into 4 quadrants of the sphincter (Brisinda et al. 2014). Twelve patients (67%) improved for at least 1 month; improvement was sustained beyond 6 months in five patients. These investigators advocated BT, not only as an alternative to myectomy in such cases but also as a diagnostic trial, with persistent symptoms after injection, despite a decrease in sphincter pressure, suggesting another etiology for the constipation.

A total of 33 children with surgically treated Hirschsprung's disease treated with intrasphincteric BT injection for obstructive symptoms were analyzed in a recent study (Minkes and Langer 2000). The median time of follow-up was 7.3 years. A median of two injections was given. Initial improvement was achieved in 76%, with a median duration of 4.1 months. Proportion of children hospitalized for enterocolitis decreased after treatment from 19 to 7. A good long-term response was found in 49%. Basson and coworkers have studied 43 patients with idiopathic constipation,

Hirschsprung's disease, anorectal malformation, and GIT dysmotility (Han-Geurts et al. 2014). A total dose of 200 Ona-A BT units has been injected. Successful outcomes occurred in 72% patients after the first BT treatment, and 25% required further surgical management of their symptoms.

Pain after hemorrhoidectomy appears to be multifactorial and dependent on individual pain tolerance, mode of anesthesia, postoperative analgesia, and surgical technique. IAS spasm is believed to play an important role (Basson et al. 2014). The BT role in reducing pain after hemorrhoidectomy has been assessed in a double-blind study on 50 consecutive patients undergoing Morgan hemorrhoidectomy and assigned to an IAS injection of 0.4 mL of solution containing either 20 Ona-A BT units or normal saline (Patti et al. 2006). Those patients who had BT had significantly less pain toward the end of the 1st week after surgery. Reduction in IAS spasm is the presumed mechanism of action.

Disclosure The authors did not receive any financial support or commercial sponsorship. All authors were involved in drafting the manuscript and revising it critically for important intellectual content and have given final approval of the version to be published. Furthermore, all authors have participated sufficiently in the work to take public responsibility for its content.

References

Achem SR, Gerson LB (2013) Distal esophageal spasm: an update. Curr Gastroenterol Rep 15:325

Ahmadi J, Azary S, Ashjaei B, Paragomi P, Khalifeh-Soltani A (2013) Intrasphincteric botulinum toxin injection in treatment of chronic idiopathic constipation in children. Iran J Pediatr 23:574–578

Akaike N, Shin MC, Wakita M, Torii Y, Harakawa T, Ginnaga A, Kato K, Kaji R, Kozaki S (2013) Transsynaptic inhibition of spinal transmission by A2 botulinum toxin. J Physiol 591:1031–1043

Albanese A, Maria G, Bentivoglio AR, Brisinda G, Cassetta E, Tonali P (1997) Severe constipation in Parkinson's disease relieved by botulinum toxin. Mov Disord 12:764–766

Albanese A, Brisinda G, Mathias CJ (2000) The autonomic nervous system and gastrointestinal disorders. In: Appenzeller O, Vinken PJ, Bruyn GW (eds) Handbook of clinical neurology, vol 75. Elsevier, Amsterdam, pp 613–663

Albanese A, Brisinda G, Bentivoglio AR, Maria G (2003) Treatment of outlet obstruction constipation in Parkinson's disease with botulinum neurotoxin A. Am J Gastroenterol 98:1439–1440

Albani G, Petroni ML, Mauro A, Liuzzi A, Lezzi G, Verti B, Marzullo P, Cattani L (2005) Safety and efficacy of therapy with botulinum toxin in obesity: a pilot study. J Gastroenterol 40:833–835

Alberty J, Oelerich M, Ludwig K, Hartmann S, Stoll W (2000) Efficacy of botulinum toxin A for treatment of upper esophageal sphincter dysfunction. Laryngoscope 110:1151–1156

Alfonsi E, Merlo IM, Ponzio M, Montomoli C, Tassorelli C, Biancardi C, Lozza A, Martignoni E (2010) An electrophysiological approach to the diagnosis of neurogenic dysphagia: implications for botulinum toxin treatment. J Neurol Neurosurg Psychiatry 81:54–60

Alfonsi E et al (2017) Botulinum toxin is effective in the management of neurogenic dysphagia. Clinical-electrophysiological findings and tips on safety in different neurological disorders. Front Pharmacol 8:80

Allen J, White CJ, Leonard R, Belafsky PC (2010) Effect of cricopharyngeus muscle surgery on the pharynx. Laryngoscope 120:1498–1503

Allescher HD, Storr M, Seige M, Gonzales-Donoso R, Ott R, Born P, Frimberger E, Weigert N, Stier A, Kurjak M et al (2001) Treatment of achalasia: Botulinum toxin injection vs. pneumatic balloon dilation. A prospective study with long-term follow-up. Endoscopy 33:1007–1017

Annese V, Basciani M, Perri F, Lombardi G, Frusciante V, Simone P, Andriulli A, Vantrappen G (1996) Controlled trial of botulinum toxin injection versus placebo and pneumatic dilation in achalasia. Gastroenterology 111:1418–1424

Annese V, Bassotti G, Coccia G, D'Onofrio V, Gatto G, Repici A, Andriulli A (1999) Comparison of two different formulations of botulinum toxin A for the treatment of oesophageal achalasia. The Gismad Achalasia Study Group. Aliment Pharmacol Ther 13:1347–1350

Annese V, Bassotti G, Coccia G, Dinelli M, D'Onofrio V, Gatto G, Leandro G, Repici A, Testoni PA, Andriulli A (2000) A multicentre randomised study of intrasphincteric botulinum toxin in patients with oesophageal achalasia. Gismad achalasia study group. Gut 46:597–600

Arroyo A, Perez F, Serrano P, Candela F, Lacueva J, Calpena R (2005a) Surgical versus chemical (botulinum toxin) sphincterotomy for chronic anal fissure: long-term results of a prospective randomized clinical and manometric study. Am J Surg 189:429–434

Arroyo A, Perez F, Serrano P, Candela F, Calpena R (2005b) Long-term results of botulinum toxin for the treatment of chronic anal fissure: prospective clinical and manometric study. Int J Color Dis 20:267–271

Asim M, Lowrie N, Stewart J, Lolohea S, Van Dalen R (2014) Botulinum toxin versus botulinum toxin with low-dose glyceryltrinitrate for healing of chronic anal fissure: a prospective, randomised trial. N Z Med J 127:80–86

Atkinson SI, Rees J (1997) Botulinum toxin for cricopharyngeal dysphagia: case reports of ct-guided injection. J Otolaryngol 26:273–276

Bagheri R, Fattahi SH, Haghi SZ, Aryana K, Aryanniya A, Akhlaghi S, Riyabi FN, Sheibani S (2013) Botulinum toxin for prevention of delayed gastric emptying after esophagectomy. Asian Cardiovasc Thorac Ann 21:689–692

Bai Y, Xu MJ, Yang X, Xu C, Gao J, Zou DW, Li ZS (2010) A systematic review on intrapyloric botulinum toxin injection for gastroparesis. Digestion 81:27–34

Bakhshipour A, Rabbani R, Shirani S, Soleimani HA, Mikaeli J (2010) Comparison of pneumatic dilation with pneumatic dilation plus botulinum toxin for treatment of achalasia. Acta Med Iran 48:107–110

Ballal MA, Sanford PA (2000) Physiology of the sphincter of Oddi--the present and the future?--Part 1. Saudi J Gastroenterol 6:129–146

Bansal R, Nostrant TT, Scheiman JM, Koshy S, Barnett JL, Elta GH, Chey WD (2003) Intrasphincteric botulinum toxin versus pneumatic balloon dilation for treatment of primary achalasia. J Clin Gastroenterol 36:209–214

Basson S, Charlesworth P, Healy C, Phelps S, Cleeve S (2014) Botulinum toxin use in paediatric colorectal surgery. Pediatr Surg Int 30:833–838

Bassotti G, D'Onofrio V, Battaglia E, Fiorella S, Dughera L, Iaquinto G, Mazzocchi A, Morelli A, Annese V (2006) Treatment with botulinum toxin of octo-nonagerians with oesophageal achalasia: a two-year follow-up study. Aliment Pharmacol Ther 23:1615–1619

Berkel AE, Rosman C, Koop R, van Duijvendijk P, van der Palen J, Klaase JM (2014) Isosorbide dinitrate ointment vs botulinum toxin A (dysport) as the primary treatment for chronic anal fissure: a randomized multicentre study. Colorectal Dis 16:O360–O366

Blitzer A, Brin MF (1997) Use of botulinum toxin for diagnosis and management of cricopharyngeal achalasia. Otolaryngol Head Neck Surg 116:328–330

Brant CQ, Siqueira ES, Ferrari AP Jr (1999a) Botulinum toxin for oropharyngeal dysphagia: case report of flexible endoscope-guided injection. Dis Esophagus 12:68–73

Brant CQ, Nakao F, Ardengh JC, Nasi A, Ferrari AP Jr (1999b) Echoendoscopic evaluation of botulinum toxin intrasphincteric injections in Chagas' disease achalasia. Dis Esophagus 12:37–40

Brant C, Moraes-Filho JP, Siqueira E, Nasi A, Libera E, Morais M, Rohr M, Macedo EP, Alonso G, Ferrari AP (2003) Intrasphincteric botulinum toxin injection in the treatment of chagasic achalasia. Dis Esophagus 16:33–38

Brisinda G, Maria G, Bentivoglio AR, Cassetta E, Gui D, Albanese A (1999) A comparison of injections of botulinum toxin and topical nitroglycerin ointment for the treatment of chronic anal fissure. N Engl J Med 341:65–69

Brisinda G, Maria G, Sganga G, Bentivoglio AR, Albanese A, Castagneto M (2002) Effectiveness of higher doses of botulinum toxin to induce healing in patients with chronic anal fissures. Surgery 131:179–184

Brisinda G, Civello IM, Albanese A, Maria G (2003a) Gastrointestinal smooth muscles and sphincters spasms: treatment with botulinum neurotoxin. Curr Med Chem 10:603–623

Brisinda D, Maria G, Fenici R, Civello IM, Brisinda G (2003b) Safety of botulinum neurotoxin treatment in patients with chronic anal fissure. Dis Colon Rectum 46:419–420

Brisinda G, Albanese A, Cadeddu F, Bentivoglio AR, Mabisombi A, Marniga G, Maria G (2004a) Botulinum neurotoxin to treat chronic anal fissure: results of a randomized "Botox vs. Dysport" controlled trial. Aliment Pharmacol Ther 19:695–701

Brisinda G, Cadeddu F, Brandara F, Brisinda D, Maria G (2004b) Treating chronic anal fissure with botulinum neurotoxin. Nat Clin Pract Gastroenterol Hepatol 1:82–89

Brisinda G, Maria G, Bentivoglio AR, Cadeddu F, Marniga G, Brandara F, Albanese A (2006) Management of bladder, prostatic and pelvic floor disorders. Neurotox Res 9:161–172

Brisinda G, Cadeddu F, Brandara F, Marniga G, Maria G (2007a) Randomized clinical trial comparing botulinum toxin injections with 0.2 per cent nitroglycerin ointment for chronic anal fissure. Br J Surg 94:162–167

Brisinda G, Cadeddu F, Mazzeo P, Maria G (2007b) Botulinum toxin A for the treatment of chronic anal fissure. Expert Rev Gastroenterol Hepatol 1:219–228

Brisinda G, Cadeddu F, Brandara F, Marniga G, Vanella S, Nigro C, Maria G (2008) Botulinum toxin for recurrent anal fissure following lateral internal sphincterotomy. Br J Surg 95:774–778

Brisinda G, Bianco G, Silvestrini N, Maria G (2014) Cost considerations in the treatment of anal fissures. Expert Rev Pharmacoecon Outcomes Res 14:511–525

Bromer MQ, Friedenberg F, Miller LS, Fisher RS, Swartz K, Parkman HP (2005) Endoscopic pyloric injection of botulinum toxin A for the treatment of refractory gastroparesis. Gastrointest Endosc 61:833–839

Burmeister S (2013) Review of current diagnosis and management of diffuse esophageal spasm, nutcracker esophagus/spastic nutcracker and hypertensive lower esophageal sphincter. Curr Opin Otolaryngol Head Neck Surg 21:543–547

Cadeddu F, Bentivoglio AR, Brandara F, Marniga G, Brisinda G, Maria G (2005) Outlet type constipation in Parkinson's disease: results of botulinum toxin treatment. Aliment Pharmacol Ther 22:997–1003

Cai XB, Dai YM, Wan XJ, Zeng Y, Liu F, Wang D, Zhou H (2013) Comparison between botulinum injection and removable covered self-expanding metal stents for the treatment of achalasia. Dig Dis Sci 58:1960–1966

Cardoso ES, Rodrigues BM, Barroso M, Menezes CJ, Lucena RS, Nora DB, Melo A (2006) Botulinum toxin type a for the treatment of the spastic equinus foot in cerebral palsy. Pediatr Neurol 34:106–109

Chen HL, Woo XB, Wang HS, Lin YJ, Luo HX, Chen YH, Chen CQ, Peng JS (2014) Botulinum toxin injection versus lateral internal sphincterotomy for chronic anal fissure: a meta-analysis of randomized control trials. Tech Coloproctol 18:693–698

Chiu MJ, Chang YC, Hsiao TY (2004) Prolonged effect of botulinum toxin injection in the treatment of cricopharyngeal dysphagia: case report and literature review. Dysphagia 19:52–57

Christiansen J, Bruun E, Skjoldbye B, Hagen K (2001) Chronic idiopathic anal pain: analysis of ultrasonography, pathology, and treatment. Dis Colon Rectum 44:661–665

Ciulla A, Cremona F, Genova G, Maiorana AM (2013) Echo-guided injection of botulinum toxin versus blind endoscopic injection in patients with achalasia: final report. Minerva Gastroenterol Dietol 59:237–240

Civelek M, Ainslie K, Garanich JS, Tarbell JM (1985) Smooth muscle cells contract in response to fluid flow via a Ca2+-independent signaling mechanism. J Appl Physiol 2002(93):1907–1917

Coskun H, Duran Y, Dilege E, Mihmanli M, Seymen H, Demirkol MO (2005) Effect on gastric emptying and weight reduction of botulinum toxin-A injection into the gastric antral layer: an experimental study in the obese rat model. Obes Surg 15:1137–1143

Cuilliere C, Ducrotte P, Zerbib F, Metman EH, de Looze D, Guillemot F, Hudziak H, Lamouliatte H, Grimaud JC, Ropert A et al (1997) Achalasia: outcome of patients treated with intrasphincteric injection of botulinum toxin. Gut 41:87–92

D'Onofrio V, Miletto P, Leandro G, Iaquinto G (2002) Long-term follow-up of achalasia patients treated with botulinum toxin. Dig Liver Dis 34:105–110

De Nardi P, Ortolano E, Radaelli G, Staudacher C (2006) Comparison of glyceryl trinitrate and botulinum toxin-A for the treatment of chronic anal fissure: long-term results. Dis Colon Rectum 49:427–432

Drendel M, Carmel E, Kerimis P, Wolf M, Finkelstein Y (2013) Cricopharyngeal achalasia in children: surgical and medical treatment. Isr Med Assoc J 15:430–433

Dughera L, Battaglia E, Maggio D, Cassolino P, Mioli PR, Morelli A, Emanuelli G, Bassotti G (2005) Botulinum toxin treatment of oesophageal achalasia in the old old and oldest old: a 1-year follow-up study. Drugs Aging 22:779–783

Emile SH et al (2016) Efficacy and safety of botulinum toxin in treatment of anismus: a systematic review. World J Gastrointest Pharmacol Ther 7(3):453–462

Ezzeddine D, Jit R, Katz N, Gopalswamy N, Bhutani MS (2002) Pyloric injection of botulinum toxin for treatment of diabetic gastroparesis. Gastrointest Endosc 55:920–923

Farid M, El Monem HA, Omar W, El Nakeeb A, Fikry A, Youssef T, Yousef M, Ghazy H, Fouda E, El Metwally T et al (2009a) Comparative study between biofeedback retraining and botulinum neurotoxin in the treatment of anismus patients. Int J Colorectal Dis 24:115–120

Farid M, Youssef T, Mahdy T, Omar W, Moneim HA, El Nakeeb A, Youssef M (2009b) Comparative study between botulinum toxin injection and partial division of puborectalis for treating anismus. Int J Colorectal Dis 24:327–334

Farid M, El Nakeeb A, Youssef M, Omar W, Fouda E, Youssef T, Thabet W, Elmoneum HA, Khafagy W (2009c) Idiopathic hypertensive anal canal: a place of internal sphincterotomy. J Gastrointest Surg 13:1607–1613

Farouk R (2014) Sphincter-preserving therapy for treating a chronic anal fissure: long-term outcomes. Ann Coloproctol 30:132–134

Fernandez Lopez F, Conde Freire R, Rios Rios A, Garcia Iglesias J, Cainzos Fernandez M, Potel Lesquereux J (1999) Botulinum toxin for the treatment of anal fissure. Dig Surg 16:515–518

Festen S, Gisbertz SS, van Schaagen F, Gerhards MF (2009) Blinded randomized clinical trial of botulinum toxin versus isosorbide dinitrate ointment for treatment of anal fissure. Br J Surg 96:1393–1399

Fiorini A, Corti RE, Valero JL, Bai JC, Boerr L (1996) Botulinum toxin is effective in the short-term treatment of esophageal achalasia. Preliminary results of a randomized trial. Acta Gastroenterol Latinoam 26:155–157

Fishman VM, Parkman HP, Schiano TD, Hills C, Dabezies MA, Cohen S, Fisher RS, Miller LS (1996) Symptomatic improvement in achalasia after botulinum toxin injection of the lower esophageal sphincter. Am J Gastroenterol 91:1724–1730

Foschi D, Corsi F, Lazzaroni M, Sangaletti O, Riva P, La Tartara G, Bevilacqua M, Osio M, Alciati A, Bianchi Porro G et al (2007) Treatment of morbid obesity by intraparietogastric administration of botulinum toxin: a randomized, double-blind, controlled study. Int J Obes 31:707–712

Foschi D, Lazzaroni M, Sangaletti O, Corsi F, Trabucchi E, Bianchi Porro G (2008) Effects of intramural administration of botulinum toxin A on gastric emptying and eating capacity in obese patients. Dig Liver Dis 40:667–672

Friedenberg F, Gollamudi S, Parkman HP (2004) The use of botulinum toxin for the treatment of gastrointestinal motility disorders. Dig Dis Sci 49:165–175

Gandomkar H, Zeinoddini A, Heidari R, Amoli HA (2015) Partial lateral internal sphincterotomy versus combined botulinum toxin A injection and topical diltiazem in the treatment of chronic anal fissure: a randomized clinical trial. Dis Colon Rectum 58:228–234

Garcia-Compean D, Mendoza-Fuerte E, Martinez JA, Villarreal I, Maldonado H (2005) Endoscopic injection of botulinum toxin in the gastric antrum for the treatment of obesity. Results of a pilot study. Gastroenterol Clin Biol 29:789–791

Ghoshal UC, Chaudhuri S, Pal BB, Dhar K, Ray G, Banerjee PK (2001) Randomized controlled trial of intrasphincteric botulinum toxin A injection versus balloon dilatation in treatment of achalasia cardia. Dis Esophagus 14:227–231

Giral A, Memisoglu K, Gultekin Y, Imeryuz N, Kalayci C, Ulusoy NB, Tozun N (2004) Botulinum toxin injection versus lateral internal sphincterotomy in the treatment of chronic anal fissure: a non-randomized controlled trial. BMC Gastroenterol 4:7

Gorelick A, Barnett J, Chey W, Anderson M, Elta G (2004) Botulinum toxin injection after biliary sphincterotomy. Endoscopy 36:170–173

Greaves RR, Mulcahy HE, Patchett SE, Gorard DA, Fairclough PD, Alstead EM, Farthing MJ (1999) Early experience with intrasphincteric botulinum toxin in the treatment of achalasia. Aliment Pharmacol Ther 13:1221–1225

Grumelli C, Corradini I, Matteoli M, Verderio C (2010) Intrinsic calcium dynamics control botulinum toxin A susceptibility in distinct neuronal populations. Cell Calcium 47:419–424

Gui D, Cassetta E, Anastasio G, Bentivoglio AR, Maria G, Albanese A (1994) Botulinum toxin for chronic anal fissure. Lancet 344:1127–1128

Gui D, De Gaetano A, Spada PL, Viggiano A, Cassetta E, Albanese A (2000) Botulinum toxin injected in the gastric wall reduces body weight and food intake in rats. Aliment Pharmacol Ther 14:829–834

Gutschow CA, Tox U, Leers J, Schafer H, Prenzel KL, Holscher AH (2010) Botox, dilation, or myotomy? Clinical outcome of interventional and surgical therapies for achalasia. Langenbecks Arch Surg 395:1093–1099

Haapaniemi JJ, Laurikainen EA, Pulkkinen J, Marttila RJ (2001) Botulinum toxin in the treatment of cricopharyngeal dysphagia. Dysphagia 16:171–175

Hackert T et al (2017) Sphincter of Oddi botulinum toxin injection to prevent pancreatic fistula after distal pancreatectomy. Surgery 161(5):1444–1450

Halahakoon VC, Pitt JP (2014) Anal advancement flap and botulinum toxin A (BT) for chronic anal fissure (CAF). Int J Color Dis 29:1175–1177

Hallan RI, Williams NS, Melling J, Waldron DJ, Womack NR, Morrison JF (1988) Treatment of anismus in intractable constipation with botulinum a toxin. Lancet 2:714–717

Han-Geurts IJ, Hendrix VC, de Blaauw I, Wijnen MH, van Heurn EL (2014) Outcome after anal intrasphincteric botox injection in children with surgically treated Hirschsprung disease. J Pediatr Gastroenterol Nutr 59:604–607

Hansen MB (2003) The enteric nervous system III: a target for pharmacological treatment. Pharmacol Toxicol 93:1–13

Hep A, Dolina J, Dite P, Plottova Z, Valek V, Kala Z, Prasek J (2000) Restoration of propulsive peristalsis of the esophagus in achalasia. Hepatogastroenterology 47:1203–1204

Huoh KC, Messner AH (2013) Cricopharyngeal achalasia in children: indications for treatment and management options. Curr Opin Otolaryngol Head Neck Surg 21:576–580

Hurwitz M, Bahar RJ, Ament ME, Tolia V, Molleston J, Reinstein LJ, Walton JM, Erhart N, Wasserman D, Justinich C et al (2000) Evaluation of the use of botulinum toxin in children with achalasia. J Pediatr Gastroenterol Nutr 30:509–514

Hussain SZ, Thomas R, Tolia V (2002) A review of achalasia in 33 children. Dig Dis Sci 47:2538–2543
Ip KS, Cameron DJ, Catto-Smith AG, Hardikar W (2000) Botulinum toxin for achalasia in children. J Gastroenterol Hepatol 15:1100–1104
Irani K, Rodriguez L, Doody DP, Goldstein AM (2008) Botulinum toxin for the treatment of chronic constipation in children with internal anal sphincter dysfunction. Pediatr Surg Int 24:779–783
Joo JS, Agachan F, Wolff B, Nogueras JJ, Wexner SD (1996) Initial north american experience with botulinum toxin type a for treatment of anismus. Dis Colon Rectum 39:1107–1111
Jost WH (1997) One hundred cases of anal fissure treated with botulin toxin: early and long-term results. Dis Colon Rectum 40:1029–1032
Jost WH (2001) Botulinum toxin type b in the treatment of anal fissures: first preliminary results. Dis Colon Rectum 44:1721–1722
Jost WH, Schimrigk K (1994) Therapy of anal fissure using botulin toxin. Dis Colon Rectum 37:1321–1324
Jost WH, Schimrigk K (1995) Botulinum toxin in therapy of anal fissure. Lancet 345:188–189
Jost WH, Schrank B (1999) Repeat botulin toxin injections in anal fissure: in patients with relapse and after insufficient effect of first treatment. Dig Dis Sci 44:1588–1589
Jung HE, Lee JS, Lee TH, Kim JN, Hong SJ, Kim JO, Kim HG, Jeon SR, Cho JY (2014) Long-term outcomes of balloon dilation versus botulinum toxin injection in patients with primary achalasia. Korean J Intern Med 29:738–745
Kelly EA, Koszewski IJ, Jaradeh SS, Merati AL, Blumin JH, Bock JM (2013) Botulinum toxin injection for the treatment of upper esophageal sphincter dysfunction. Ann Otol Rhinol Laryngol 122:100–108
Kent MS, Pennathur A, Fabian T, McKelvey A, Schuchert MJ, Luketich JD, Landreneau RJ (2007) A pilot study of botulinum toxin injection for the treatment of delayed gastric emptying following esophagectomy. Surg Endosc 21:754–757
Keshtgar AS, Ward HC, Sanei A, Clayden GS (2007) Botulinum toxin, a new treatment modality for chronic idiopathic constipation in children: long-term follow-up of a double-blind randomized trial. J Pediatr Surg 42:672–680
Keshtgar AS, Ward HC, Clayden GS (2009) Transcutaneous needle-free injection of botulinum toxin: a novel treatment of childhood constipation and anal fissure. J Pediatr Surg 44:1791–1798
Kim do H, Jung HY (2014) The long-term outcome of balloon dilation versus botulinum toxin injection in patients with primary achalasia. Korean J Intern Med 29:727–729
Kim DY, Park CI, Ohn SH, Moon JY, Chang WH, Park SW (2006) Botulinum toxin type a for poststroke cricopharyngeal muscle dysfunction. Arch Phys Med Rehabil 87:1346–1351
Kolbasnik J, Waterfall WE, Fachnie B, Chen Y, Tougas G (1999) Long-term efficacy of botulinum toxin in classical achalasia: a prospective study. Am J Gastroenterol 94:3434–3439
Krause E, Schirra J, Gurkov R (2008) Botulinum toxin A treatment of cricopharyngeal dysphagia after subarachnoid hemorrhage. Dysphagia 23:406–410
Kroupa R, Hep A, Dolina J, Valek V, Matyasova Z, Prokesova J, Mrazova J, Sedmik J, Novotny I (2010) Combined treatment of achalasia – botulinum toxin injection followed by pneumatic dilatation: long-term results. Dis Esophagus 23:100–105
Kuhn MA, Belafsky PC (2013) Management of cricopharyngeus muscle dysfunction. Otolaryngol Clin North Am 46:1087–1099
Lacy BE, Crowell MD, Schettler-Duncan A, Mathis C, Pasricha PJ (2004) The treatment of diabetic gastroparesis with botulinum toxin injection of the pylorus. Diabetes Care 27:2341–2347
Lepiarczyk E, Bossowska A, Majewski M (2015) Changes in chemical coding of sympathetic chain ganglia (SChG) neurons supplying porcine urinary bladder after botulinum toxin (BTX) treatment. Cell Tissue Res 360(2):263–272
Leyden JE, Moss AC, MacMathuna P (2014) Endoscopic pneumatic dilation versus botulinum toxin injection in the management of primary achalasia. Cochrane Database Syst Rev 12: CD005046

Lindsey I, Jones OM, Cunningham C, Mortensen NJ (2004a) Chronic anal fissure. Br J Surg 91:27179

Lindsey I, Cunningham C, Jones OM, Francis C, Mortensen NJ (2004b) Fissurectomy-botulinum toxin: a novel sphincter-sparing procedure for medically resistant chronic anal fissure. Dis Colon Rectum 47:1947–1952

Liu LW, Tarnopolsky M, Armstrong D (2004) Injection of botulinum toxin A to the upper esophageal sphincter for oropharyngeal dysphagia in two patients with inclusion body myositis. Can J Gastroenterol 18:397–399

Lourenssen S, Miller KG, Blennerhassett MG (2009) Discrete responses of myenteric neurons to structural and functional damage by neurotoxins in vitro. Am J Physiol Gastrointest Liver Physiol 297:G228–G239

Lund JN, Scholefield JH (1996) Aetiology and treatment of anal fissure. Br J Surg 83:1335–1344

Lysy J, Israelit-Yatzkan Y, Sestiery-Ittah M, Weksler-Zangen S, Keret D, Goldin E (2001) Topical nitrates potentiate the effect of botulinum toxin in the treatment of patients with refractory anal fissure. Gut 48:221–224

Mabvuure NT, Hey SY, Forshaw M (2014) Recurrent respiratory distress and cardiopulmonary arrest caused by megaoesophagus secondary to achalasia. Int J Surg Case Rep 5:628–632

Madalinski M, Jagiello K, Labon M, Adrich Z, Kryszewski A (1999) Botulinum toxin injection into only one point in the external anal sphincter: a modification of the treatment for chronic anal fissure. Endoscopy 31:S63

Madalinski MH, Slawek J, Zbytek B, Duzynski W, Adrich Z, Jagiello K, Kryszewski A (2001) Topical nitrates and the higher doses of botulinum toxin for chronic anal fissure. Hepatogastroenterology 48:977–979

Madalinski MH, Slawek J, Duzynski W, Zbytek B, Jagiello K, Adrich Z, Kryszewski A (2002) Side effects of botulinum toxin injection for benign anal disorders. Eur J Gastroenterol Hepatol 14:853–856

Madoff RD, Fleshman JW (2003) Aga technical review on the diagnosis and care of patients with anal fissure. Gastroenterology 124:235–245

Mandal A, Robinson RJ (2001) Indications and efficacy of botulinum toxin in disorders of the gastrointestinal tract. Eur J Gastroenterol Hepatol 13:603–609

Maradey-Romero C, Gabbard S, Fass R (2014) Treatment of esophageal motility disorders based on the chicago classification. Curr Treat Options Gastroenterol 12:441–455

Maria G, Cassetta E, Gui D, Brisinda G, Bentivoglio AR, Albanese A (1998a) A comparison of botulinum toxin and saline for the treatment of chronic anal fissure. N Engl J Med 338:217–220

Maria G, Brisinda G, Bentivoglio AR, Cassetta E, Gui D, Albanese A (1998b) Botulinum toxin injections in the internal anal sphincter for the treatment of chronic anal fissure: long-term results after two different dosage regimens. Ann Surg 228:664–669

Maria G, Brisinda D, Ruggieri MP, Civello IM, Brisinda G (1999) Identification of anti-endothelial cell antibodies in patients with chronic anal fissure. Surgery 126:535–540

Maria G, Brisinda G, Bentivoglio AR, Cassetta E, Albanese A (2000a) Botulinum toxin in the treatment of outlet obstruction constipation caused by puborectalis syndrome. Dis Colon Rectum 43:376–380

Maria G, Brisinda G, Bentivoglio AR, Cassetta E, Gui D, Albanese A (2000b) Influence of botulinum toxin site of injections on healing rate in patients with chronic anal fissure. Am J Surg 179:46–50

Maria G, Brisinda G, Bentivoglio AR, Albanese A, Sganga G, Castagneto M (2001) Anterior rectocele due to obstructed defecation relieved by botulinum toxin. Surgery 129:524–529

Maria G, Sganga G, Civello IM, Brisinda G (2002) Botulinum neurotoxin and other treatments for fissure-in-ano and pelvic floor disorders. Br J Surg 89:950–961

Maria G, Cadeddu F, Brandara F, Marniga G, Brisinda G (2006) Experience with type a botulinum toxin for treatment of outlet-type constipation. Am J Gastroenterol 101:2570–2575

Mariotti R, Bentivoglio M (1996) Botulinum toxin induces nitric oxide synthase activity in motoneurons. Neurosci Lett 219:25–28

Marjoux S, Pioche M, Benet T, Lanne JS, Roman S, Ponchon T, Mion F (2013) Fatal mediastinitis following botulinum toxin injection for esophageal spasm. Endoscopy 45(Suppl 2 UCTN): E405–E406
Marjoux S, Brochard C, Roman S, Gincul R, Pagenault M, Ponchon T, Ropert A, Mion F (2014) Botulinum toxin injection for hypercontractile or spastic esophageal motility disorders: may high-resolution manometry help to select cases? Dis Esophagus 28(8):735–741
Martin JT, Federico JA, McKelvey AA, Kent MS, Fabian T (2009) Prevention of delayed gastric emptying after esophagectomy: a single center's experience with botulinum toxin. Ann Thorac Surg 87:1708–1713; discussion 1713–1704
Martinek J, Spicak J (2003) A modified method of botulinum toxin injection in patients with achalasia: a pilot trial. Endoscopy 35:841–844
Martinek J, Siroky M, Plottova Z, Bures J, Hep A, Spicak J (2003) Treatment of patients with achalasia with botulinum toxin: a multicenter prospective cohort study. Dis Esophagus 16:204–209
Masiero S, Briani C, Marchese-Ragona R, Giacometti P, Costantini M, Zaninotto G (2006) Successful treatment of long-standing post-stroke dysphagia with botulinum toxin and rehabilitation. J Rehabil Med 38:201–203
Mason PF, Watkins MJ, Hall HS, Hall AW (1996) The management of chronic fissure in-ano with botulinum toxin. J R Coll Surg Edinb 41:235–238
Mentes BB, Irkorucu O, Akin M, Leventoglu S, Tatlicioglu E (2003) Comparison of botulinum toxin injection and lateral internal sphincterotomy for the treatment of chronic anal fissure. Dis Colon Rectum 46:232–237
Miftakhov RN, Wingate DL (1993) Mathematical modelling of the enteric nervous network. Facilitation and inhibition of the cholinergic transmission. J Biomed Eng 15:311–318
Mikaeli J, Fazel A, Montazeri G, Yaghoobi M, Malekzadeh R (2001) Randomized controlled trial comparing botulinum toxin injection to pneumatic dilatation for the treatment of achalasia. Aliment Pharmacol Ther 15:1389–1396
Mikaeli J, Yaghoobi M, Montazeri G, Ansari R, Bishehsari F, Malekzadeh R (2004) Efficacy of botulinum toxin injection before pneumatic dilatation in patients with idiopathic achalasia. Dis Esophagus 17:213–217
Mikaeli J, Bishehsari F, Montazeri G, Mahdavinia M, Yaghoobi M, Darvish-Moghadam S, Farrokhi F, Shirani S, Estakhri A, Malekzadeh R (2006) Injection of botulinum toxin before pneumatic dilatation in achalasia treatment: a randomized-controlled trial. Aliment Pharmacol Ther 24:983–989
Miller LS, Pullela SV, Parkman HP, Schiano TD, Cassidy MJ, Cohen S, Fisher RS (2002a) Treatment of chest pain in patients with noncardiac, nonreflux, nonachalasia spastic esophageal motor disorders using botulinum toxin injection into the gastroesophageal junction. Am J Gastroenterol 97:1640–1646
Miller LS, Szych GA, Kantor SB, Bromer MQ, Knight LC, Maurer AH, Fisher RS, Parkman HP (2002b) Treatment of idiopathic gastroparesis with injection of botulinum toxin into the pyloric sphincter muscle. Am J Gastroenterol 97:1653–1660
Minguez M, Melo F, Espi A, Garcia-Granero E, Mora F, Lledo S, Benages A (1999) Therapeutic effects of different doses of botulinum toxin in chronic anal fissure. Dis Colon Rectum 42:1016–1021
Minkes RK, Langer JC (2000) A prospective study of botulinum toxin for internal anal sphincter hypertonicity in children with Hirschsprung's disease. J Pediatr Surg 35:1733–1736
Mishra R, Thomas S, Maan MS, Hadke NS (2005) Topical nitroglycerin versus lateral internal sphincterotomy for chronic anal fissure: prospective, randomized trial. ANZ J Surg 75:1032–1035
Mittermair R, Keller C, Geibel J (2007) Intragastric injection of botulinum toxin A for the treatment of obesity. Obes Surg 17:732–736
Moerman MB (2006) Cricopharyngeal botox injection: indications and technique. Curr Opin Otolaryngol Head Neck Surg 14:431–436

Moerman M, Callier Y, Dick C, Vermeersch H (2002) Botulinum toxin for dysphagia due to cricopharyngeal dysfunction. Eur Arch Otorhinolaryngol 259:1–3

Muehldorfer SM, Schneider TH, Hochberger J, Martus P, Hahn EG, Ell C (1999) Esophageal achalasia: intrasphincteric injection of botulinum toxin A versus balloon dilation. Endoscopy 31:517–521

Murray WR (2011) Botulinum toxin-induced relaxation of the sphincter of Oddi may select patients with acalculous biliary pain who will benefit from cholecystectomy. Surg Endosc 25:813–816

Murry T, Wasserman T, Carrau RL, Castillo B (2005) Injection of botulinum toxin A for the treatment of dysfunction of the upper esophageal sphincter. Am J Otolaryngol 26:157–162

Nasr M, Ezzat H, Elsebae M (2010) Botulinum toxin injection versus lateral internal sphincterotomy in the treatment of chronic anal fissure: a randomized controlled trial. World J Surg 34:2730–2734

Nelson RL, Thomas K, Morgan J, Jones A (2012) Non surgical therapy for anal fissure. Cochrane Database Syst Rev 2:CD003431

Neubrand M, Scheurlen C, Schepke M, Sauerbruch T (2002) Long-term results and prognostic factors in the treatment of achalasia with botulinum toxin. Endoscopy 34:519–523

Panaccione R, Gregor JC, Reynolds RP, Preiksaitis HG (1999) Intrasphincteric botulinum toxin versus pneumatic dilatation for achalasia: a cost minimization analysis. Gastrointest Endosc 50:492–498

Parameswaran MS, Soliman AM (2002) Endoscopic botulinum toxin injection for cricopharyngeal dysphagia. Ann Otol Rhinol Laryngol 111:871–874

Pasricha PJ, Ravich WJ, Hendrix TR, Sostre S, Jones B, Kalloo AN (1995) Intrasphincteric botulinum toxin for the treatment of achalasia. N Engl J Med 332:774–778

Pasricha PJ, Rai R, Ravich WJ, Hendrix TR, Kalloo AN (1996) Botulinum toxin for achalasia: long-term outcome and predictors of response. Gastroenterology 110:1410–1415

Patti MG, Fisichella PM (2014) Controversies in management of achalasia. J Gastrointest Surg 18:1705–1709

Patti R, Almasio PL, Arcara M, Sammartano S, Romano P, Fede C, Di Vita G (2006) Botulinum toxin vs. topical glyceryl trinitrate ointment for pain control in patients undergoing hemorrhoidectomy: a randomized trial. Dis Colon Rectum 49:1741–1748

Porter RF, Gyawali CP (2011) Botulinum toxin injection in dysphagia syndromes with preserved esophageal peristalsis and incomplete lower esophageal sphincter relaxation. Neurogastroenterol Motil 23:139–144, e127–138

Poulain B, Tauc L, Maisey EA, Wadsworth JD, Mohan PM, Dolly JO (1988) Neurotransmitter release is blocked intracellularly by botulinum neurotoxin, and this requires uptake of both toxin polypeptides by a process mediated by the larger chain. Proc Natl Acad Sci U S A 85:4090–4094

Rayner CK, Horowitz M (2005) New management approaches for gastroparesis. Nat Clin Pract Gastroenterol Hepatol 2:454–462; quiz 493

Regan J, Murphy A, Chiang M, McMahon BP, Coughlan T, Walshe M (2014) Botulinum toxin for upper oesophageal sphincter dysfunction in neurological swallowing disorders. Cochrane Database Syst Rev 5:CD009968

Restivo DA, Marchese-Ragona R, Lauria G, Squatrito S, Gullo D, Vigneri R (2006) Botulinum toxin treatment for oropharyngeal dysphagia associated with diabetic neuropathy. Diabetes Care 29:2650–2653

Restivo DA, Marchese-Ragona R, Patti F, Solaro C, Maimone D, Zappala G, Pavone A (2011) Botulinum toxin improves dysphagia associated with multiple sclerosis. Eur J Neurol 18:486–490

Rollnik JD, Meier PN, Manns MP, Goke M (2003) Antral injections of botulinum a toxin for the treatment of obesity. Ann Intern Med 138:359–360

Roman S, Kahrilas PJ (2013) Management of spastic disorders of the esophagus. Gastroenterol Clin North Am 42:27–43

Ron Y, Avni Y, Lukovetski A, Wardi J, Geva D, Birkenfeld S, Halpern Z (2001) Botulinum toxin type-A in therapy of patients with anismus. Dis Colon Rectum 44:1821–1826

Sajid MS, Hunte S, Hippolyte S, Kiri VA, Maringe C, Baig MK (2008) Comparison of surgical vs chemical sphincterotomy using botulinum toxin for the treatment of chronic anal fissure: a meta-analysis. Colorectal Dis 10:547–552

Saliakellis E, Fotoulaki M (2013) Gastroparesis in children. Ann Gastroenterol 26:204–211

Samim M, Twigt B, Stoker L, Pronk A (2012) Topical diltiazem cream versus botulinum toxin A for the treatment of chronic anal fissure: a double-blind randomized clinical trial. Ann Surg 255:18–22

Schneider I, Thumfart WF, Pototschnig C, Eckel HE (1994) Treatment of dysfunction of the cricopharyngeal muscle with botulinum a toxin: introduction of a new, noninvasive method. Ann Otol Rhinol Laryngol 103:31–35

Scholes MA, McEvoy T, Mousa H, Wiet GJ (2014) Cricopharyngeal achalasia in children: botulinum toxin injection as a tool for diagnosis and treatment. Laryngoscope 124:1475–1480

Scholz T, Hetzer FH, Dindo D, Demartines N, Clavien PA, Hahnloser D (2007) Long-term follow-up after combined fissurectomy and botox injection for chronic anal fissures. Int J Colorectal Dis 22:1077–1081

Shafik A, El-Sibai O (1998) Botulin toxin in the treatment of nonrelaxing puborectalis syndrome. Dig Surg 15:347–351

Sharata A, Kurian AA, Dunst CM, Bhayani NH, Reavis KM, Swanstrom LL (2013) Peroral endoscopic myotomy (POEM) is safe and effective in the setting of prior endoscopic intervention. J Gastrointest Surg 17:1188–1192

Shaw GY, Searl JP (2001) Botulinum toxin treatment for cricopharyngeal dysfunction. Dysphagia 16:161–167

Shawki S, Costedio M (2013) Anal fissure and stenosis. Gastroenterol Clin North Am 42:729–758

Shogan AN, Rogers DJ, Hartnick CJ, Kerschner JE (2014) Use of botulinum toxin in pediatric otolaryngology and laryngology. Int J Pediatr Otorhinolaryngol 78:1423–1425

Shrestha S, Pasricha PJ (2001) Chronic visceral right upper quadrant pain without gallstones. Curr Treat Options Gastroenterol 4:123–131

Simms HN, McCallion K, Wallace W, Campbell WJ, Calvert H, Moorehead RJ (2004) Efficacy of botulinum toxin in chronic anal fissure. Ir J Med Sci 173:188–190

Siproudhis L, Sebille V, Pigot F, Hemery P, Juguet F, Bellissant E (2003) Lack of effficacy of botulinum toxin in chronic anal fissure. Aliment Pharmacol Ther 18:515–524

Spector ME, Callaway E, McKean EL, Prince ME (2013) Videofluoroscopic-guided botulinum toxin injections for pharyngoesophageal spasm after total laryngectomy. Laryngoscope 123:394–397

Sterling JL et al (2018) The role of botulinum toxin injections for esophageal motility. Curr Treat Options Gastroenterol 16(4):528–540

Storr M, Allescher HD, Rosch T, Born P, Weigert N, Classen M (2001a) Treatment of symptomatic diffuse esophageal spasm by endoscopic injection of botulinum toxin: a prospective study with long term follow-up. Gastrointest Endosc 54:18A

Storr M, Allescher HD, Rosch T, Born P, Weigert N, Classen M (2001b) Treatment of symptomatic diffuse esophageal spasm by endoscopic injections of botulinum toxin: a prospective study with long-term follow-up. Gastrointest Endosc 54:754–759

Suzukia Y, Sano N, Shinonaga C, Fukuda M, Hyodo M, Morimoto T (2007) Successful botulinum toxin treatment of dysphagia in a spinal muscular atrophy type 2 patient. Brain and Development 29:662–665

Tack J, Zaninotto G (2015) Therapeutic options in oesophageal dysphagia. Nat Rev Gastroenterol Hepatol 12(6):332–341

Thomas A et al (2018) Botulinum toxin as a treatment for refractory gastroparesis: a literature review. Curr Treat Options Gastroenterol 16(4):479–488

Tilney HS, Heriot AG, Cripps NP (2001) Complication of botulinum toxin injections for anal fissure. Dis Colon Rectum 44:1721–1724

Topazian M, Camilleri M, De La Mora-Levy J, Enders FB, Foxx-Orenstein AE, Levy MJ, Nehra V, Talley NJ (2008) Endoscopic ultrasound-guided gastric botulinum toxin injections in obese subjects: a pilot study. Obes Surg 18:401–407

Topazian M, Camilleri M, Enders FT, Clain JE, Gleeson FC, Levy MJ, Rajan E, Nehra V, Dierkhising RA, Collazo-Clavell ML et al (2013) Gastric antral injections of botulinum toxin delay gastric emptying but do not reduce body weight. Clin Gastroenterol Hepatol 11:145–150 e141

Tranqui P, Trottier DC, Victor C, Freeman JB (2006) Nonsurgical treatment of chronic anal fissure: nitroglycerin and dilatation versus nifedipine and botulinum toxin. Can J Surg 49:41–45

Vaezi MF (2013) Botulinum toxin in nonachalasia motility disorders: a welcomed therapy in an area with limited therapeutic options. Clin Gastroenterol Hepatol 11:1122–1124

Valizadeh N, Jalaly NY, Hassanzadeh M, Kamani F, Dadvar Z, Azizi S, Salehimarzijarani B (2012) Botulinum toxin injection versus lateral internal sphincterotomy for the treatment of chronic anal fissure: randomized prospective controlled trial. Langenbecks Arch Surg 397:1093–1098

Vanuytsel T, Bisschops R, Farre R, Pauwels A, Holvoet L, Arts J, Caenepeel P, De Wulf D, Mimidis K, Rommel N et al (2013) Botulinum toxin reduces dysphagia in patients with nonachalasia primary esophageal motility disorders. Clin Gastroenterol Hepatol 11:1115–1121 e1112

Vela MF (2014) Management strategies for achalasia. Neurogastroenterol Motil 26:1215–1221

Vela MF, Richter JE, Wachsberger D, Connor J, Rice TW (2004) Complexities of managing achalasia at a tertiary referral center: use of pneumatic dilatation, heller myotomy, and botulinum toxin injection. Am J Gastroenterol 99:1029–1036

Walzer N, Hirano I (2008) Achalasia. Gastroenterol Clin N Am 37:807–825

Wehrmann T, Kokabpick H, Jacobi V, Seifert H, Lembcke B, Caspary WF (1999) Long-term results of endoscopic injection of botulinum toxin in elderly achalasic patients with tortuous megaesophagus or epiphrenic diverticulum. Endoscopy 31:352–358

Wehrmann T, Schmitt TH, Arndt A, Lembcke B, Caspary WF, Seifert H (2000) Endoscopic injection of botulinum toxin in patients with recurrent acute pancreatitis due to pancreatic sphincter of Oddi dysfunction. Aliment Pharmacol Ther 14:1469–1477

Wiesel PH, Schneider R, Dorta G, Blum AL, Gillet M, Michetti P (1997) Botulinum toxin for refractory postoperative pyloric spasm. Endoscopy 29:132

Witte ME, Klaase JM (2007) Botulinum toxin A injection in isdn ointment-resistant chronic anal fissures. Dig Surg 24:197–201

Woltman TA, Pellegrini CA, Oelschlager BK (2005) Achalasia. Surg Clin North Am 85:483–493

Zaninotto G, Marchese Ragona R, Briani C, Costantini M, Rizzetto C, Portale G, Zanetti L, Masiero S, Costantino M, Nicoletti L et al (2004a) The role of botulinum toxin injection and upper esophageal sphincter myotomy in treating oropharyngeal dysphagia. J Gastrointest Surg 8:997–1006

Zaninotto G, Annese V, Costantini M, Del Genio A, Costantino M, Epifani M, Gatto G, D'Onofrio V, Benini L, Contini S et al (2004b) Randomized controlled trial of botulinum toxin versus laparoscopic heller myotomy for esophageal achalasia. Ann Surg 239:364–370

Zaninotto G, Leusink A, Markar SR (2019) Management of achalasia in 2019. Curr Opin Gastroenterol. https://doi.org/10.1097/MOG.0000000000000544

Zarate N, Mearin F, Baldovino F, Armengol JR, Malagelada JR (2002) Achalasia treatment in the elderly: is botulinum toxin injection the best option? Eur J Gastroenterol Hepatol 14:285–290

Zhang Y, Wang ZN, He L, Gao G, Zhai Q, Yin ZT, Zeng XD (2014) Botulinum toxin type-A injection to treat patients with intractable anismus unresponsive to simple biofeedback training. World J Gastroenterol 20:12602–12607

Zhu Q, Liu J, Yang C (2009) Clinical study on combined therapy of botulinum toxin injection and small balloon dilation in patients with esophageal achalasia. Dig Surg 26:493–498

The Use of Botulinum Toxin in the Management of Headache Disorders

Hsiangkuo Yuan and Stephen D. Silberstein

Contents

Abstract

Tremendous progress has been made in the past decades for the treatment of headache disorders. Chronic migraine is the most disabling type of headache and requires the use of acute and preventive medications, many of which are associated with adverse events that limit patient adherence. Botulinum toxin (BoNT) serotype A, a neurotoxin derived from certain strains of *Clostridium*, disrupts neuropeptide secretion and receptor translocation related to trigeminal nociception, thereby preventing pain sensitization through peripheral and possibly central mechanisms. Ever since the first randomized controlled trial on onabotulinumtoxinA (onabotA) for migraine was published two decades ago, onabotA has been the only BoNT formulation approved for use in the prevention

The original version of this chapter was revised. A correction to this chapter can be found at https://doi.org/10.1007/164_2020_416

H. Yuan · S. D. Silberstein (✉)
Jefferson Headache Center, Thomas Jefferson University, Philadelphia, PA, USA
e-mail: Stephen.silberstein@jefferson.edu

S. M. Whitcup, M. Hallett (eds.), *Botulinum Toxin Therapy*,
Handbook of Experimental Pharmacology 263, https://doi.org/10.1007/164_2020_365

of chronic migraine. Superior tolerability and efficacy have been demonstrated on multiple migraine endpoints in many controlled trials and real-life studies. OnabotA is a safe and efficacious treatment for chronic migraine and possibly high-frequency episodic migraine. Further research is still needed to understand its mechanism of action to fully develop its therapeutic potential.

Keywords

Botulinum toxin · Calcitonin gene-related peptide · Migraine

1 Introduction

Headache is one of the most common neurological disorders. According to the Global Burden of Diseases, Injuries, and Risk Factors Study 2016 (GBD 2016), in individuals 15–49 years of age, headache is the number one global cause (level 3) of years lived with disability and number two of disability-adjusted life years (GBD 2016 DALYs Hale Collaborators 2016, 2017). The international classification of headache disorder (ICHD-3) classifies headache into primary headaches, secondary headaches, neuropathies and facial pains, and other headaches (Headache Classification Committee of the International Headache Society (IHS) 2018). Primary headaches include migraine, tension-type headache (TTH), trigeminal autonomic cephalalgia (TAC), and others. While TTH is the most prevalent headache, migraine is associated with greater disability and poorer quality of life.

Migraine is a complex neurovascular disorder characterized by pulsatile, disabling headaches associated with neurologic, gastric, and autonomic symptoms, typically lasting 4 to 72 h (Headache Classification Committee of the International Headache Society (IHS) 2018). It affects roughly 1 billion people worldwide, posing a significant socioeconomic impact (Stokes et al. 2011; Bloudek et al. 2012). The ICHD-3 lists criteria for six subtypes of migraine (Headache Classification Committee of the International Headache Society (IHS) 2018). Based on its attack frequency, migraine can be categorized as episodic migraine (EM) or chronic migraine (CM). CM affects roughly 1–3% of the population, and about 2.5% of EM patients progress yearly to CM. Compared to EM, CM is associated with increased headache-related disability, cardiopulmonary and psychiatric comorbidities, and greater financial and occupational burdens (Adams et al. 2015; Buse et al. 2010). CM is characterized by recurring headaches $\geq$15 days per month with $\geq$8 headaches being migrainous for $\geq$3 months (Headache Classification Committee of the International Headache Society (IHS) 2018). These headaches can phenomenologically resemble a mixture of TTH and migraine but still respond to triptans, a migraine-specific medication. CM is often associated with cutaneous allodynia as a result of central pain sensitization. CM is frequently complicated by acute medication overuse but does not always improve after drug withdrawal. Less than 5% of CM patients received a correct diagnosis and treatment (Dodick et al. 2016), and only a quarter of them achieved remission (headache $<$10 days per month) in 2 years (Manack et al. 2011). CM patients are usually sensitive to adverse events (AEs) and are thus more likely to withdraw from treatment. The adherence to preventive medication drops significantly by day 30 of treatment, with only 25% adherence by 6 months (Hepp et al. 2017). This is likely due to drug-related AEs, inadequate response to treatment, nocebo effect, patient preference, and loss to follow-up. There remains a strong demand for a CM preventive agent with greater efficacy and lower AEs compared to currently available treatment.

Driven by the research on migraine pathophysiology, new therapeutic entities have been developed. OnabotulinumtoxinA (onabotA; Botox®, Allergan plc, Dublin, Ireland) injection was approved by the US Federal Drug Administration (FDA) specifically for CM prevention in 2010. In addition, four therapeutic monoclonal antibodies (mAbs) targeting calcitonin gene-related peptide (eptinezumab, fremanezumab, galcanezumab) or its receptor (erenumab) were approved for migraine prophylaxis in the past 2 years. Galcanezumab also received approval for prophylactic use in episodic cluster headache. With the clinical success of onabotA for use in CM, it is important to understand the role of botulinum neurotoxin A (BoNTA) in migraine pathophysiology and management.

2 Pathophysiology of Primary Headache Disorders

Migraine, TTH, and TAC likely share a common pathophysiology: the trigeminovascular pathway (Vollesen et al. 2018). Abnormal activation of the trigeminovascular system by cortical spreading depression (CSD) or other activators is believed to cause not only vasodilation and neurogenic inflammation but also peripheral and central pain sensitization that leads to sustained headache (Goadsby et al. 2017). Trigeminal sensory fibers, along with high cervical sensory fibers, innervate pain-sensitive structures in the head (e.g., dura, falx, face, scalp, blood vessels) and upper neck. Some intracranial meningeal afferents innervate extracranial tissues (e.g., periosteum, muscle) via collateral fibers through the skull (Schueler et al. 2014; Kosaras et al. 2009). These extra- and intracranial sensory nerve fibers and their overlapping collaterals transmit sensory signals to the trigeminal cervical complex (TCC), which then projects to multiple central nuclei (e.g., thalamus, hypothalamus, parabrachial, locus coeruleus, dorsal raphe, periaqueductal gray, basal ganglia) involved in the modulation and perception of head pain and associated symptoms. The trigeminal nerve also connects with the parasympathetic system through the superior salivatory nucleus and sphenopalatine ganglion, which are responsible for autonomic symptoms (e.g., conjunctival injection, tearing, and rhinorrhea) in migraine and TACs.

Trigeminal nociceptive afferents are pseudounipolar neurons with cell bodies primarily located in the trigeminal ganglion (TG). While trigeminal somata were traditionally considered free of synaptic contacts, they are now believed to interact with neighboring neurons, satellite glial cells, and other cell types (Goadsby et al. 2017); cross-talks in the TG (and TCC) likely modulate pain sensitization (Messlinger et al. 2020). Trigeminal pain afferents are mainly thinly myelinated Aδ and unmyelinated C fibers utilizing a spectrum of neurotransmitters (e.g., glutamate, gamma-aminobutyric acid [GABA], serotonin, histamine, nitric oxide, etc.) and neuropeptides (calcitonin gene-related peptide [CGRP], substance P [SP], somatostatin, cholecystokinin, etc.) as their signaling messengers (Lazarov 2002). Upon activation, a subgroup of C-fibers release CGRP, SP, and glutamate, which play a role in neurogenic inflammation, nociception modulation, and pain amplification. They often coexist in large dense-core vesicles at synapses on nerve terminals or in varicosities along axons. CGRP, one of the most potent cerebral vasodilators, plays a significant role in migraine pathogenesis. CGRP interacts with its receptors on trigeminal Aδ neurons, satellite glial cells, endothelial cells, immune cells, the endocannabinoid system, and blood vessel smooth muscle cells. It induces vasodilation, neurogenic inflammation, pain receptors upregulation, and subsequently peripheral/central sensitization, leading to a dysfunctional activation of the trigeminovascular system. CGRP serum levels are elevated during migraine attacks, and CGRP functional blockade aborts and prevents migraine attacks. Migraine pain most likely results from activation of pain-producing structures and reduction in endogenous pain inhibition (Silberstein et al. 2007). Factors such as stress, diet, environmental change, extracranial noxious activation, and hormonal fluctuation also interact with the pain formation network. More details regarding migraine pathophysiology are beyond the scope of this paper. Here, we will focus on migraine and its relationship with BoNTA.

3 Treatment of Migraine

Headache management can be acute, preventive, or both. Many patients with migraine self-medicate with simple analgesics (e.g., aspirin, acetaminophen, nonsteroidal anti-inflammatory drugs), while others require prescription medication (e.g., triptan, dihydroergotamine, and neuroleptics) for acute treatment. The level of evidence for acute migraine pharmacotherapies was recently published by the American Headache Society (AHS) (Marmura et al. 2015). Since overusing acute medications often leads to medication overuse (i.e., rebound) headaches (Solomon et al. 2011), a preventive treatment regimen is required.

The AHS guidelines recommend starting preventive treatment if the patient has $\geq$3 headache days per month with severe disability, $\geq$4 headache days per month with mild disability, or $\geq$6 headache days per month with no disability (American Headache Society 2019). In the USA, the FDA has approved antiepileptic medications (divalproex, valproate, topiramate), beta-adrenergic blockers (propranolol, timolol), and CGRP functional blocking mAbs for migraine prevention and onabotA only for CM prevention. Other medications such as tricyclic antidepressants, serotonin norepinephrine reuptake inhibitors, antihistamines, angiotensin-converting enzyme inhibitors, and angiotensin receptor blockers are used off-label for migraine prevention. Per AHS recommendation, if the patient demonstrates an inability to tolerate (due to side effects) or an inadequate response to a 6-week trial of at least 2 classes of medication listed above or a 6-month trial of onabotA, a CGRP functional blocking mAb injection can be used (American Headache Society 2019).

4 Botulinum Toxin Formulations

Botulinum toxins (BoNT) are produced by *Clostridium botulinum*, *C. butyricum*, and other subspecies. Traditionally, there are seven BoNT serotypes (A–G) separated by their antigenicity; newer serotypes continue to be discovered. As of now, more than 40 genetic BoNT variants have been identified using DNA sequencing technology. The serotype designation is defined by the absence of cross-neutralization in an animal bioassay by type-specific monovalent botulinum antitoxin (Rummel 2015; Zhang et al. 2017; Barash and Arnon 2014). Each serotype has specific molecular targets on the soluble *N*-ethylmaleimide-sensitive factor attachment protein receptor (SNARE) proteins, which mediate Ca^{2+}-dependent vesicular release machinery. BoNT serotypes A, C, and E catalyze proteolysis of distinct sites on synaptosome-associated proteins of 25 kDa (SNAP-25). Serotypes B, D, F, and G cleave different locations on vesicle-associated membrane proteins (VAMP). BoNTC also cleaves syntaxin 1. All BoNT serotypes inhibit vesicle release, but their intracellular target proteins and metabolic pathways vary, resulting in different potencies and durations of action. Serotypes A and B are used clinically for neuromuscular blockade due to their longer-lasting effect. Serotype A is also used for aesthetics and pain treatment. Although the amino acid sequence varies between different serotypes, they are structurally similar, consisting of a single inactive polypeptide chain (150 kDa) and neurotoxin accessory proteins (NAPs). The 150 kDa polypeptide can be activated by proteolytic cleavage into 100 KDa heavy chain (H_C) and 50 kDa light chain (L_C), connected by a disulfide bond. The NAPs have no direct therapeutic effect but are responsible for neurotoxin structure stability, protection from proteolysis, and transportation across epithelial barriers (Gu and Jin 2013; Ghosal et al. 2018).

To date, three BoNTA and one BoNTB formulations are available for clinical use in the USA: onabotA, abobotulinumtoxinA (abobotA; Dysport®, Ipsen S.A., Paris, France), incobotulinumtoxinA (incobotA; XEOMIN®, Merz Pharma GmbH & Co, Frankfurt, Germany), and rimabotulinumtoxinB (rimabotB; Myobloc®/ Neurobloc®, US WorldMeds, LLC; Louisville, KY). Each has its own indications and dosage with a noninterchangeable BoNT unit. Their potency used to be determined by the median lethal dose (LD_{50}) in mice. However, this lethality assay is time-consuming and expensive, so it has been largely replaced by cell-based assays uniquely developed by each manufacturer. While vacuum-dried onabotA and lyophilized abobotA and incobotA require reconstitution before use, rimabotB is formulated as a ready-to-use solution. IncobotA, even without NAPs, can be stored at room temperature, whereas onbotA and abobotA must be stored refrigerated. Reconstituted BoNTA should be refrigerated and administered within 24 h. Among them, onabotA has the highest LD_{50}/ED_{50} ratio (19.8 ± 3.38) and the lowest off-target migration (Ferrari et al. 2018), likely due to its larger molecular weight and the use of a different excipient. Since only onabotA is approved for CM use, we will focus on the discussion of BoNTA.

5 Uptake and Trafficking of BoNTA

Upon injection, BoNTA dissociates from NAP, diffuses, and binds to unmyelinated regions on nerve fibers. Via a dual receptor strategy, the BoNTA H_C domain binds sequentially to the polysialogangliosides (GT1b > GD1a = GD1b > GM1) and then to the luminal L4 region of synaptic vesicle glycoprotein 2 (SV2) (Rummel 2017). The initial low-affinity binding to gangliosides facilitates the accumulation of BoNTA to specific microdomains on the unmyelinated area of the nerve fiber followed by the high-affinity interaction with exposed luminal SV2C. H_C binds to SV2C L4 and enters the endosome via clathrin-mediated endocytosis rather than pinocytosis (Pellett et al. 2015; Black and Dolly 1986). The endocytosis occurs in minutes, taking in a few toxin molecules per vesicle (Colasante et al. 2013). Since the L4 domain is transiently exposed to the membrane surface during exocytosis, BoNTA is taken up in a higher amount in more active and sensitized neurons. Keep in mind that neurons can still take up BoNTA independent of SNARE dysfunction; BoNTA significantly reduces fast depolarization-dependent endocytosis but not slow depolarization-independent endocytosis (Pellett et al. 2015). Inside the endosome, the oxidized and acidic environment maintains the disulfide bridge between H_C and L_C, thereby facilitating the insertion of the translocation domain (H_N) onto the vesicle membrane. H_N helps translocate L_C from the endosome to the cytosol (Connan and Popoff 2017). Once outside the endosome, L_C is then released by the thioredoxin system on the extrinsic surface of the vesicle (Pirazzini et al. 2013) and functions as a zinc-dependent endopeptidase that cleaves the host SNAP-25 as long as Lc remains active (usually for a few months in vivo) (Blasi et al. 1993). L_C's prolonged stability is determined by many factors, including deubiquitination (Tsai et al. 2017), tyrosine phosphorylation (Toth et al. 2012), and a dileucine motif (Wang et al. 2011). Although SNAP-25 synthesis increases, newly formed SNAP-25 continues to be cleaved by active Lc. Cleaved SNAP-25 exhibits reduced affinity to intracellular Ca^{2+}-sensing synaptotagmin, limiting the fusion of vesicles with the cell membrane. Such neurosecretory dysfunction is dependent on Ca^{2+} and thus can be restored by higher Ca^{2+} concentrations (Gerona et al. 2000). The faulty SNARE complexes have a prolonged existence on the synaptic membrane; BoNTA-cleaved SNAP-25 can be found in the nerve terminal for months. Recovery of function is associated with a decrease in the cleaved to intact SNAP-25 ratio (Jurasinski et al. 2001). Typically, paresis occurs 2–5 days after injection in skeletal muscle and lasts 2–3 months before wearing off (longer for smooth muscle or sweat glands). The actual onset and duration vary depending on the target neuron treated, injection dose, and formulation used.

The trafficking of the BoNTA has been a research interest for decades. It has been known that axonal transport is cargo-specific and essential for neuronal function. While cytoskeletal polymers are delivered via slow axonal transport, organelles and vesicles are transported rapidly (250–400 mm/day) with specific directions (anterograde, retrograde, or bidirectional) depending on the cargo type (Maday et al. 2014). Whether BoNTA undergoes axonal trafficking to central terminals remains an issue of debate. Early studies using high-dose radiolabeled BoNTA demonstrated its retrograde transport centrally. Recent studies further suggested the transport of catalytic-active BoNTA to distal nerve terminals (Restani et al. 2011, 2012). However, higher doses of BoNTA employed beyond the therapeutic dose (>10 U/kg) may cause systemic non-axonal spread, thus limiting the study validity. Due to challenges in the direct measurement of BoNTA, most studies rely on the measurement of BoNTA's function (e.g., cleaved SNAP-25, receptor trafficking) as indirect biomarkers. Upon peripheral BoNTA injection, cleaved SNAP-25 can be found in the sensory region of the brainstem, sensory ganglion, or spinal cord segment associated with the peripherally injected area (Matak et al. 2019). Decreased numbers of α-amino-3-hydroxy-5-methyl-4-isoxazolepropionic acid (AMPA) receptors were also found in spinal cord dorsal horn neurons after low-dose BoNTA injection (6 U/kg) (Hong et al. 2017). However, there are concerns regarding the lack of specificity on commercially available anti-cleaved SNAP-25 antibodies, rendering a low target specificity particularly on immunohistochemistry (Rheaume et al. 2015). Colocalization of cleaved SNAP-25 signal within synapses using 2D rather than 3D imaging may also lead to erroneous conclusions (Cai et al. 2017). Although conflicting data exist, growing evidence suggests that BoNTA possibly exerts its function throughout the affected neuron via retrograde axonal trafficking. It remains to be determined whether the full BoNTA, cleaved SNAP-25, or both get transported. This phenomenon likely occurs within the first few hours after injection and is facilitated by neutral pH vesicles intended for retrograde trafficking (Harper et al. 2016). Furthermore, whether there is a true transsynaptic transport to upstream neurons remains conflicted (Cai et al. 2017; Caleo et al. 2018). Bilateral effects after unilateral BoNTA injection (without evidence of contralateral toxin transportation) were demonstrated in several pain hypersensitivity animal models (Matak et al. 2019). Keep in mind that preclinical findings are not always translatable to humans. In human studies, peripheral BoNTA injection elicits gray matter volume changes and partial restoration of connectivity on fMRI in cervical dystonia patients (Delnooz et al. 2013, 2015). Whether the central effect of BoNTA originated from transported toxins directly or from cerebral plasticity secondary to peripheral modulations remains to be explored (Hallett 2018).

6 Mechanism of Action of BoNTA in Migraine

BoNTA cleaves SNAP-25, interrupts SNARE function, alters calcium-triggered fast exocytosis, and possibly affects some other SNAP-25 functions (Antonucci et al. 2016). In vitro studies have demonstrated that by interfering with vesicle release, BoNTA not only inhibits the release of neuropeptides and neurotransmitters but also disrupts the expression of pain-sensing receptors on the plasma membrane. For instance, BoNTA has been shown to inhibit the release of CGRP (Durham et al. 2004), substance P (Welch et al. 2000), and glutamate from sensory peptidergic neurons (Cui et al. 2004). BoNTA impaired the release of glutamate more than GABA (Verderio et al. 2007); the latter is likely essential for BoNTA's antinociceptive action (Drinovac et al. 2014). Pericranially injected BoNTA prevented inflammatory cell infiltration and inhibited the increase of CGRP levels in the dura, with cleaved SNAP-25 colocalizing with CGRP in intracranial dural nerve endings (Lackovic et al. 2016). BoNTA altered the trafficking and expression of pain-related receptors (e.g., transient receptor potential cation channel vanilloid subfamily member 1 [TRPV1], transient receptor potential cation channel ankyrin subfamily member 1 [TRPA1], ATP-gated P2X receptor cation channel family 3 [$P2X_3$], AMPA) (Hong et al. 2017; Shimizu et al. 2012; Zhang et al. 2016). It influenced immune cell (e.g., microglia, monocytes) activation and modulated neuroimmune balance in cytokine secretion (Mika et al. 2011; Zychowska et al. 2016). BoNTA's action can also be affected by the endogenous opioid system (involving μ-opioid receptor) (Drinovac et al. 2013).

In animal models, BoNTA attenuated pain sensitizations (e.g., mechanical allodynia, thermal hyperalgesia) (Lackovic et al. 2016; Mika et al. 2011) but did not change pain threshold or tolerance to heat and electrical stimuli (Blersch et al. 2002; Voller et al. 2003). Pre-treatment with BoNTA inhibited formalin-induced nociceptive behavior in the absence of apparent muscle weakness in a dose-dependent manner (Cui et al. 2004). It reduced the c-Fos activation in the trigeminal nucleus caudalis (TNC), locus coeruleus, and periaqueductal gray (Matak et al. 2014). BoNTA had no immediate effect on the spontaneous activity of nonsensitized meningeal nociceptor units in the TG, but it inhibited suprathreshold (not threshold) mechanical response (more on C-units than Aδ-units) when administered 3 h prior. Pre-treatment with BoNTA prevented inflammatory soup (IS)-induced increase in spontaneous activity and suprathreshold mechanical sensitivity in C- but not A-δ-units (Burstein et al. 2014). Pre-treatment with BoNTA reduced the prolonged firing of the nociceptors on C fibers but not the probability of their response to mechanical CSD (Melo-Carrillo et al. 2019). This is in contrast with the effect from CGRP mAb that prevents CSD-induced activation of Aδ-fibers (Melo-Carrillo et al. 2017). These findings suggest that BoNTA does not directly inhibit pain but prevents pain sensitization. But how does extracranially administered BoNTA alter the intracranial nociceptors' response?

The antinociceptive action of BoNTA takes place peripherally and perhaps centrally. The extracranial administration of BoNTA (~30 U/kg) suppressed the response of meningeal nociceptors to the stimulation of their intracranial dural receptive fields (Zhang et al. 2016). Although direct transport through suture lines is possible, other transsynaptic pathways likely exist. Since injection of the axonal transport blocker (colchicine) into the TG prevented the formation of cleaved SNAP-25 in dura, it is plausible that extracranial BoNTA (5 U/kg) was axonally transported to the TG and then underwent transsynaptic transport to bilateral dural afferents (Lackovic et al. 2016). Such bilateral effects of BoNTA and dependence on retrograde axonal transport suggest a central site of action (Filipovic et al. 2012). A few days after injection of BoNTA (15 U/kg) into rat whisker pads, cleaved SNAP-25 were found in the primary afferent terminal within the TNC but not in higher central nuclei (thalamus, hypothalamus, sensory cortex, locus coeruleus, periaqueductal gray, etc.). Despite this, the pain-invoked activity in the locus coeruleus and periaqueductal gray was reduced, suggesting a central indirect plasticity effect

(Matak et al. 2014). Chemical denervation of either the TG or sensory nerves prevented both the appearance of cleaved SNAP-25 in TNC and BoNTA's (3.5 U/kg) antinociceptive activity in formalin-induced orofacial pain (Matak et al. 2011). This suggests that BoNTA's antinociceptive activity is associated with capsaicin-sensitive neurons. BoNTA's bilateral antinociceptive effect occurred after peripheral ipsilateral (5 U/kg) and intrathecal (1 U/kg) but not intracisternal application (Drinovac Vlah et al. 2016). These findings are suggestive of central involvement at the level of the brainstem and spinal cord but not the cerebrum or cerebellum. Again, the lack of antibody specificity to cleaved SNAP-25 and higher BoNTA injection dosage causing nonspecific spread remain a methodological concern.

Studies in healthy volunteers highlighted BoNTA's role in preventing pain sensitization. BoNTA produced a marked and specific decrease in noxious mechanical pain sensitivity, whereas sensitivity to low-threshold mechanical and thermal stimuli as well as cutaneous innervation remained unchanged (Paterson et al. 2014). BoNTA reduced capsaicin-induced trigeminal pain and heat pain threshold but not electrical or pressure pain threshold (Gazerani et al. 2009). The antinociceptive effect of BoNTA was found to be independent of the muscular effect in patients with cervical dystonia and post-stroke spasticity (Brashear et al. 2002; Naumann et al. 2002). Pain improvement in patients with cervical dystonia occurred before motor improvement, and the pain relief lasted longer than muscle weakness (Relja and Miletic 2017). These findings suggest that there is a modulation of afferent sensory input (e.g., muscle spindles) to reduce pain sensitization (Weise et al. 2019). Furthermore, since BoNTA induces no central nervous system AEs in humans, there is likely no true supratentorial transport or mechanism of action. Its central involvement is likely restricted to the brainstem/cord or secondary to neuronal network plasticity from peripheral modulation of SNAP-25 (Antonucci et al. 2016).

7 Efficacy of BoNTA in Migraine

The first randomized controlled trial of the use of BoNTA in migraine was published in 2000 (Silberstein et al. 2000). Since then, there have been many new developments in the clinical use of BoNTA (mostly onabotA) for headache disorders. The two major studies conducted on patients with CM in the 2010s, Phase III Research Evaluating Migraine Prophylaxis Therapy (PREEMPT) I and II (Aurora et al. 2010; Diener et al. 2010), resulted in the FDA approval of the use of intramuscular 155 U of onabotA administered to 31 injection sites across 7 head and neck muscles using a fixed-site, fixed-dose injection paradigm every 12 weeks (Blumenfeld et al. 2010). Up to 40 U of additional onabotA can be administered to 8 injection sites across 3 head and neck muscle groups using a modified follow-the-pain approach with a maximum dose of 195 U. The PREEMPT results showed a significant improvement in multiple headache endpoints as well as patients' productivity, vitality, psychological distress, and overall quality of life (Dodick et al. 2010). Those who did not improve after the first cycle often become responders after additional cycles (Silberstein et al. 2015). Some studies have shown that the effects of onabotA may wear off about 10 weeks after the initial injection (Masters-Israilov and Robbins 2019; Zidan et al. 2019). Higher doses may thus be beneficial (Masters-Israilov and Robbins 2019; Negro et al. 2015a), especially if there is an insufficient response or duration of effect. OnabotA seems as effective as amitriptyline, divalproex, or topiramate for CM prophylaxis but with greater tolerability (Cady et al. 2011; Magalhaes et al. 2010; Blumenfeld et al. 2008). It most likely is effective in high-frequency EM (8–15 headache days per month) but not low-frequency EM or chronic TTH (Alpuente et al. 2019; Freund and Rao 2019). Based on recent two meta-analyses, onabotA treatment compared with placebo resulted in a reduction of 2.0 (95% CI 1.1 to 2.8; $n = 1{,}384$) or 1.6 (95% CI 0.1 to 3.1; $n = 1{,}546$) migraine days per month in patients with CM. The relative risk of treatment-related AEs were 2.2 (95% CI 1.7 to 2.8; $n = 2{,}839$) or 1.2 (95% CI 1.08 to 1.32, $n = 829$) when compared to placebo and 0.76 (95% CI 0.59 to 0.98; $n = 73$) when compared to oral prophylaxis (Herd et al. 2019; Bruloy et al. 2019). In most of these controlled onabotA trials, the placebo responses were high. Blinding may be an issue, as 60% correctly guessed their treatment after the third injection in the phase 2 study (Mathew et al. 2005). Despite the methodological concerns in these controlled trials, many long-term real-life (prospective and retrospective) studies confirmed onabotA's efficacy in improving headache frequency, disability, analgesic overuse, associated psychiatric symptoms, and quality of life (Table 1). It did not interact with oral preventives but could lead to their discontinuation. The majority of patients (70–80%) responded to onabotA within the first year of treatment, and many may continue to improve afterwards. The AE rate was less than 20%, including neck pain, ptosis, musculoskeletal stiffness, headache, and injection site discomfort (all were < 5%). Discontinuation rates were 15–48%, and 3–27% reported a lack of efficacy in these studies. The American Academy of Neurology practice guideline (2016) recommends that onabotA should be offered to patients with CM (to increase headache-free days) but not to those with EM (Level A). It is probably ineffective for treating chronic TTH (level B) (Simpson et al. 2016).

Some factors may be predictive of onabotA's treatment response. Exploding headaches (buildup of pressure inside the head) were more common (92%) in onabotA non-responders. In contrast, imploding headaches (perceiving the head to be crushed, clamped, or stubbed by an external force) were more common (74%) in responders (Jakubowski et al. 2006). Ocular headache was also associated with a higher response rate than nonocular headache (55% vs. 39%) (Lin et al. 2014). Allodynia has been associated with onabotA efficacy in peripheral neuropathic pain (Attal et al. 2016), but cephalic allodynia did not predict onabotA's clinical outcome (Sandrini et al. 2011; Young et al. 2019). Pericranial muscle tenderness, shorter duration of migraine, and onabotA use within the first 12 months of CM diagnosis were also associated with greater efficacy (Sandrini et al. 2011; Dominguez et al. 2018; Eross et al. 2005). Headache frequency, wearing off of effect, certain psychiatric symptoms, and medication overuse have been reported as predictors but may require further validation (Alpuente et al. 2020; Schiano di Cola et al. 2019; Ahmed et al. 2019b, 2019c). Interictal CGRP and, to a lesser degree, VIP levels measured in peripheral blood were predictive of the response to onabotA (Cernuda-Morollon

Table 1 Long-term (>1 year) studies on onabotulinumtoxinA for use in chronic migraine

Study	Subjects #	Duration	Effectiveness
Negro et al. 2015b P, single-center	132 (108F), 100% MOH	8 cycles	MHD: pre 22.3 ± 4.1, post 7.3 ± 2.1.* MMD: pre 21.4 ± 4.3, post 6.8 ± 2.3.* MAD: Pre 20.8 ± 4.5, post 5.3 ± 1.7*. 15% discontinued. 5% not effective
Negro et al. 2015a P, single-center	143 (114F), 100% MOH	8 cycles	195 U significantly more effective than 155 U in reducing MHD, MMD, MAD at every time point post-treatment. 17% discontinued. 2.9% not effective. TEAE: 17.5% for 155 U, 20.3% 195 U (nonsignificant)
Cernuda-Morollon et al. 2015 R, single-center	132 (119F), 41% MOH	7.7 (2–29) cycles	81.8% and 74.2% ≥50% responders at 1 and 2 years, respectively. 27% no response. 40%, injection interval extended to 4 months after 1 year
Aicua-Rapun et al. 2016 R, multi-center	115 (98F), 80% MOH	7.6 (5–13) cycles	78.5% remitted to EM. 61.9% no more MO. 45.2% preventive retired. 15.7% discontinued due to lack of efficacy. 36.5% injected >155 U
Kollewe et al. 2016 P, single-center	27 (25F), 61% MOH	6.5 (4–13) cycles	MHD: −10.2 ± 5.1*, MMD: −9.4 ± 5.8*, MAD: −5.9 ± 6.2*. 19% discontinued, 26% partial responders, 11% non-responders. Significant improvement on SF-36, MSQ, HIT-6, and BDI
Guerzoni et al. 2017 R, single-center	90 (76F), 100% MOH	Up to 13 cycles	CM (all MOH) 98% before treatment, 63% ($n = 37/59$) after 1 year; 67% ($n = 14/21$) after 2 years, 54% ($n = 7/13$) after 3 years. MOH 59% ($n = 35/59$) after 1 year; 62% ($n = 13/21$) after 2 years; 54% ($n = 7/13$) after 3 years. 8 patients used 195 U
Blumenfeld et al. 2018, 2019 P, multicenter NCT01516892	716 (607F), 64% MOH	9 cycles	MHD: −10.7* after 9 treatments ($n = 402$). TEAE 10.5%. HIT-6, PHQ-9, GAD-7 significantly improved. 56% completed and 48% withdrew from study. 6% lack of efficacy
Stark et al. 2019 R, multi-center	211 (187F), 61% MOH	2–14 cycles	MHD: −16.9 ± 9.0* ($n = 137$), MMD: −10.0 ± 8.4* ($n = 129$), MAD: −12.7 ± 8.1* ($n = 103$). 26% <50% reduction after 2 cycles
Ahmed et al. 2019a P, multicenter NCT01686581	633 (540F), 36% MOH	5.5 (1–13) cycles	MHD: −13.1* after 8 cycles ($n = 200$). 36% >155 U. Significant improvement in MSQ, EQ-5D. 23% discontinued. 14% lack of efficacy. 79% interval > 13 weeks

The list is sorted by publication year. *$P < 0.05$

P prospective, *R* retrospective, *MOH* medication overuse headache, *F* female, *MHD* mean monthly headache days, *MMD* mean monthly migraine days, *MAD* mean acute medication use days, *TEAE* treatment emergent adverse event. *HIT-6™* headache impact test, *SF-36* 36-item short form survey, *MSQ* Migraine-specific quality of life questionnaire, *BDI* Beck's depression inventory, *PHQ-9* patient health questionnaire-9, *GAD-7* general anxiety disorder-7, *EQ-5D* EuroQol 5D et al. 2014). The ratio of the mean velocity of the middle cerebral artery to that of the ipsilateral internal carotid artery (MCA/ICA index) was significantly higher in responders than non-responders (Lee et al. 2016). While white-matter lesions on MRIs were of no predictive value (Bumb et al. 2013), onabotA responders may have certain distinct morphometric (e.g., cortical thickness) and functional (e.g., connectivity) features that differ from non-responders (Hubbard et al. 2016). To date, there is no definitive biomarker for predicting the benefit of onabotA, but the abovementioned features may help clinicians predict the treatment response for their patients.

With the advent of CGRP functional blocking mAbs, it is of great interest whether onabotA works synergistically with these mAbs. CGRP mAb blocks CGRP's function on Aδ trigeminal sensory fibers. In contrast, onabotA acts on unmyelinated C fibers. These complementary mechanisms of action strongly suggest that functional blockade via CGRP has a synergistic effect with onabotA. In a retrospective analysis of 67 patients with CM, Armanious et al. showed that the addition of erenumab 70 mg and 140 mg to patients on onabotA resulted in a reduction of 3.1 and 11.5 mean monthly migraine days, respectively, after 60 days (Armanious and Jimenez-Sanders 2019). Yuan et al. reported in their retrospective analysis that following erenumab addition to onabotA therapy, 18/34 patients had further decreases in headache days while 6/34 had not (Yuan et al. 2019). In a prospective observational study of 158 patients, Boudreau et al. found that in patients with CM who had failed more than three preventive drugs, the addition of erenumab (after 4 injections) to those who were on onabotA or oral preventives had better outcome than those without concomitant preventive therapy; erenumab and onabotA combined was the most effective (Boudreau and Catherine 2019). Combination therapy thus appears effective and perhaps synergistic, but more studies are needed to confirm this.

8 Treatment Guidelines

OnabotA therapy is indicated for the preventive treatment of patients with CM. It may also be effective for patients with high-frequency EM (Alpuente et al. 2019). Currently, the FDA has approved onabotA but no other BoNTA formulation for use in CM. OnabotA use is contraindicated in patients with sensitivity to any BoNT. It must be used with caution in patients with neuromuscular disorders, such as myasthenia gravis (Blumenfeld et al. 2003). If there is insufficient response or wearing off, a higher dose (up to 195 U) administered using a follow-the-pain approach can be considered. For stable responders >1 year, the injection interval can be extended or even completely tapered off. At least three cycles of onabotA should be administered before considering treatment failure.

9 OnabotA Injection Techniques

Sterile technique should be observed for the entire onabotA injection procedure. Injections do not have to be intramuscular; the muscles are just used as reference sites for injections, which are most commonly administered in the glabellar and frontal regions, the temporalis muscle, the occipitalis muscle, and the cervical paraspinal region.

The injection protocols commonly used are: (1) the fixed-site approach, which uses fixed, symmetrical injection sites and a range of predetermined doses; (2) the follow-the-pain approach, which often employs asymmetrical injections and adjusts the sites and doses depending on where the patient feels pain and where the examiner can elicit pain and tenderness on palpation of the muscle; and a combination approach, which uses injections at fixed frontal sites, supplemented with follow-the-pain injections (this approach typically uses higher doses of onabotA). More procedural detail can be seen in references from Blumenfeld et al. (Blumenfeld et al. 2010, 2017).

Figures 1, 2, 3, 4, 5, and 6 list the recommended anatomical sites of onabotA injection for headache and the onabotA dose per site used in the PREEMPT trials (mean dose used in PREEMPT is 165 U). OnabotA (155 U) is administered via 31 fixed-site, fixed-dose injections across 7 specific head and neck muscle areas. A sterile 1 ml Luer-Lok syringe with a 30-gauge 0.5 in needle is used. Each injection is 0.1 mL, which contains 5 U of onabotA. Up to 40 U of additional onabotA can be administered, using a follow-the-pain approach, into the temporalis, occipitalis, and/or trapezius muscles, with a maximum dose of 195 U administered to 39 sites. In early studies, there was more neck pain if the trapezius muscles were injected, likely due to the use of a longer needle and different injection angles. When deciding on dose and location for additional onabotA, the location of the patient's

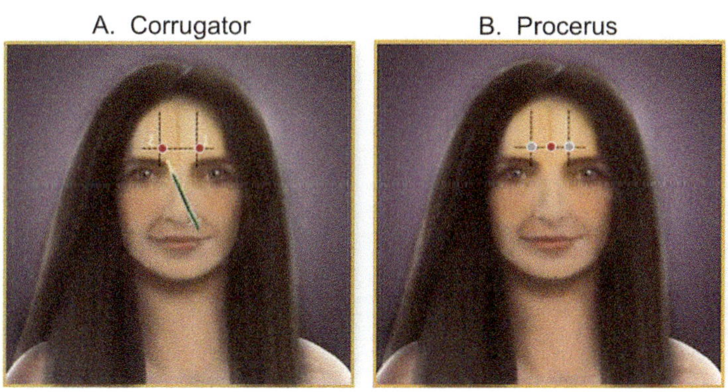

Fig. 1 (**a**) The corrugator injection sites (bilateral) are above the medial superior edge of the orbital ridge (bony landmark). (**b**) The procerus site is above and midline to the medial superior aspect of the orbital ridge (bony landmark) of each eye

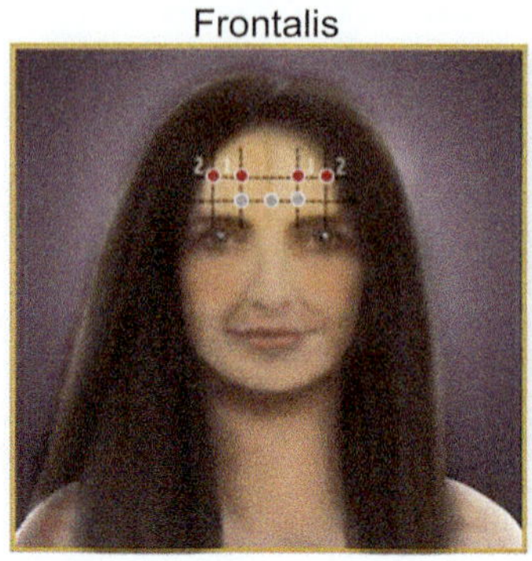

Fig. 2 The frontalis injection sites (bilateral) are located just beneath the skin surface of the central and forehead regions

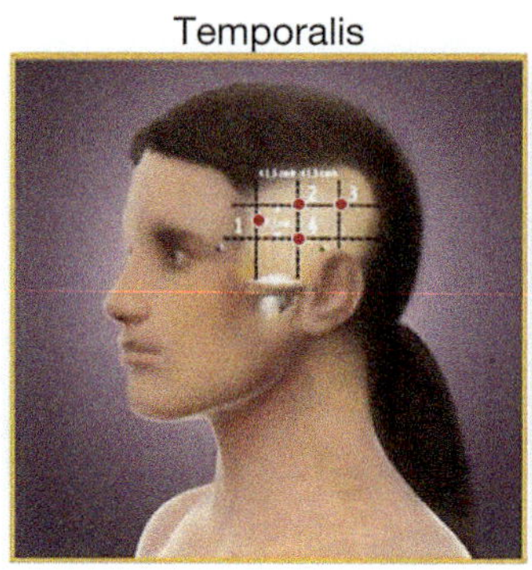

Fig. 3 The first injection site is located in the anterior aspect of the temporalis muscle. The second and fourth sites are within the medial aspect, and the third site is located in the posterior aspect of this muscle. These injections should be repeated on the left side for a total of eight injections into the temporalis muscle. Additional injections can be distributed between the right and left temporalis muscles in areas of maximal tenderness and/or pain

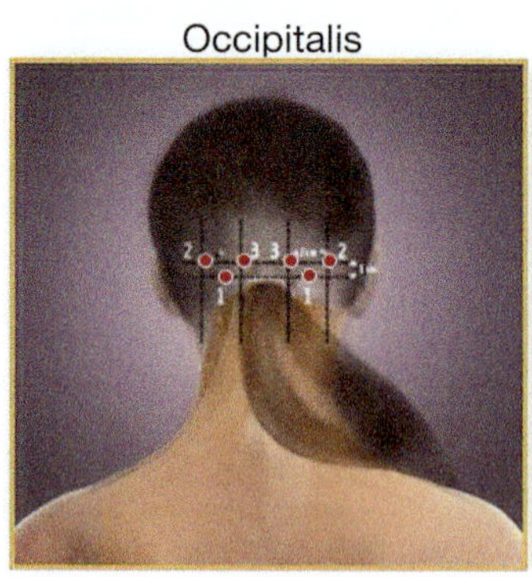

Fig. 4 The six occipitalis muscle injection sites are located superior to the supranuchal ridge on either side of the occipital protuberance. In the areas of maximal tenderness and/or pain, up to two additional injections can be distributed across the right and left occipitalis muscles

Cervical Paraspinal

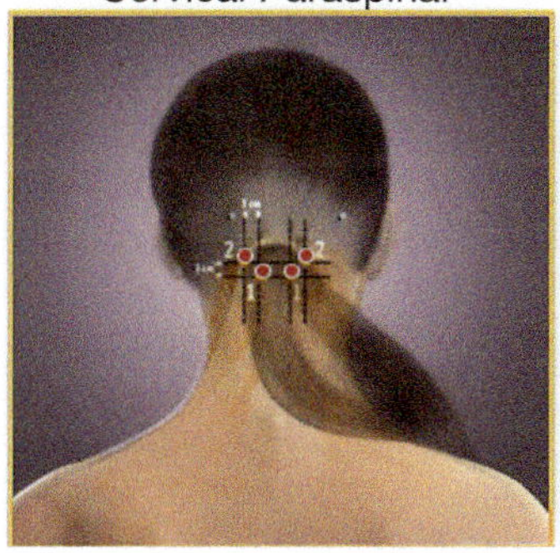

Fig. 5 The first cervical paraspinal injection site is lateral to the midline and inferior to the occipital protuberance. The second site is lateral and superior to the first injection. These injections should be repeated symmetrically on the contralateral side for a total of four injections

Trapezius

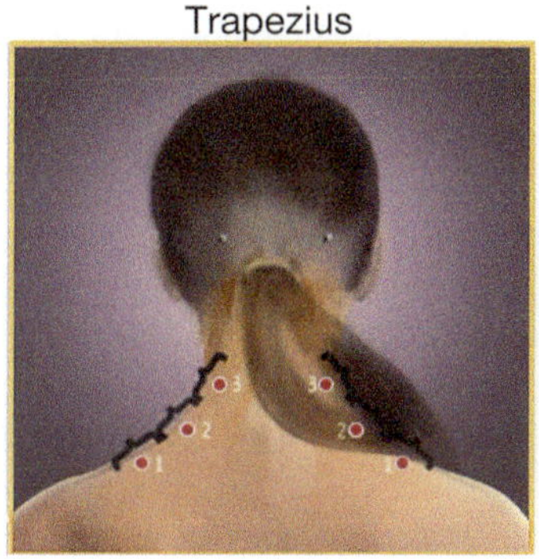

Fig. 6 The first of the three trapezius muscle injection sites is located in the lateral aspect of the muscle. The second site is within the mid-portion of the muscle, and the third site is within the superior aspect of the muscle. Symmetrical injections should be repeated on the contralateral side for a total of six injections. Up to four additional injections can be distributed between the right and left trapezius muscles, in the areas identified as having maximal tenderness

predominant pain and the severity of palpable muscle tenderness should be considered. Proper understanding of the anatomy behind each injection and variability between each individual not only optimizes efficacy but also minimizes unwanted outcomes and AEs (Blumenfeld et al. 2017).

10 Adverse Events Associated with OnabotA Use

More than two decades of clinical use have established the safety of onabotA. The most frequently reported AEs were neck pain, muscle weakness, musculoskeletal stiffness, ptosis, injection site pain, and headache. Rash and flu-like symptoms can rarely occur as a result of an allergic reaction. However, serious allergic reactions have never been reported. Injection of anterior neck muscles can cause dysphagia (swallowing difficulties) in some patients. The most common side effects when treating facial muscles are cosmetic and include ptosis or asymmetry of the position of the eyebrows. Another possible but rare side effect is difficulty in holding the head erect because of neck muscle weakness. Frontotemporal muscle atrophy has been reported after treatment for >5 years. Headache patients occasionally develop a headache following the injection procedure, although some have immediate relief of an acute attack; the latter is most likely due to a trigger point injection effect. Worsening of headaches and neck pain can occur and last for several days or, rarely, weeks after the injections because of the irritating effect of the needling and delay in the muscle relaxing effect of onabotA.

11 Summary

Migraine, especially CM, is a common debilitating disorder that profoundly impacts patient's quality of life. Existing preventive and acute nonbiologic treatments vary in efficacy and may be associated with intolerable AEs. The biologic onabotA is a safe and effective treatment for the prevention of CM, and perhaps high-frequency EM. BoNTA cleaves SNAP-25 and exerts its action on peptidergic trigeminal sensory fibers, interfering with neuropeptide release and pain-related receptors insertion into the plasma membrane. Through retrograde transport, BoNTA itself or cleaved SNAP-25 may act in the TCC and contralateral dural fibers to block peripheral and central pain sensitizations. Further research is still needed to understand the mechanism of action of BoNTA in headache, identify predictive biomarkers for its efficacy, establish its potential synergy with CGRP functional blocking agents, and fully develop its therapeutic potential.

Acknowledgments *Disclosure:* Dr. Yuan received honoraria from Supernus Pharmaceuticals, Inc. Dr. Silberstein is a consultant and/or advisory panel member for and has receives honoraria from Alder (Lundbeck) Biopharmaceuticals; Allergan, Inc.; Amgen; Biohaven; Cefaly; Curelator; eNeura Inc.; electroCore Medical, LLC; Impel NeuroPharma; Medscape, LLC; Novartis; Ipsen Biopharmaceuticals; Eli Lilly, MedscapevLLC; Satsuma; Supernus Pharmaceuticals, Inc.; Teva Pharmaceuticals and Trigemina, Inc. No funding was received for this manuscript.

References

Adams AM, Serrano D, Buse DC, Reed ML, Marske V, Fanning KM, Lipton RB (2015) The impact of chronic migraine: the chronic migraine epidemiology and outcomes (CaMEO) study methods and baseline results. Cephalalgia 35:563–578

Ahmed F, Gaul C, Garcia-Monco JC, Sommer K, Martelletti P, Investigators RP (2019a) An open-label prospective study of the real-life use of on a botulinum toxin A for the treatment of chronic migraine: the REPOSE study. J Headache Pain 20:26

Ahmed F, Buture A, Tanvir T, Khalil M (2019b) Medication overuse does not affect response to OnabotulinumtoxinA treatment in patients with Chronic Migraine; update from real-life data from Hull Headache Clinic. In: Paper presented at the international headache congress. Dublin, Ireland, September 5–8, pp IHC-PO-397

Ahmed F, Buture A, Tanvir T, Khalil M (2019c) Analysis of patterns of response to OnabotulinumtoxinA in Chronic Migraine in predicting long-term outcome. In: Paper presented at the International Headache Congress. Dublin, Ireland, September 5–8, pp IHC-PO-398

Aicua-Rapun I, Martinez-Velasco E, Rojo A, Hernando A, Ruiz M, Carreres A, Porqueres E, Herrero S, Iglesias F, Guerrero AL (2016) Real-life data in 115 chronic migraine patients treated with Onabotulinumtoxin a during more than one year. J Headache Pain 17:112

Alpuente A, Gallardo VJ, Torres-Ferrus M, Alvarez-Sabin J, Pozo-Rosich P (2019) Early efficacy and late gain in chronic and high-frequency episodic migraine with onabotulinumtoxinA. Eur J Neurol 26:1464–1470

Alpuente A, Gallardo VJ, Torres-Ferrus M, Alvarez-Sabin J, Pozo-Rosich P (2020) Short and mid-term predictors of response to OnabotulinumtoxinA: real-life experience observational study. Headache 60:677–685

American Headache Society (2019) The American headache society position statement on integrating new migraine treatments into clinical practice. Headache 59:1–18

Antonucci F, Corradini I, Fossati G, Tomasoni R, Menna E, Matteoli M (2016) SNAP-25, a known presynaptic protein with emerging postsynaptic functions. Front Synaptic Neurosci 8:7

Armanious M, Jimenez-Sanders R (2019) A retrospective analysis to evaluate the response of the addition of erenumab to onabotulinumtoxinA for the prevention of intractable chronic migraine without aura. In: Paper presented at the AHS 61st annual scientific meeting. Philadelphia, USA, July 11–14, pp P226LB

Attal N, de Andrade DC, Adam F, Ranoux D, Teixeira MJ, Galhardoni R, Raicher I, Uceyler N, Sommer C, Bouhassira D (2016) Safety and efficacy of repeated injections of botulinum toxin in peripheral neuropathic pain (BOTNEP): a randomised, double-blind, placebo-controlled trial. Lancet Neurol 15:555–565

Aurora SK, Dodick DW, Turkel CC, DeGryse RE, Silberstein SD, Lipton RB, Diener HC, Brin MF, PREEMPT Chronic Migraine Study Group (2010) OnabotulinumtoxinA for treatment of chronic migraine: results from the double-blind, randomized, placebo-controlled phase of the PREEMPT 1 trial. Cephalalgia 30:793–803

Barash JR, Arnon SS (2014) A novel strain of Clostridium botulinum that produces type B and type H botulinum toxins. J Infect Dis 209:183–191

Black JD, Dolly JO (1986) Interaction of 125I-labeled botulinum neurotoxins with nerve terminals. II. Autoradiographic evidence for its uptake into motor nerves by acceptor-mediated endocytosis. J Cell Biol 103:535–544

Blasi J, Chapman ER, Link E, Binz T, Yamasaki S, Camilli PD, Südhof TC, Niemann H, Jahn R (1993) Botulinum neurotoxin A selectively cleaves the synaptic protein SNAP-25. Nature 365:160–163

Blersch W, Schulte-Mattler WJ, Przywara S, May A, Bigalke H, Wohlfarth K (2002) Botulinum toxin A and the cutaneous nociception in humans: a prospective, double-blind, placebo-controlled, randomized study. J Neurol Sci 205:59–63

Bloudek LM, Stokes M, Buse DC, Wilcox TK, Lipton RB, Goadsby PJ, Varon SF, Blumenfeld AM, Katsarava Z, Pascual J, Lanteri-Minet M, Cortelli P, Martelletti P (2012) Cost of healthcare for patients with migraine in five European countries: results from the International Burden of Migraine Study (IBMS). J Headache Pain 13:361–378

Blumenfeld AM, Binder W, Silberstein SD, Blitzer A (2003) Procedures for administering botulinum toxin type A for migraine and tension-type headache. Headache 43:884–891

Blumenfeld AM, Schim JD, Chippendale TJ (2008) Botulinum toxin type A and divalproex sodium for prophylactic treatment of episodic or chronic migraine. Headache 48:210–220

Blumenfeld A, Silberstein SD, Dodick DW, Aurora SK, Turkel CC, Binder WJ (2010) Method of injection of onabotulinumtoxinA for chronic migraine: a safe, well-tolerated, and effective treatment paradigm based on the PREEMPT clinical program. Headache 50:1406–1418

Blumenfeld AM, Silberstein SD, Dodick DW, Aurora SK, Brin MF, Binder WJ (2017) Insights into the functional anatomy behind the PREEMPT injection paradigm: guidance on achieving optimal outcomes. Headache 57:766–777

Blumenfeld AM, Stark RJ, Freeman MC, Orejudos A, Manack AA (2018) Long-term study of the efficacy and safety of OnabotulinumtoxinA for the prevention of chronic migraine: COMPEL study. J Headache Pain 19:13

Blumenfeld AM, Tepper SJ, Robbins LD, Manack Adams A, Buse DC, Orejudos A, Silberstein DS (2019) Effects of onabotulinumtoxinA treatment for chronic migraine on common comorbidities including depression and anxiety. J Neurol Neurosurg Psychiatry 90:353–360

Boudreau GP, Catherine D (2019) Treatment of chronic migraine with erenumab alone or as an add on therapy; a real world prospective observational study. Paper presented at the International Headache Congress. Dublin, Ireland, September 5–8, pp IHC-OR-025

Brashear A, Gordon MF, Elovic E, Kassicieh VD, Marciniak C, Do M, Lee CH, Jenkins S, Turkel C (2002) Botox post-stroke spasticity study G. intramuscular injection of botulinum toxin for the treatment of wrist and finger spasticity after a stroke. N Engl J Med 347:395–400

Bruloy E, Sinna R, Grolleau JL, Bout-Roumazeilles A, Berard E, Chaput B (2019) Botulinum toxin versus placebo: a meta-analysis of prophylactic treatment for migraine. Plast Reconstr Surg 143:239–250

Bumb A, Seifert B, Wetzel S, Agosti R (2013) Patients profiling for Botox(R) (onabotulinum toxin A) treatment for migraine: a look at white matter lesions in the MRI as a potential marker. Springerplus 2:377

Burstein R, Zhang X, Levy D, Aoki KR, Brin MF (2014) Selective inhibition of meningeal nociceptors by botulinum neurotoxin type A: therapeutic implications for migraine and other pains. Cephalalgia 34:853–869

Buse DC, Manack A, Serrano D, Turkel C, Lipton RB (2010) Sociodemographic and comorbidity profiles of chronic migraine and episodic migraine sufferers. J Neurol Neurosurg Psychiatry 81:428–432

Cady RK, Schreiber CP, Porter JA, Blumenfeld AM, Farmer KU (2011) A multi-center double-blind pilot comparison of onabotulinumtoxinA and topiramate for the prophylactic treatment of chronic migraine. Headache 51:21–32

Cai BB, Francis J, Brin MF, Broide RS (2017) Botulinum neurotoxin type A-cleaved SNAP25 is confined to primary motor neurons and localized on the plasma membrane following intramuscular toxin injection. Neuroscience 352:155–169

Caleo M, Spinelli M, Colosimo F, Matak I, Rossetto O, Lackovic Z, Restani L (2018) Transynaptic action of botulinum neurotoxin type a at central cholinergic Boutons. J Neurosci 38:10329–10337

Cernuda-Morollon E, Martinez-Camblor P, Ramon C, Larrosa D, Serrano-Pertierra E, Pascual J (2014) CGRP and VIP levels as predictors of efficacy of Onabotulinumtoxin type A in chronic migraine. Headache 54:987–995

Cernuda-Morollon E, Ramon C, Larrosa D, Alvarez R, Riesco N, Pascual J (2015) Long-term experience with onabotulinumtoxinA in the treatment of chronic migraine: what happens after one year? Cephalalgia 35:864–868

Colasante C, Rossetto O, Morbiato L, Pirazzini M, Molgo J, Montecucco C (2013) Botulinum neurotoxin type A is internalized and translocated from small synaptic vesicles at the neuromuscular junction. Mol Neurobiol 48:120–127

Connan C, Popoff MR (2017) Uptake of Clostridial neurotoxins into cells and dissemination. In: Barth H (ed) Uptake and trafficking of protein toxins. Springer, Basel, pp 39–78

Cui M, Khanijou S, Rubino J, Aoki KR (2004) Subcutaneous administration of botulinum toxin A reduces formalin-induced pain. Pain 107:125–133

Delnooz CC, Pasman JW, Beckmann CF, van de Warrenburg BP (2013) Task-free functional MRI in cervical dystonia reveals multi-network changes that partially normalize with botulinum toxin. PLoS One 8:e62877

Delnooz CC, Pasman JW, van de Warrenburg BP (2015) Dynamic cortical gray matter volume changes after botulinum toxin in cervical dystonia. Neurobiol Dis 73:327–333

Diener HC, Dodick DW, Aurora SK, Turkel CC, DeGryse RE, Lipton RB, Silberstein SD, Brin MF, Group PCMS (2010) OnabotulinumtoxinA for treatment of chronic migraine: results from the double-blind, randomized, placebo-controlled phase of the PREEMPT 2 trial. Cephalalgia 30:804–814

Dodick DW, Turkel CC, DeGryse RE, Aurora SK, Silberstein SD, Lipton RB, Diener HC, Brin MF, Group PCMS (2010) OnabotulinumtoxinA for treatment of chronic migraine: pooled results from the double-blind, randomized, placebo-controlled phases of the PREEMPT clinical program. Headache 50:921–936

Dodick DW, Loder EW, Manack Adams A, Buse DC, Fanning KM, Reed ML, Lipton RB (2016) Assessing barriers to chronic migraine consultation, diagnosis, and treatment: results from the chronic migraine epidemiology and outcomes (CaMEO) study. Headache 56:821–834

Dominguez C, Pozo-Rosich P, Torres-Ferrus M, Hernandez-Beltran N, Jurado-Cobo C, Gonzalez-Oria C, Santos S, Monzon MJ, Latorre G, Alvaro LC, Gago A, Gallego M, Medrano V, Huerta M, Garcia-Alhama J, Belvis R, Leira Y, Leira R (2018) On a botulinum toxin A in chronic migraine: predictors of response. A prospective multicentre descriptive study. Eur J Neurol 25:411–416

Drinovac Vlah V, Bach-Rojecky L, Lackovic Z (2016) Antinociceptive action of botulinum toxin type A in carrageenan-induced mirror pain. J Neural Transm (Vienna) 123:1403–1413

Drinovac V, Bach-Rojecky L, Matak I, Lackovic Z (2013) Involvement of mu-opioid receptors in antinociceptive action of botulinum toxin type A. Neuropharmacology 70:331–337

Drinovac V, Bach-Rojecky L, Lackovic Z (2014) Association of antinociceptive action of botulinum toxin type A with GABA-A receptor. J Neural Transm (Vienna) 121:665–669

Durham PL, Cady R, Cady R (2004) Regulation of calcitonin gene-related peptide secretion from trigeminal nerve cells by botulinum toxin type A: implications for migraine therapy. Headache 44:35–42; discussion 42-33

Eross EJ, Gladstone JP, Lewis S, Rogers R, Dodick DW (2005) Duration of migraine is a predictor for response to botulinum toxin type A. Headache 45:308–314

Ferrari A, Manca M, Tugnoli V, Alberto L (2018) Pharmacological differences and clinical implications of various botulinum toxin preparations: a critical appraisal. Funct Neurol 33:7–18

Filipovic B, Matak I, Bach-Rojecky L, Lackovic Z (2012) Central action of peripherally applied botulinum toxin type A on pain and dural protein extravasation in rat model of trigeminal neuropathy. PLoS One 7:e29803

Freund B, Rao A (2019) Efficacy of botulinum toxin in tension-type headaches: a systematic review of the literature. Pain Pract 19:541–551

Gazerani P, Pedersen NS, Staahl C, Drewes AM, Arendt-Nielsen L (2009) Subcutaneous botulinum toxin type A reduces capsaicin-induced trigeminal pain and vasomotor reactions in human skin. Pain 141:60–69

GBD 2016 DALYs Hale Collaborators (2017) Global, regional, and national disability-adjusted life-years (DALYs) for 333 diseases and injuries and healthy life expectancy (HALE) for 195 countries and territories, 1990–2016: a systematic analysis for the Global Burden of Disease Study 2016. Lancet 390:1260–1344

GBD 2016 Disease and Injury Incidence and Prevalence Collaborators (2017) Global, regional, and national incidence, prevalence, and years lived with disability for 328 diseases and injuries for 195 countries, 1990–2016: a systematic analysis for the Global Burden of Disease Study 2016. Lancet 390:1211–1259

Gerona RR, Larsen EC, Kowalchyk JA, Martin TF (2000) The C terminus of SNAP25 is essential for Ca(2+)-dependent binding of synaptotagmin to SNARE complexes. J Biol Chem 275:6328–6336

Ghosal KJ, Patel K, Singh BR, Hale ML (2018) Role of critical elements in botulinum neurotoxin complex in toxin routing across intestinal and bronchial barriers. PLoS One 13:e0199524

Goadsby PJ, Holland PR, Martins-Oliveira M, Hoffmann J, Schankin C, Akerman S (2017) Pathophysiology of migraine: a disorder of sensory processing. Physiol Rev 97:553–622
Gu S, Jin R (2013) Assembly and function of the botulinum neurotoxin progenitor complex. Curr Top Microbiol Immunol 364:21–44
Guerzoni S, Pellesi L, Baraldi C, Cainazzo MM, Negro A, Martelletti P, Pini LA (2017) Long-term treatment benefits and prolonged efficacy of OnabotulinumtoxinA in patients affected by chronic migraine and medication overuse headache over 3 years of therapy. Front Neurol 8:586
Hallett M (2018) Mechanism of action of botulinum neurotoxin: unexpected consequences. Toxicon 147:73–76
Harper CB, Papadopulos A, Martin S, Matthews DR, Morgan GP, Nguyen TH, Wang T, Nair D, Choquet D, Meunier FA (2016) Botulinum neurotoxin type-A enters a non-recycling pool of synaptic vesicles. Sci Rep 6:19654
Headache Classification Committee of the International Headache Society (IHS) (2018) The international classification of headache disorders, 3rd edition. Cephalalgia 38:1–211
Hepp Z, Dodick DW, Varon SF, Chia J, Matthew N, Gillard P, Hansen RN, Devine EB (2017) Persistence and switching patterns of oral migraine prophylactic medications among patients with chronic migraine: a retrospective claims analysis. Cephalalgia 37:470–485
Herd CP, Tomlinson CL, Rick C, Scotton WJ, Edwards J, Ives NJ, Clarke CE, Sinclair AJ (2019) Cochrane systematic review and meta-analysis of botulinum toxin for the prevention of migraine. BMJ Open 9:e027953
Hong B, Yao L, Ni L, Wang L, Hu X (2017) Antinociceptive effect of botulinum toxin A involves alterations in AMPA receptor expression and glutamate release in spinal dorsal horn neurons. Neuroscience 357:197–207
Hubbard CS, Becerra L, Smith JH, DeLange JM, Smith RM, Black DF, Welker KM, Burstein R, Cutrer FM, Borsook D (2016) Brain changes in responders vs. non-responders in chronic migraine: markers of disease reversal. Front Hum Neurosci 10:497
Jakubowski M, McAllister PJ, Bajwa ZH, Ward TN, Smith P, Burstein R (2006) Exploding vs. imploding headache in migraine prophylaxis with botulinum toxin A. Pain 125:286–295
Jurasinski CV, Lieth E, Dang Do AN, Schengrund C-L (2001) Correlation of cleavage of SNAP-25 with muscle function in a rat model of Botulinum neurotoxin type A induced paralysis. Toxicon 39:1309–1315
Kollewe K, Escher CM, Wulff DU, Fathi D, Paracka L, Mohammadi B, Karst M, Dressler D (2016) Long-term treatment of chronic migraine with OnabotulinumtoxinA: efficacy, quality of life and tolerability in a real-life setting. J Neural Transm (Vienna) 123:533–540
Kosaras B, Jakubowski M, Kainz V, Burstein R (2009) Sensory innervation of the calvarial bones of the mouse. J Comp Neurol 515:331–348
Lackovic Z, Filipovic B, Matak I, Helyes Z (2016) Activity of botulinum toxin type A in cranial dura: implications for treatment of migraine and other headaches. Br J Pharmacol 173:279–291
Lazarov NE (2002) Comparative analysis of the chemical neuroanatomy of the mammalian trigeminal ganglion and mesencephalic trigeminal nucleus. Prog Neurobiol 66:19–59
Lee MJ, Lee C, Choi H, Chung CS (2016) Factors associated with favorable outcome in botulinum toxin A treatment for chronic migraine: a clinic-based prospective study. J Neurol Sci 363:51–54
Lin KH, Chen SP, Fuh JL, Wang YF, Wang SJ (2014) Efficacy, safety, and predictors of response to botulinum toxin type A in refractory chronic migraine: a retrospective study. J Chin Med Assoc 77:10–15
Maday S, Twelvetrees AE, Moughamian AJ, Holzbaur EL (2014) Axonal transport: cargo-specific mechanisms of motility and regulation. Neuron 84:292–309
Magalhaes E, Menezes C, Cardeal M, Melo A (2010) Botulinum toxin type A versus amitriptyline for the treatment of chronic daily migraine. Clin Neurol Neurosurg 112:463–466
Manack A, Buse DC, Serrano D, Turkel CC, Lipton RB (2011) Rates, predictors, and consequences of remission from chronic migraine to episodic migraine. Neurology 76:711–718

Marmura MJ, Silberstein SD, Schwedt TJ (2015) The acute treatment of migraine in adults: the American headache society evidence assessment of migraine pharmacotherapies. Headache 55:3–20

Masters-Israilov A, Robbins MS (2019) OnabotulinumtoxinA wear-off phenomenon in the treatment of chronic migraine. Headache 59:1753–1761

Matak I, Bach-Rojecky L, Filipovic B, Lackovic Z (2011) Behavioral and immunohistochemical evidence for central antinociceptive activity of botulinum toxin A. Neuroscience 186:201–207

Matak I, Rossetto O, Lackovic Z (2014) Botulinum toxin type A selectivity for certain types of pain is associated with capsaicin-sensitive neurons. Pain 155:1516–1526

Matak I, Bolcskei K, Bach-Rojecky L, Helyes Z (2019) Mechanisms of botulinum toxin type A action on pain. Toxins (Basel) 11:459

Mathew NT, Frishberg BM, Gawel M, Dimitrova R, Gibson J, Turkel C, Group BCS (2005) Botulinum toxin type A (BOTOX) for the prophylactic treatment of chronic daily headache: a randomized, double-blind, placebo-controlled trial. Headache 45:293–307

Melo-Carrillo A, Strassman AM, Nir RR, Schain AJ, Noseda R, Stratton J, Burstein R (2017) Fremanezumab-A humanized monoclonal anti-CGRP antibody-inhibits thinly Myelinated (Adelta) but not unmyelinated (C) meningeal Nociceptors. J Neurosci 37:10587–10596

Melo-Carrillo A, Strassman AM, Schain AJ, Noseda R, Ashina S, Adams A, Brin MF, Burstein R (2019) Exploring the effects of extracranial injections of botulinum toxin type A on prolonged intracranial meningeal nociceptors responses to cortical spreading depression in female rats. Cephalalgia 39:1358–1365

Messlinger K, Balcziak LK, Russo AF (2020) Cross-talk signaling in the trigeminal ganglion: role of neuropeptides and other mediators. J Neural Transm (Vienna) 127:431–444

Mika J, Rojewska E, Makuch W, Korostynski M, Luvisetto S, Marinelli S, Pavone F, Przewlocka B (2011) The effect of botulinum neurotoxin A on sciatic nerve injury-induced neuroimmunological changes in rat dorsal root ganglia and spinal cord. Neuroscience 175:358–366

Naumann M, Yakovleff A, Durif F, Group BCDPS (2002) A randomized, double-masked, crossover comparison of the efficacy and safety of botulinum toxin type A produced from the original bulk toxin source and current bulk toxin source for the treatment of cervical dystonia. J Neurol 249:57–63

Negro A, Curto M, Lionetto L, Martelletti P (2015a) A two years open-label prospective study of OnabotulinumtoxinA 195 U in medication overuse headache: a real-world experience. J Headache Pain 17:1

Negro A, Curto M, Lionetto L, Crialesi D, Martelletti P (2015b) OnabotulinumtoxinA 155 U in medication overuse headache: a two years prospective study. Springerplus 4:826

Paterson K, Lolignier S, Wood JN, McMahon SB, Bennett DL (2014) Botulinum toxin-A treatment reduces human mechanical pain sensitivity and mechanotransduction. Ann Neurol 75:591–596

Pellett S, Tepp WH, Scherf JM, Johnson EA (2015) Botulinum neurotoxins can enter cultured neurons independent of synaptic vesicle recycling. PLoS One 10:e0133737

Pirazzini M, Rossetto O, Bertasio C, Bordin F, Shone CC, Binz T, Montecucco C (2013) Time course and temperature dependence of the membrane translocation of tetanus and botulinum neurotoxins C and D in neurons. Biochem Biophys Res Commun 430:38–42

Relja M, Miletic V (2017) When movement disorders hurt: addressing pain in hyperkinetic disorders. Parkinsonism Relat Disord 44:110–113

Restani L, Antonucci F, Gianfranceschi L, Rossi C, Rossetto O, Caleo M (2011) Evidence for anterograde transport and transcytosis of botulinum neurotoxin A (BoNT/A). J Neurosci 31:15650–15659

Restani L, Giribaldi F, Manich M, Bercsenyi K, Menendez G, Rossetto O, Caleo M, Schiavo G (2012) Botulinum neurotoxins A and E undergo retrograde axonal transport in primary motor neurons. PLoS Pathog 8:e1003087

Rheaume C, Cai BB, Wang J, Fernandez-Salas E, Aoki KR, Francis J, Broide RS (2015) A highly specific monoclonal antibody for botulinum neurotoxin type A-cleaved SNAP25. Toxins (Basel) 7:2354–2370

Rummel A (2015) The long journey of botulinum neurotoxins into the synapse. Toxicon 107:9–24

Rummel A (2017) Two feet on the membrane: uptake of clostridial neurotoxins. In: Barth H (ed) Uptake and trafficking of protein toxins. Springer, Basel, pp 1–37

Sandrini G, Perrotta A, Tassorelli C, Torelli P, Brighina F, Sances G, Nappi G (2011) Botulinum toxin type-A in the prophylactic treatment of medication-overuse headache: a multicenter, double-blind, randomized, placebo-controlled, parallel group study. J Headache Pain 12:427–433

Schiano di Cola F, Caratozzolo S, Liberini P, Rao R, Padovani A (2019) Response predictors in chronic migraine: medication overuse and depressive symptoms negatively impact Onabotulinumtoxin-A treatment. Front Neurol 10:678

Schueler M, Neuhuber WL, De Col R, Messlinger K (2014) Innervation of rat and human dura mater and pericranial tissues in the parieto-temporal region by meningeal afferents. Headache 54:996–1009

Shimizu T, Shibata M, Toriumi H, Iwashita T, Funakubo M, Sato H, Kuroi T, Ebine T, Koizumi K, Suzuki N (2012) Reduction of TRPV1 expression in the trigeminal system by botulinum neurotoxin type-A. Neurobiol Dis 48:367–378

Silberstein S, Mathew N, Saper J, Jenkins S (2000) Botulinum toxin type A as a migraine preventive treatment. For the BOTOX migraine clinical research group. Headache 40:445–450

Silberstein SD, Dodick DW, Aurora SK, Diener HC, DeGryse RE, Lipton RB, Turkel CC (2015) Per cent of patients with chronic migraine who responded per onabotulinumtoxinA treatment cycle: PREEMPT. J Neurol Neurosurg Psychiatry 86:996–1001

Simpson DM, Hallett M, Ashman EJ, Comella CL, Green MW, Gronseth GS, Armstrong MJ, Gloss D, Potrebic S, Jankovic J, Karp BP, Naumann M, So YT, Yablon SA (2016) Practice guideline update summary: botulinum neurotoxin for the treatment of blepharospasm, cervical dystonia, adult spasticity, and headache: report of the guideline development Subcommittee of the American Academy of neurology. Neurology 86:1818–1826

Solomon M, Nahas SJ, Segal JZ, Young WB (2011) Medication adaptation headache. Cephalalgia 31:515–517

Stark C, Stark R, Limberg N, Rodrigues J, Cordato D, Schwartz R, Jukic R (2019) Real-world effectiveness of on a botulinum toxin A treatment for the prevention of headaches in adults with chronic migraine in Australia: a retrospective study. J Headache Pain 20:81

Stokes M, Becker WJ, Lipton RB, Sullivan SD, Wilcox TK, Wells L, Manack A, Proskorovsky I, Gladstone J, Buse DC, Varon SF, Goadsby PJ, Blumenfeld AM (2011) Cost of health care among patients with chronic and episodic migraine in Canada and the USA: results from the International Burden of Migraine Study (IBMS). Headache 51:1058–1077

Toth S, Brueggmann EE, Oyler GA, Smith LA, Hines HB, Ahmed SA (2012) Tyrosine phosphorylation of botulinum neurotoxin protease domains. Front Pharmacol 3:102

Tsai YC, Kotiya A, Kiris E, Yang M, Bavari S, Tessarollo L, Oyler GA, Weissman AM (2017) Deubiquitinating enzyme VCIP135 dictates the duration of botulinum neurotoxin type A intoxication. Proc Natl Acad Sci USA 114:E5158–E5166

Verderio C, Grumelli C, Raiteri L, Coco S, Paluzzi S, Caccin P, Rossetto O, Bonanno G, Montecucco C, Matteoli M (2007) Traffic of botulinum toxins A and E in excitatory and inhibitory neurons. Traffic 8:142–153

Voller B, Sycha T, Gustorff B, Schmetterer L, Lehr S, Eichler HG, Auff E, Schnider P (2003) A randomized, double-blind, placebo controlled study on analgesic effects of botulinum toxin A. Neurology 61:940–944

Vollesen AL, Benemei S, Cortese F, Labastida-Ramirez A, Marchese F, Pellesi L, Romoli M, Ashina M, Lampl C (2018) School of advanced studies of the European headache F. Migraine and cluster headache - the common link. J Headache Pain 19:89

Wang J, Zurawski TH, Meng J, Lawrence G, Olango WM, Finn DP, Wheeler L, Dolly JO (2011) A dileucine in the protease of botulinum toxin A underlies its long-lived neuroparalysis: transfer of longevity to a novel potential therapeutic. J Biol Chem 286:6375–6385

Weise D, Weise CM, Naumann M (2019) Central effects of botulinum neurotoxin-evidence from human studies. Toxins (Basel):11

Welch MJ, Purkiss JR, Foster KA (2000) Sensitivity of embryonic rat dorsal root ganglia neurons to Clostridium botulinum neurotoxins. Toxicon 38:245–258

Silberstein SD, Lipton RB, Dodick DW (eds) (2007) Wolff's headache and other head pain, 8th edn. Oxford University Press, New York

Young WB, Ivan Lopez J, Rothrock JF, Orejudos A, Manack Adams A, Lipton RB, Blumenfeld AM (2019) Effects of on a botulinum toxin A treatment in patients with and without allodynia: results of the COMPEL study. J Headache Pain 20:10

Yuan H, Baggaley S, Ozudogru S, Digre K (2019) CGRP antibodies as adjunctive prophylactic therapy for prolonging the therapeutic effect of OnabotulinumtoxinA injections among chronic migraine patients. In: Paper presented at the AHS 61st annual scientific meeting. Philadelphia, USA, July 11–14, pp P09

Zhang X, Strassman AM, Novack V, Brin MF, Burstein R (2016) Extracranial injections of botulinum neurotoxin type A inhibit intracranial meningeal nociceptors' responses to stimulation of TRPV1 and TRPA1 channels: are we getting closer to solving this puzzle? Cephalalgia 36:875–886

Zhang S, Masuyer G, Zhang J, Shen Y, Lundin D, Henriksson L, Miyashita SI, Martinez-Carranza-M, Dong M, Stenmark P (2017) Identification and characterization of a novel botulinum neurotoxin. Nat Commun 8:14130

Zidan A, Roe C, Burke D, Mejico L (2019) OnabotulinumtoxinA wear-off in chronic migraine, observational cohort study. J Clin Neurosci 69:237–240

Zychowska M, Rojewska E, Makuch W, Luvisetto S, Pavone F, Marinelli S, Przewlocka B, Mika J (2016) Participation of pro- and anti-nociceptive interleukins in botulinum toxin A-induced analgesia in a rat model of neuropathic pain. Eur J Pharmacol 791:377–388

Botulinum Toxin and Pain

Zdravko Lacković

Contents

Abstract

This chapter is focused on analgesic mechanism of action of botulinum toxin type A (BoNT-A) including the action beyond peripheral nerve endings. With the exception of the meninges and possibly urinary bladder, the presence of BoNT-A activity in the periphery, cleaving SNAP25 as a target molecule, up to now was not convincingly shown. In contrast many reports demonstrated BoNT-A activity and the presence of cleaved SNAP25 in the brain and spinal cord. In a model of mirror pain BoNT-A analgesic effect can be achieved even without participation of peripheral nerve ending. Thus generalized hypothesis central or peripheral mechanism of action belongs to history, and there is a need to confirm or dispute the results with meninges, urinary bladder, and possibly with other, especially visceral organs.

The original version of this chapter was revised. A correction to this chapter can be found at https://doi.org/10.1007/164_2020_415

Z. Lacković (✉)
Department of Pharmacology, University of Zagreb School of Medicine, Zagreb, Croatia
e-mail: zdravko.lackovic@mef.hr

S. M. Whitcup, M. Hallett (eds.), *Botulinum Toxin Therapy*,
Handbook of Experimental Pharmacology 263, https://doi.org/10.1007/164_2019_348

There are two general options for the central actions of BoNT-A:

1. The activity ends by silencing primary sensory neuron thereby stopping the pain information further in the CNS.
2. Or thereafter, indirectly or transsynaptically, BoNT-A triggers smaller or larger neural loops, forming memory of pain in the CNS that could explain the bilateral effects after unilateral peripheral administration, similar effect in mirror image allodynia and the like

Intensive research has shown that peripherally administered BoNT-A reaches the CNS by axonal transport. There is increasing evidence that BoNT-A is preventing pain in a growing range of disorders. In the absence of unexpected findings, or an increase in the uncontrolled use of illicit preparations by uneducated persons, BoNT-A is emerging as a new long-lasting and relatively safe analgesic.

Keywords
Analgesia · Axonal transport · Botulinum toxin type A · Botulinum toxin type B · CNS · Pain · SNAP25 · SNARE · Transsynaptic transport

1 Introduction

There is enormous advancement in medical science and practice in this twenty-first century: artificial organs, gene therapy, robotic surgery, brain-computer interface, and more. One area is lagging behind, and it is pain, especially chronic pain, the greatest source of human misery and suffering. Hundreds of potential analgesics are being investigated, including opioids and nonsteroidal anti-inflammatory drugs. However, finding analgesics that will work that are strong enough, long enough, and free from serious side effects is a long-standing hope. From the multitude of substances under study, botulinum toxins, especially botulinum toxin type A (BoNT-A), and to a lesser extent BoNT-B, emerge as medicine that might have a special place. This review focuses mostly on BoNT-A.

2 How Neurotoxin Became an Analgesic

The beneficial effect of botulinum toxin serotype A (BoNT-A) in pain was first observed in the patients with painful cervical dystonia in 1985 (Tsui et al. 1985) and considered to be a consequence of reduced muscular contraction. However, eventually it becomes evident that the analgesic activity occurs before and lasts longer than the antispastic activity (Freund and Schwartz 2003). From the first observation in 1987, BoNT-A analgesic potency did not attract major attention for a long time. Several large trials, 33 years later, showed that BoNT-A is beneficial in chronic but not episodic migraine (Diener et al. 2000; Dodick et al. 2010). In 2010 FDA approved BoNT-A for chronic migraine.

3 BoNT-A Analgesia

A common classification is nociceptive, inflammatory, and chronic pain. However in many cases, this is a continuum. Touching something painfully hot, we remove the hand from the heat; this is the reflex present in life forms even without a brain. If the burn is severe enough, there will be inflammation lasting several days or longer. Especially if something went wrong, acute pain can become chronic, and for a long period we feel an unpleasant sensation; in some cases *traces of memory of pain* develop, and the pain and allodynia can continue after tissue damage is already repaired. Moreover in rare cases, where an amputation of burned body part was necessary, some patients still feel the pain in the extremity which does not exist anymore (*phantom pain*). As hippocampal and other neurons are developing plastic changes underlying the mechanism of memory and learning, nociceptive neurons are also developing plastic changes that can last longer than nociceptive stimuli: memory of the pain. Accordingly chronic pain could be considered a CNS disorder (Tracey and Bushnell 2009; Ji et al. 2013).

The first experimental evidence in vivo of antinociceptive effect of BoNT-A was published in 2004. It was shown that in the rat formalin test, BoNT-A diminished only the second inflammatory phase (Cui et al. 2004). The formalin test consists of an injection of dilute formaldehyde (1% in saline) usually in dorsal surface of one hindpaw, but could be any part of the body. The response is the amount of time the animals spend by movement pointing to painful place, usually licking the injection place. There are two distinct periods of high licking activity, an early phase lasting the first 5 min and a late phase lasting from 20 to 30 min after the injection of formalin.

On the basis of (1) this experiment in which analgesic effect of BoNT-A in vivo was shown only in the second, inflammatory phase of formalin test, (2) different in vitro studies showing inhibitory action of BoNT-A, (3) first evidence of analgesic effect of BoNT-A in patients, and (4) knowledge that BoNT-A inhibits the release of the acetylcholine at the neuromuscular junction, it was suggested that the analgesic mechanism in the sensory nerve is the same as it is in motor nerve: enzymatic blockade of neurotransmitter release (Aoki 2003, 2015). This idea was that both inflammation and pain is associated with peripheral release of inflammatory mediators and neurotransmitters like glutamate. However BoNT-A has antinociceptive effect in pain induced by intraplantar injection of capsaicin or carrageenan but no visible effect on inflammation. Thus, it was concluded that the *mechanism of the antinociceptive action of BoNT-A might be much more complex than the suggested inhibition of transmitter release in the periphery* (Bach-Rojecky and Lacković 2005). Contrary to BoNT-A, the skeletal muscle relaxant dantrolene produced more motor impairment than analgesia (Favre-Guilmard et al. 2009). Apparently, release of inflammatory substances and neurotransmitters involved in inflammation is separate from BoNT-A antinociceptive effect.

The most common model to study chronic pain in experimental animals is chronic constriction injury (CCI) of a nerve that results in mononeuropathy with long-lasting pain hypersensitivity and allodynia. There are a number of reports showing that BoNT-A reduces pain in CCI (Bach-Rojecky et al. 2005, 2010; Shinoda et al. 2007; Kitamura et al. 2009; Filipović et al. 2012).

In the first of those report, a peripheral application of BoNT-A (7 U/kg) significantly reduced thermal and mechanical hypersensitivity in rats with the partial sciatic nerve transection (Bach-Rojecky et al. 2005). Treatment with high dose of BoNT-A was associated with faster nerve regeneration (Marinelli et al. 2010).

4 Axonal Transport of BoNT-A

BoNT-B and tetanus toxin at molecular level cleave the same SNARE protein: VAMP/synaptobrevin. However BoNT-B produces flaccid paralysis, while tetanus toxin has clinically opposite effect, spastic paralysis. The basic difference is the existence of retrograde axonal transport of tetanus toxin in contrast to BoNT-s. This shows fundamental importance of the existence of axonal transport of BoNT-s.

First publication of central effect of peripherally applied BoNT dates to 1956 and was published in *Byulleten Eksperimental'noi Biologii i Meditsiny* (USSR). Soviet scientist V. V. Michailov found that administration of BoNT to experimental animals causes defect in brain stem reflexes (Michailov 1956, cit. by Tyler 1963). In 1963 Tyler published a case report in *Science* reporting that electrical stimulation of multiple peripheral nerves elicited "H" reflexes in a patient, 61 years old, with botulism. The author emphasized that *this "central" action of botulinum toxin is similar to that suggested for tetanus toxin* (Tyler 1963). Two years later Polley et al. (1965) found that BoNT-A has a depressant effect on the cortical electrical activity of monkeys.

To evaluate if peripherally applied BoNT-A can reach the CNS, the radioiodinated BoNT-A was prepared and observed in the spinal cord after peripheral injection (Habermann 1974; Wiegand et al. 1976). After unilaterally injected sublethal doses of ^{125}I-BoNT-A into gastrocnemius muscle of the cat, most radioactivity was localized in the spinal cord ipsilateral to radioactivity injection, as well as ventral roots innervating the injected muscle (Wiegand et al. 1976). Nearly 30 years later, distribution of radiolabeled holotoxin and only the active 150 KDa free toxin was investigated at the same experiment. However, almost no radioactivity was found in the brain (Tang-Liu et al. 2003).

Evidence for axonal transport was investigated in vitro. Using rat hippocampal neurons cultured in microfluidic devices, Koizumi et al. (2014) studied uptake and transfer of BoNT-A heavy chain in compartmentalized platforms. The simple system consists of two tiny chambers connected by a narrow passage. In one chamber neuronal soma is placed, while in another nerve terminals will grow. Passage between these chambers is so narrow that cell soma cannot pass through and it is retained in soma chamber. In such system Koizumi found activity-dependent uptake of BoNT-A Hc in the terminal chamber that led to a significant increase in SNAP25 cleavage detected in the soma chamber. Blocking autophagosome formation or acidification with wortmannin or bafilomycin A1, respectively, inhibited the activity-dependent retrograde transport of BoNT-A-Hc.

In another study in compartmented cultures of sympathetic neurons from rat superior cervical ganglion, neurons were examined after focal application of BoNT-A. A majority of cleaved SNAP-25 was seen locally, but some appeared along neurites and accumulated in the soma over several weeks. Neurite transection prevented movement of BoNT-A. However, spontaneous or evoked transmission to cell bodies was not inhibited by retrogradely migrated BoNT-A except with high doses. This was interpreted as the lack of evidence for a direct central action of BoNT-A (Lawrence et al. 2012). Opposite results in those two experiments might be addressed to the methodological differences; however, hippocampal (Koizumi et al. 2014) or sympathetic neurons (Lawrence et al. 2012) are not pain-transmitting sensory neurons.

Quantity of BoNT-A used in experimental animals or in human studies is very low, in picograms or low nanograms range that is not possible to trace to individual neurons. Development of antibody against BoNT-A enzymatic product cl-BoNT-A enabled tracing augmented signal and presence cl-SNAP25 in spinal cord and nuclei of cranial nerves (Antonucci et al. 2008; Matak et al. 2011, 2012, 2019).

Percutaneously, injected *formalin* into trigeminal ganglia destroyed all nerves, and after that BoNT-A had no behavioral effect, and no cl-SAP25 was detected upstream of the ganglia. Although rough this experiment excludes the possibility that BoNT-A can bypass trigeminal ganglia and reach trigeminal nucleus by some alternative route.

Cleaved SNAP25 in caudal trigeminal nuclei disappeared after sensory denervation, induced by transcutaneous application of *capsaicin* into trigeminal ganglia (Matak et al. 2014). Finally, the analgesic effect of BoNT-A in the formalin test, in bilateral pain as well, as in CC test, and cl-SNAP25 in the brain was prevented with intraneuronal application of axonal transport blocker colchicine (Bach-Rojecky and Lacković 2009; Filipović et al. 2014). These observations exclude passive spreading of BoNT-A along axons, establishing existence of axonal transport and partial overlapping with the capsaicin sensitive, i.e., vanilloid receptor.

5 BoNT-A Target Molecule

Most peripheral terminals of sensory neurons in the skin, viscera, and autonomic ganglia of guinea pig and mice lack immunoreactivity for SV2, SNARE (including SNAP25), or glutamate transporters. In dorsal root ganglia, most small neurons with immunoreactivity for both substance P and CGRP lacked immunoreactivity for SNAP25. Thus, molecular machinery considered essential for vesicular uptake and exocytotic release of glutamate or other neurotransmitters is not expressed at detectable levels by most peripheral sensory neurons containing SP and CGRP in rodents and guinea pig (Morris et al. 2005).

Marinelli et al. (2012) after peripheral administration of BoNT-A, together with the behavioral effects on CC neuropathic pain, found immunofluorescence of the cl-SNAP-25 in all tissues examined, from the peripheral nerve endings, sensory ganglia. Interestingly in the skin sections of naive mice intraplantarly injected only with saline, there was almost undetectable staining of cl-SNAP25. However in naïve mice injected with BoNT-A, intense GFAP staining in hindpaw nerve endings was accompanied by a diffuse staining of cl-SNAP25. High magnification images show a punctuate staining, interpreted as localization of cl-SNAP25 in the peripheral nerve terminals. Whether those peripheral nerves were sensory or autonomic was not identified. Appearance of cl-SNAP25 is expected effect of BoNT-A protease. On the contrary "punctate staining" as well as presence of any cl-SNAP25 in BoNT-A naïve mice raises the question about the specificity of anti cl-SNAP25 antibody (Marinelli et al. 2012).

Examination of cl-SNAP25 immunohistochemistry in guinea pig bladder after in vivo intramural injection of a toxin showed SNAP25 immunoreactive fibers abundant throughout the bladder tissue in the mucosa and muscular layer. Double labeling showed that toxin cleaves the SNAP25 protein mainly in cholinergic (parasympathetic) but also in adrenergic and sensory fibers (Coelho et al. 2012).

6 Transsynaptic, Cell-to-Cell Transport of BoNT-A

Bomba-Warczak et al. (2016) investigated the potential distal effects of BoNT-A, BoNT-D, and tetanus toxin, using hippocampal neurons grown in compartmentalized microfluidic devices. Neurons are placed in soma chamber. After 2 weeks axons were found in the opposing axon chamber. When the axon chamber was incubated with BoNT-A, intensive cleavage was observed in the soma chamber. Using axotomy of cultured neurons and specific antibody against BoNT-A, it was found that all three toxins are taken up, via two separate pathways: (1) usual synaptic vesicle recycling pathway that leads to local effects and (2) a distinct secondary uptake pathway that directs these toxins into non-acidified organelles that mediated retrograde transport to the soma chamber. Toxins were then released into the media, where they exerted their effects upon upstream neurons. These discoveries reveal that BoNT-A and -E similar to tetanus toxin undergo interneuronal transfer and transcytosis in an active form producing long-distance effects (Bomba-Warczak et al. 2016).

In vivo evidence for transsynaptic transport of BoNT-a was found in the CNS or motoric system (Antonucci et al. 2008; Caleo et al. 2018). Possibility of transsynaptic transport in sensory system is discussed together with mirror pain.

7 Bilateral Effect of BoNT-A Following Unilateral Injection

Effect of BoNT-A was studied in polyneuropathic pain caused by experimental streptozotocin diabetes (Bach-Rojecky et al. 2010; Favre-Guilmard et al. 2017) and paclitaxel-induced peripheral neuropathy (Favre-Guilmard et al. 2009). In both conditions neuropathic pain develops in both legs, and BoNT-A applied unilaterally reduced pain on both side.

Mirror pain or mirror-image allodynia occurs in the healthy body region contralateral to the site of nociceptive stimuli. This is still a mysterious phenomenon that occurs in association with many clinical pain syndromes and in different animal models of pathological pain.

Sluka et al. (2001) developed model of mirror pain induced by repeated intramuscular administration of acidic saline. Two unilateral injections of low pH saline, 5 days apart, caused a pH-dependent bilateral mechanical, but not heat, hyperalgesia lasting 30 days. Histopathological changes were minimal showing that such chronic muscle-induced pain is unrelated to tissue damage. Lidocaine injection into the gastrocnemius muscle or unilateral dorsal rhizotomy had no effect on the contralateral mechanical hyperalgesia. Apparently after second injection, some memory of the pain has been developed (Sluka et al. 2001).

Injection of 3% carrageenan in the muscle or knee produced hyperalgesia to mechanical and heat stimuli ipsilaterally, which lasted 7–8 weeks and spread to the contralateral side 1–2 weeks after injection. Histologically acute inflammation after 1 week transforms to chronic. Interestingly hyperalgesia that spreads to the contralateral side appeared at the same time period as the inflammation transforms from acute to chronic (Radhakrishnan et al. 2003).

In acidic saline, mirror pain ipsilateral injection of BoNT-A had a bilateral effect, while contralateral injection diminished pain only on that side. Injection of colchicine into the ipsilateral sciatic nerve bilaterally prevented antinociceptive activity of the BoNT-A. However, when colchicine was injected into the sciatic nerve opposite to the site of pain induction and BoNT-A injection, it did not prevent the BoNT-A antinociceptive effect on either side. This observation eliminated possible contribution of the contralateral peripheral nerve endings to the BoNT-A effect. After sciatic nerve was transected, BoNT-A in a dose as low as 0.5 U/kg was injected into the proximal part of a distally cut sciatic nerve, which reduced mirror pain hypersensitivity on the contralateral side. This observation surprisingly demonstrates that BoNT-A antinociceptive effect is independent from peripheral nerve endings (Bach-Rojecky and Lacković 2009).

Bilateral effect of BoNT-A was also demonstrated in mirror pain induced by carrageenan (Favre-Guilmard et al. 2017). In all models tested, BoNT-A alleviates the pain bilaterally.

8 Convergence Point of Pain in Trigeminal Region

Different types of pain in trigeminal region caused by formalin injection in a whisker pad, temporomandibular inflammation caused by injection of CFA, or infraorbital nerve constriction injury, much less occipital nerve constriction injury results in neurogenic inflammation of cranial dura (Filipović et al. 2012, 2014; Lacković et al. 2016) characterized by dural extravasation measured by appearance of proinflammatory cells and plasma protein extravasation in meningeal tissue. This phenomenon accompanies selectively only pain in extracranial trigeminal region and cannot be induced by pain in other parts of the body, as well as it is absent in spinal meninges. Apparently neurogenic inflammation of cranial meninges is a common, convergence point of different types of pain in trigeminal region. Application of BoNT-A abolishes pain behavior and in parallel abolishes the dural inflammation. Immunohistochemically, cl-SNAP25 was found in nerve elements of cranial dura, where it was colocalized with CGRP (Lacković et al. 2016). Those observations create the intriguing question how peripherally applied BoNT-A arrived to dura. BoNT-A effect can be prevented by colchicine injected into the trigeminal ganglion, indicating toxin's axonal transport (Filipović et al. 2012). Still the question remains how BoNT-A crosses from trigeminal extracranial nerve endings to trigeminal nerve endings in dura. Namely, meninges and extracranial trigeminal regions are innervated by separate sensory neurons (Shimizu et al. 2012). The logical conclusion seems that there is transsynaptic transport of retrogradely transported BoNT-A through peripheral branch of trigeminal nerves to trigeminal branches of the same nerve innervating dura. Transcytosis within the trigeminal ganglion after its peripheral injection has been suggested (Kitamura et al. 2009; Shimizu et al. 2012). However, transcytosis in the trigeminal sensory nuclei cannot be excluded (Matak and Lacković 2015; Ramachandran and Yaksh 2014). The third option is extracranial extensions of the nerves innervating the meninges. It is relatively less known that some nerves from dura have extracranial projections through sutures of the skull bones (Kosaras et al. 2009). Using electrophysiological techniques applied on those extracranial nerve terminals of experimental animals, it was found that BoNT-A achieves electrophysiological effects consistent with antimigraine effect (Burstein et al. 2017). Retrograde transport of BoNT-A through those dural extracranial nerves, passing to dura and trigeminal ganglia, seems as a third possibility. However, appearance of BoNT-A activity (cl-SNAP25) in dura after single injection in rat temporomandibular joint (Lacković et al. 2016), or vibrissal pad (Filipović et al. 2012), which is far away from skull sutures, indicates that extracranial extensions of dural nerves could not be only source of BoNT-A activity on meningeal nociceptors. In vitro spontaneous cholinergic neurotransmission is blocked over 80% by 1 pM BoNT-A despite cleaving only less than 20% of the SNAP25 (Lawrence et al. 2013). Clearly only a portion of SNAP25 needs to be cleaved to induce near-complete synaptic silencing.

The effect of BoNT-A beyond first, peripheral sensory neuron has been only fragmentarily investigated. Administration of BoNT-A into the rat whisker pad was without effect on 10 brain regions related to sensation of pain. The only significant effect was increase of concentration of noradrenaline in striatum and serotonin in hypothalamus (Ibragić et al. 2016). Whether this can play a role in reported BoNT-A efficacy for the treatment of depression remains to be investigated (Stearns et al. 2018).

Antinociceptive effects of BoNT-A in formalin and sciatic CC pain were abolished by low dose of intrathecal naltrexone or selective μ-antagonist naloxonazine. Additionally BoNT-A-induced decrease in dorsal horn c-Fos expression was prevented by naltrexone. Apparently this is a central effect because naltrexone abolished the effect of BoNT-A on pain and dural plasma protein extravasation, whereas peripherally acting methylnaltrexone did not. However, methylnaltrexone decreased the antinociceptive effect of morphine only partially in the second phase of the formalin test and had no significant effect on morphine-mediated reduction in dural neurogenic inflammation (Drinovac Vlah et al. 2018). BoNT-A enhances the analgesic effects of morphine on inflammatory pain and antagonizes tolerance induced by morphine in mice. Since the effects of BoNT-a on the opioid system were prevented by antagonist and augmented by agonist (morphine) (Vacca et al. 2012), it is clear that normal tone of endogenous opioid system, involving central μ-opioid receptor, is required for antinociceptive activity of BoNT-A.

Cl-SNAP25 has been identified in parasympathetic (pre- and postganglionic), sympathetic, and afferent fibers in the urinary bladder. BoNT-A reduces the release of acetylcholine from parasympathetic, norepinephrine from sympathetic, and glutamate and neuropeptides from sensory neurons (Cruz 2014).

Chronic pain is associated with glial activation: hypertrophy, proliferation, and upregulation of glial markers/mediators that modulate excitatory and inhibitory synaptic transmission (Rojewska et al. 2018). In vivo in sciatic nerve and dorsal root ganglia application of BoNT-A diminished neuroimmunological changes, activation of microglia/macrophages. The title of one publication emphasized the importance of glia: *Glia and pain: is chronic pain a gliopathy* (Ji et al. 2013)?

9 Emerging New Analgesic

Preclinical studies on experimental animals suggest specific pharmacological and pharmacotherapeutic characteristics:

A. A unique characteristic of BoNT-A is a long-lasting analgesic effect. In experimental animal analgesic, effect lasts up to 3 weeks or longer, in human 3 months and more.
B. BoNT-A has no effect on acute, reflexive, painful stimuli regardless of the cause (thermal, mechanical, chemical) that has important warning function and is vital to the life of the organism. In preclinical research first phase of formal test is the best known example.
C. Antinociceptive activity of BoNT-A is achieved at a lower dose than neuroparalytic activity and occurs usually with delay of 3–5 days after peripheral administration. The lowest effective antinociceptive dose of BoNT-A in rats is 3.5 U/kg after peripheral intraplantar administration (Bach-Rojecky et al. 2005), while a dose of 30 U/kg cause muscular weakness (Cui et al. 2004). Because BoNT-A has both analgesic and muscle relaxing/paralytic activity, the behavioral outcome, which is measured in experimental animals, is the balance between the two. This is probably the reason that up to now, there is no any reliable dose response of BoNT-A analgesic activity.
D. BoNT-A has a long-lasting analgesic effect in chronic pain of different origin like inflammatory pain, including neurogenic pain of the meninges, neuropathic pain, chronic visceral pain arising from inflammation, benign and malignant tumors, etc.

E. Regardless of the cause of chronic pain, up to now there are no negative results published. Accordingly, in experimental animals, it seems that BoNT-A has beneficial effect on all types of chronic pain, hyperalgesia, and allodynia.

In experimental animal beneficial effect of BoNT-A was reported in a model of trigeminal neuropathy (Filipović et al. 2012), trigeminal mandibular disorder (Lacković et al. 2016) or streptozotocin diabetes (Bach-Rojecky et al. 2010), and dozens more.

After registration for treatment of chronic migraine, analgesic effects of BoNT-A have been studied in many human disorders. In a short survey of PubMed in the 5-year period (Dec 2012–Oct 2019), we found 23 clinical trials in 14 painful disorders and 30 review-type publication (review, systemic review, meta-analysis, Cochrane, etc.) focused on 18 indications.

According to the American Academy of Neurology (AAN), the quality and risk of bias of clinical trials are evaluated and are categorized into Classes I, II, III, or IV. Quality and risk of bias decrease and increase, respectively, from Class I to Class IV studies. Moreover, recommendations for treatments can be formed after evaluation and classification of the evidence from strong to weak (level A to C), based on the quality and quantity of all the available scientific evidence.

Level of evidence for efficacy of BoNT-s in different pain syndromes using the recommended efficacy criteria from the Assessment and Therapeutic Subcommittee of the American Academy of Neurology is as follows (Safarpour and Jabbari 2018):

There is a level A evidence (effective) for BoNT therapy in

- Post-herpetic neuralgia
- Trigeminal neuralgia
- Posttraumatic neuralgia

There is a level B evidence (probably effective) for:

- Diabetic neuropathy
- Plantar fasciitis
- Piriformis syndrome
- Pain associated with total knee arthroplasty
- Male pelvic pain syndrome
- Chronic low back pain and male pelvic pain
- Neuropathic pain secondary to traumatic spinal cord injury

BoNT-s are possibly effective (Level C – one class II study):

- For female pelvic pain
- Painful knee osteoarthritis
- Post-operative pain in children with cerebral palsy after adductor release surgery
- Anterior knee pain with vastus lateralis imbalance

10 Conclusion

This review is focused on mechanism of action of BoNT-A including the action beyond peripheral nerve endings. With the exception of meninges and possibly urinary bladder, the inactivation of peripheral SNAP25, as a target molecule, is not convincingly demonstrated. In a model of mirror pain, BoNT-A analgesic effect can be achieved even without participation of peripheral nerve ending. Finding of axonal transport of BoNT-A from periphery to the CNS opens the way for new discoveries of its action beyond first sensory neuron, as well as new discoveries about chronic pain and formation of the memory of pain. There is increasing evidence that BoNT-A is preventing pain in a growing range of disorders. In the absence of unexpected findings, or an increase in the uncontrolled use of illicit preparations by uneducated persons, BoNT-A is emerging as a new long-lasting and relatively safe analgesic.

References

Antonucci F, Rossi C, Gianfranceschi L, Rossetto O, Caleo M (2008) Long-distance retrograde effects of botulinum neurotoxin A. J Neurosci 28(14):3689–3696. https://doi.org/10.1523/JNEUROSCI.0375-08.2008

Aoki KR (2003) Evidence for antinociceptive activity of botulinum toxin type A in pain management. Headache 43(Suppl 1):S9–S15. https://doi.org/10.1046/j.1526-4610.43.7s.3.x

Aoki KR (2015) Review of a proposed mechanism for the antinociceptive action of botulinum toxin type A. Neurotoxicology 26(5):785–793. https://doi.org/10.1016/j.neuro.2005.01.017

Bach-Rojecky L, Lacković Z (2005) Antinociceptive effect of botulinum toxin type a in rat model of carrageenan and capsaicin induced pain. Croat Med J 46(2):201–208

Bach-Rojecky L, Lacković Z (2009) Central origin of the antinociceptive action of botulinum toxin type A. Pharmacol Biochem Behav 94(2):234–238. https://doi.org/10.1016/j.pbb.2009.08.012pain

Bach-Rojecky L, Relja M, Lacković Z (2005) Botulinum toxin type A in experimental neuropathic pain. J Neural Transm (Vienna) 112(2):215–219. https://doi.org/10.1007/s00702-004-0265-1

Bach-Rojecky L, Salković-Petrisić M, Lacković Z (2010) Botulinum toxin type A reduces pain supersensitivity in experimental diabetic neuropathy: bilateral effect after unilateral injection. Eur J Pharmacol 633(1–3):10–14. https://doi.org/10.1016/j.ejphar.2010.01.020

Bomba-Warczak E, Vevea JD, Brittain JM, Figueroa-Bernier A, Tepp WH, Johnson EA, Yeh FL, Chapman ER (2016) Interneuronal transfer and distal action of tetanus toxin and botulinum neurotoxins A and D in central neurons. Cell Rep 16:1974–1987. https://doi.org/10.1016/j.celrep.2016.06.104

Burstein R, Blake P, Schain A, Perry C (2017) Extracranial origin of headache. Curr Opin Neurol 30(3):263–271. https://doi.org/10.1097/WCO.0000000000000437

Caleo M, Spinelli M, Colosimo F, Matak I, Rossetto O, Lackovic Z, Restani L (2018) Transynaptic action of botulinum neurotoxin type A at central cholinergic boutons. J Neurosci 38(48):10329–10337. https://doi.org/10.1523/JNEUROSCI.0294-18.2018

Coelho A, Cruz F, Cruz CD, Avelino A (2012) Spread of onabotulinumtoxinA after bladder injection. Experimental study using the distribution of cleaved SNAP-25 as the marker of the toxin action. Eur Urol 61(6):1178–1184. https://doi.org/10.1016/j.eururo.2012.01.046

Cruz F (2014) Targets for botulinum toxin in the lower urinary tract. Neurourol Urodyn 33(1):31–38. https://doi.org/10.1002/nau.22445

Cui M, Khanijou S, Rubino J, Aoki KR (2004) Subcutaneous administration of botulinum toxin A reduces formalin-induced pain. Pain 107(1–2):125–133. https://doi.org/10.1016/j.pain.2003.10.008

Diener H, Dodick DW, Aurora S, Turkel C, DeGryse R, Lipton R, Silberstein S, Brin M (2000) OnabotulinumtoxinA for treatment of chronic migraine: results from the double-blind, randomized, placebo-controlled phase of the PREEMPT 2 trial. Cephalalgia 30(7):804–814. https://doi.org/10.1177/0333102410364677

Dodick DW, Turkel CC, DeGryse RE, Aurora SK, Silberstein SD, Lipton RB, Diener H-C, Brin MF (2010) OnabotulinumtoxinA for treatment of chronic migraine: pooled results from the double-blind, randomized, placebo-controlled phases of the PREEMPT clinical program. Headache 50(6):921–936. https://doi.org/10.1111/j.1526-4610.2010.01678.x

Drinovac Vlah V, Filipović B, Bach-Rojecky L, Lacković Z (2018) Role of central versus peripheral opioid system in antinociceptive and anti-inflammatory effect of botulinum toxin type A in trigeminal region. Eur J Pain 22(3):583–591. https://doi.org/10.1002/ejp.1146

Favre-Guilmard C, Auguet M, Chabrier PE (2009) Different antinociceptive effects of botulinum toxin type A in inflammatory and peripheral polyneuropathic rat models. Eur J Pharmacol 617 (1–3):48–53. https://doi.org/10.1016/j.ejphar.2009.06.047

Favre-Guilmard C, Chabrier PE, Kalinichev M (2017) Bilateral analgesic effects of abobotulinumtoxinA (Dysport®) following unilateral administration in the rat. Eur J Pain 21(5):927–937. https://doi.org/10.1002/ejp.995

Filipović B, Matak I, Bach-Rojecky L, Lacković Z (2012) Central action of peripherally applied botulinum toxin type A on pain and dural protein extravasation in rat model of trigeminal neuropathy. PLoS One 7(1):e29803. https://doi.org/10.1371/journal.pone.0029803.e29803

Filipović B, Matak I, Lacković Z (2014) Dural neurogenic inflammation induced by neuropathic pain is specific to cranial region. J Neural Transm (Vienna) 121(5):555–563. https://doi.org/10.1007/s00702-013-1144-4

Freund B, Schwartz M (2003) Temporal relationship of muscle weakness and pain reduction in subjects treated with botulinum toxin A. J Pain 4(3):159–165. https://doi.org/10.1054/jpai.2003.435

Habermann E (1974) 125I-labeled neurotoxin from Clostridium botulinum A: preparation, binding to synaptosomes and ascent to the spinal cord. Naunyn Schmiedeberg's Arch Pharmacol 281(1):47–56. https://doi.org/10.1007/bf00500611

Ibragić S, Matak I, Dračić A, Smajlović A, Muminović M, Sofić E, Lacković Z, Riederer P (2016) Effects of botulinum toxin type A facial injection on monoamines and their metabolites in sensory, limbic and motor brain regions in rats. Neurosci Lett 617:213–217. https://doi.org/10.1016/j.neulet.2016.02.020

Ji RR, Berta T, Nedergaard M (2013) Glia and pain: is chronic pain a gliopathy? Pain 154(Suppl 1): S10–S28. https://doi.org/10.1016/j.pain.2013.06.022

Kitamura Y, Matsuka Y, Spigelman I, Ishihara Y, Yamamoto Y, Sonoyama W, Kamioka H, Yamashiro T, Kuboki T, Oguma K (2009) Botulinum toxin type a (150 kDa) decreases exaggerated neurotransmitter release from trigeminal ganglion neurons and relieves neuropathy behaviors induced by infraorbital nerve constriction. Neuroscience 159(4):1422–1429. https://doi.org/10.1016/j.neuroscience.2009.01.066

Koizumi H, Goto S, Okita S, Morigaki R, Akaike N, Torii Y, Harakawa T, Ginnaga A, Kaji R (2014) Spinal central effects of peripherally applied botulinum neurotoxin A in comparison between its subtypes A1 and A2. Front Neurol 5:98. https://doi.org/10.3389/fneur.2014.00098

Kosaras B, Jakubowski M, Kainz V, Burstein R (2009) Sensory innervation of the calvarial bones of the mouse. J Comp Neurol 515(3):331–348

Lacković Z, Filipović B, Matak I, Helyes Z (2016) Activity of botulinum toxin type A in cranial dura: implications for treatment of migraine and other headaches. Br J Pharmacol 173 (2):279–291. https://doi.org/10.1111/bph.13366

Lawrence GW, Ovsepian SV, Wang J, Aoki KR, Dolly J (2012) Extravesicular intraneuronal migration of internalized botulinum neurotoxins without detectable inhibition of distal neurotransmission. Biochem J 441:443–452. https://doi.org/10.1042/BJ20111117

Lawrence GW, Ovsepian SV, Wang J, Aoki KR, Dolly J (2013) Therapeutic effectiveness of botulinum neurotoxin A: potent blockade of autonomic transmission by targeted cleavage of only the pertinent SNAP25. Neuropharmacology 70:287–295. https://doi.org/10.1016/j.neuropharm.2013.02.012

Marinelli S, Luvisetto S, Cobianchi S, Makuch W, Obara I, Mezzaroma E, Caruso M, Straface E, Przewlocka B, Pavone F (2010) Botulinum neurotoxin type A counteracts neuropathic pain and facilitates functional recovery after peripheral nerve injury in animal models. Neuroscience 171 (1):316–328

Marinelli S, Vacca V, Ricordy R, Uggenti C, Tata AM, Luvisetto S, Pavone F (2012) The analgesic effect on neuropathic pain of retrogradely transported botulinum neurotoxin A involves Schwann cells and astrocytes. PLoS One 7(10):e47977. https://doi.org/10.1371/journal.pone.0047977

Matak I, Lacković Z (2015) Botulinum toxin A, brain and pain. Prog Neurobiol 119–120:39–59. https://doi.org/10.1016/j.pneurobio.2014.06.00

Matak I, Bach-Rojecky L, Filipović B, Lacković Z (2011) Behavioral and immunohistochemical evidence for central antinociceptive activity of botulinum toxin A. Neuroscience 186:201–207. https://doi.org/10.1016/j.neuroscience.2011.04.026

Matak I, Riederer P, Lacković Z (2012) Botulinum toxin's axonal transport from periphery to the spinal cord. Neurochem Int 61(2):236–239. https://doi.org/10.1016/j.neuint.2012.05.001

Matak I, Rossetto O, Lacković Z (2014) Botulinum toxin type A selectivity for certain types of pain is associated with capsaicin-sensitive neurons. Pain 155(8):1516–1526. https://doi.org/10.1016/j.pain.2014.04.027

Matak I, Bölcskei K, Bach-Rojecky L, Helyes Z (2019) Mechanisms of botulinum toxin type A action on pain. Toxins (Basel) 11(8). pii: E459. https://doi.org/10.3390/toxins11080459

Michailov VV (1956) Bull Exp Biol Med. USSR 41, 45 (cit. by Tyler 1963)

Morris JL, König P, Shimizu T, Jobling P, Gibbins IL (2005) Most peptide-containing sensory neurons lack proteins for exocytotic release and vesicular transport of glutamate. J Comp Neurol 483(1):1–16. https://doi.org/10.1002/cne.20399

Polley EH, Vick JA, Ciuchta HP, Fischetti DA, MacChitelli FJ, Montanarelli N (1965) Botulinum toxin, type A: effects on central nervous system. Science 147:1036–1037. https://doi.org/10.1126/science.147.3661.1036

Radhakrishnan R, Moore SA, Sluka KA (2003) Unilateral carrageenan injection into muscle or joint induces chronic bilateral hyperalgesia in rats. Pain 104(3):567–577. https://doi.org/10.1016/s0304-3959(03)00114-3

Ramachandran R, Yaksh TL (2014) Therapeutic use of botulinum toxin in migraine: mechanisms of action. Br J Pharmacol 171(18):4177–4192. https://doi.org/10.1111/bph.12763

Rojewska E, Piotrowska A, Popiolek-Barczyk K, Mika J (2018) Botulinum toxin type A – a modulator of spinal neuron-glia interactions under neuropathic pain conditions. Toxins (Basel) 10(4). pii: E145. https://doi.org/10.3390/toxins10040145

Safarpour Y, Jabbari B (2018) Botulinum toxin treatment of pain syndromes – an evidence based review. Toxicon 147:120–128. https://doi.org/10.1016/j.toxicon.2018.01.017

Shimizu T, Shibata M, Toriumi H, Iwashita T, Funakubo M, Sato H, Kuroi T, Ebine T, Koizumi K, Suzuki N (2012) Reduction of TRPV1 expression in the trigeminal system by botulinum neurotoxin type-A. Neurobiol Dis 48(3):367–378. https://doi.org/10.1016/j.nbd.2012.07.010

Shinoda M, Kawashima K, Ozaki N, Asai H, Nagamine K, Sugiura Y (2007) P2X3 receptor mediates heat hyperalgesia in a rat model of trigeminal neuropathic pain. J Pain 8:588–597

Sluka KA, Kalra A, Moore SA (2001) Unilateral intramuscular injections of acidic saline produce a bilateral, long-lasting hyperalgesia. Muscle Nerve 24(1):37–46. https://doi.org/10.1002/1097-4598(200101)24:1<37::aid-mus4>3.0.co;2-8

Stearns TP, Shad MU, Guzman GC (2018) Glabellar botulinum toxin injections in major depressive disorder: a critical review. Prim Care Companion CNS Disord 20(5). pii: 18r02298. https://doi.org/10.4088/PCC.18r02298

Tang-Liu DD, Aoki KR, Dolly JO, de Paiva A, Houchen TL, Chasseaud LF, Webber C (2003) Intramuscular injection of 125I-botulinum neurotoxin-complex versus 125I-botulinum-free neurotoxin: time course of tissue distribution. Toxicon 42(5):461–469. https://doi.org/10.1016/s0041-0101(03)00196-x

Tracey I, Bushnell MC (2009) How neuroimaging studies have challenged us to rethink: is chronic pain a disease? J Pain 10:1113–1120. https://doi.org/10.1016/j.jpain.2009.09.001

Tsui JK, Eisen A, Mak E, Carruthers J, Scott A, Calne DB (1985) A pilot study on the use of botulinum toxin in spasmodic torticollis. Can J Neurol Sci 12(4):314–316. https://doi.org/10.1017/s031716710003540x

Tyler HR (1963) Botulinus toxin: effect on the central nervous system of man. Science 139:847–848. https://doi.org/10.1126/science.139.3557.847

Vacca V, Marinelli S, Eleuteri C, Luvisetto S, Pavone F (2012) Botulinum neurotoxin A enhances the analgesic effects on inflammatory pain and antagonizes tolerance induced by morphine in mice. Brain Behav Immun 26(3):489–499. https://doi.org/10.1016/j.bbi.2012.01.002

Wiegand H, Erdmann G, Wellhöner HH (1976) 125I-labelled botulinum A neurotoxin: pharmacokinetics in cats after intramuscular injection. Naunyn Schmiedeberg's Arch Pharmacol 292:161–165. https://doi.org/10.1007/bf00498587

The Use of Botulinum Toxin for Treatment of Depression

M. Axel Wollmer, Michelle Magid, Tillmann H. C. Kruger, and Eric Finzi

Contents

Abstract

A series of clinical studies have shown that botulinum toxin can treat major depression. Subjects suffering from unipolar depression may experience a quick, strong, and sustained improvement in the symptoms of depression after a single glabellar treatment with botulinum toxin.

The original version of this chapter was revised. A correction to this chapter can be found at https://doi.org/10.1007/164_2020_411

M. A. Wollmer (✉)
Asklepios Clinic North–Ochsenzoll, Asklepios Campus Hamburg, Medical Faculty, Semmelweis University, Hamburg, Germany
e-mail: m.wollmer@asklepios.com

M. Magid
Dell Medical School, University of Texas at Austin, Austin, TX, USA

T. H. C. Kruger
Department of Psychiatry, Social Psychiatry and Psychotherapy, Medical School Hannover, Hannover, Germany
e-mail: Krueger.Tillmann@mh-hannover.de

E. Finzi
Department of Psychiatry and Behavioral Sciences, George Washington School of Medicine, Washington, DC, USA

S. M. Whitcup, M. Hallett (eds.), *Botulinum Toxin Therapy*,
Handbook of Experimental Pharmacology 263, https://doi.org/10.1007/164_2019_272

Preliminary data suggest that botulinum toxin therapy may also be effective in the treatment of other mental disorders characterized by an excess of negative emotions, such as borderline personality disorder.

The mood-lifting effect of botulinum toxin therapy is probably mediated by the interruption of a proprioceptive feedback loop from the facial musculature to the emotional brain.

Keywords
Embodiment · Emotional proprioception · Facial feedback hypothesis

1 Background

More than 300 million people in the world suffer from depression, which is the leading global cause of disability (WHO). In spite of standard treatments with antidepressant medications and psychotherapy, a considerable proportion of patients do not respond and suffer from chronic depression. Thus, there is a need for new treatment approaches.

Injection of botulinum toxin into the glabellar region (i.e., the muscles above and between the eyebrows) may be such a new approach. Glabellar frown lines are produced by the contraction of the corrugator muscles (*musculi corrugatores superciliorum*), which are key muscles in the expression of negative emotions like anger, fear, or sadness. Charles Darwin called these muscles "grief muscles." Their activity also accounts for facial features of emotional distress like the "*omega melancholicum*" or Veraguth's folds (Greden et al. 1985; Fig. 1).

Injection of botulinum toxin in the glabellar region is probably the most popular intervention in esthetic medicine (https://www.isaps.org/wp-content/uploads/2017/10/GlobalStatistics2016-1.pdf). The wish for an emotionally more positive/less

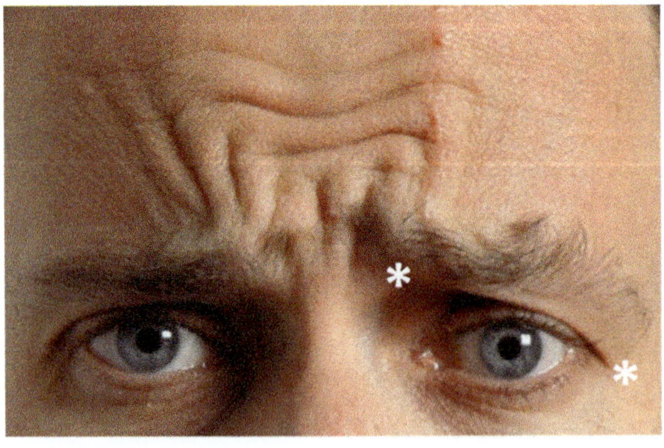

Fig. 1 The corrugator muscles are key muscles for the expression of negative emotions. Combined contraction of the corrugator muscles (FACS action unit 4) with the medial proportion of the frontalis muscle (FACS action unit 1) produces a wrinkle configuration described as "*omega melancholicum*," as it resembles the Greek letter Ω. It also results in the formation of Veraguth's folds from the lateral angle of the eye to the medial end of the eyebrow (∗)

negative facial appearance may be as important as the wish for a more beautiful and youthful look, when it comes to explaining why so many people seek removal of glabellar frown lines by the injection of botulinum toxin. The treatment can actually change facial expression in a way that makes it appear less negative or more positive, respectively (Heckmann et al. 2003).

The treatment can also influence emotional experience: It seems to enhance emotional well-being beyond a mere cosmetic benefit (Sommer et al. 2003). Moreover, it reduces irritability, as well as depressed and anxious mood (Lewis and Bowler 2009). Facial botulinum toxin treatment as it is applied in cosmetic medicine can influence the perception of visual emotional stimuli and delay the comprehension of sentences with negative emotional connotations (Davis et al. 2010; Havas et al. 2010; Baumeister et al. 2016). These behavioral effects are backed up with observations from functional studies showing that the treatment can reduce amygdala activation during viewing or imitation of an angry facial expression (Hennenlotter et al. 2009; Kim et al. 2014). Physicians in esthetic medicine are familiar with the enhanced well-being in patients they treat with botulinum toxin, and it is possible that the described effects may contribute to patients' desire for continued treatments.

2 Botulinum Toxin Therapy of Depression

There is cumulating evidence that glabellar botulinum toxin injections may have an antidepressant effect (for review, see also Kruger and Wollmer 2015; Wollmer et al. 2018). The first report of this effect was a case series published in 2006: Ten middle-aged women with moderate to severe, partly chronic, and treatment-resistant depression received a single open-label application of glabellar botulinum toxin injections. Botulinum toxin was injected according to a protocol for the cosmetic treatment of frown lines, as a sole or as an adjunctive treatment of depression (Finzi and Wasserman 2006). The treatment led to a marked improvement in the self-rated depression scores on the Beck Depression Inventory (BDI) II from before to 8 weeks after the treatment, with a high response and remission rate.

The first randomized controlled trial (RCT) of botulinum toxin therapy (BTT) for depression was published in 2012 and showed that a single treatment can lead to a quick, strong, and sustained improvement in depressive symptoms (Wollmer et al. 2012). The study included 30 middle-aged, mostly female patients, suffering from mild to moderate, partly chronic and treatment-resistant unipolar depression on stable treatment with antidepressant medication. Ability to produce moderate to severe frown lines was an inclusion criterion. The participants were randomized to a blinded treatment with either BTT or saline placebo injections. To account for the higher muscle mass, men received a higher dose of onabotulinumtoxinA than women. While the placebo group remained more or less stably depressed throughout the study, the BTT group showed a significant improvement in the symptoms of depression as early as 2 weeks after the injection, which was measurable on both the Hamilton Depression Rating Scale (HAM-D) expert rating and the BDI self-rating scales. At the primary end point 6 weeks after the baseline, the improvement had a large effect size ($d = 1.28$) and increased even further until the end of the study

16 weeks after treatment ($d = 1.87$). An improvement in Clinical Global Impressions reflected the improvement in the depression scales. Partial response (>25% reduction in HAM-D score; 87%) and response (>50% reduction in HAM-D score; 60%) rates were significantly higher in the BTT group than in the placebo group, and 33% of the botulinum toxin-treated patients attained remission. Psychomotor agitation was a predictor of response in this study (Wollmer et al. 2014).

A second RCT with a larger sample ($n = 74$) confirmed the antidepressant effect of BTT. The participants of this trial had similar baseline characteristics like those in the previous study with slightly higher depression scores (Finzi and Rosenthal 2014). After 3 weeks there was a highly significant improvement in depression measured by the BDI-II and Montgomery-Åsberg Depression Rating Scale (MADRS) depression rating scales that became even more pronounced at the primary end point 6 weeks after baseline. Improvement and response rates were comparable to those observed in the previous study, and the remission rate was significantly higher in the BTT group compared to placebo. In this trial, BTT was equally effective as a sole or adjunctive treatment. Presence of glabellar frown lines at the baseline was not an inclusion criterion and was shown not to be required for either response or remission.

A third RCT with 30 patients further corroborated and extended the previous findings (Magid et al. 2014). The crossover design of this study provided switching of the patients who were initially in the placebo arm to BTT after 12 weeks and vice versa. Given the long-lasting effect of botulinum toxin, this corresponds to a delayed start design, in which one group received BTT immediately and the other one with a delay of 12 weeks. The overall follow-up period was 24 weeks, and both groups improved significantly after botulinum toxin treatment. Remarkably, the clinical improvement in depression outlasted the muscle-relaxing effect: In the original BTT group, depression scores showed further improvement from the visit after 16 weeks to the final visit after 24, while frown line severity returned to baseline.

An individual patient data meta-analysis by the authors of the original studies and an independent conventional systematic review and meta-analysis summarize the results of these three first RCTs (Fig. 2; Magid et al. 2015; Parsaik et al. 2016). Both meta-analyses confirmed a marked reduction in the symptoms of major depression and high response and remission rates by BTT with low numbers needed to treat (2.2–2.3 and 2.9–4.9, respectively). With these meta-analyses, there is a high level of evidence for the efficacy of BTT, especially as an adjunctive treatment for women with mild to moderate unipolar depression. The treatment was very well tolerated in all three studies with no significant difference in the incidence of side effects between the BTT and the placebo groups.

Meanwhile, a fourth RCT with 28 patients with major depression further confirmed the efficacy of the treatment, with a statistically significant reduction in BDI score at week 6 in the BTT compared to the placebo group (Zamanian et al. 2017).

The Botox® manufacturer Allergan has recently completed a multicenter phase II RCT. The trial tested one-time treatment with two doses of Botox® (30 or 50 U) against saline placebo in 258 moderately to severely depressed women. In the 24-week trial, change in MADRS score from baseline to 6 weeks in the Botox®

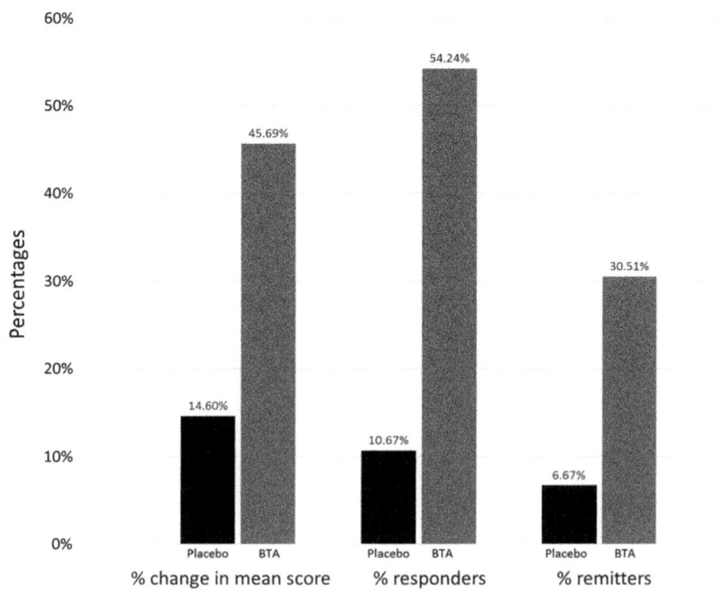

Fig. 2 In a pooled analysis of three previous randomized controlled trials on BTT in depression, there was a 45% reduction in depression scores. In 54% of the BTT recipients, there was a reduction by at least 50% (responders), and 30% of the BTT-treated patients' depression scores fell below the clinical threshold (remitters; Magid et al. 2015)

vs. the respective placebo groups was defined as the primary end point. Only the 30 U Botox® dose was superior to placebo, but the difference was not statistically significant at the primary end point. The results of the study are not yet published as a scientific report but are posted at ClinicalTrials.gov (https://clinicaltrials.gov/ct2/show/results/NCT02116361?term=botulinum&cond=depression&rank=4). Based on the results of the study, Allergan has announced to proceed with a phase III trial.

Ongoing trials investigate glabellar botulinum toxin injections as a treatment for depression in Parkinson's disease and in geriatric depression (NCT03069911, NCT03833063). Another trial compares the antidepressant effect of glabellar injections with the effect of injections into the lateral parts of the orbicularis muscles of the eyes, which is associated with the Duchenne smile (a "genuine" smile, which involves crinkling of the eyes) and the expression of positive emotions (NCT03484754).

Another case series with 42 patients suffering from severe treatment-resistant unipolar depression confirmed improvement in the symptoms of depression within 3 weeks after BTT (Chugh et al. 2018). Importantly, more than half of the patients of this case series were men and improved equally to the female participants, indicating that the antidepressant effect of the treatment is not dependent on patient gender.

A recent case series suggests that BTT may be effective in bipolar depression, too (Finzi et al. 2018).

3 Depression in Chronic Migraine

Based on the results of the two Phase 3 REsearch Evaluating Migraine Prophylaxis Therapy (PREEMPT) studies, onabotulinumtoxinA has been registered as a treatment of chronic migraine (Dodick et al. 2010). Because depression is highly prevalent among chronic migraine patients, studies in these patients may provide additional information on the antidepressant effect of botulinum toxin injections. In the injection scheme for the treatment of chronic migraine, five units of onabotulinumtoxinA are provided for injection into the corrugator muscles on each side. This dose is below the doses used for the treatment of glabellar frown lines in esthetic medicine or in the treatment of depression. However, it will lead to at least partial relaxation of these muscles.

Several studies support the antidepressant effect of botulinum toxin injections:

One study, which investigated the effect of botulinum toxin on mild depressive symptoms in 32 chronic migraine patients, found an improvement in the severity of depressive symptoms as measured by the BDI-II after 12 and 24 weeks (Boudreau et al. 2015).

In a retrospective study on treatment of migraine with botulinum toxin in 359 patients, there was an improvement in depressive symptoms measured by the Patient Health Questionnaire (PHQ-9). The improvement in depressive symptoms was correlated with the reduction in headache (Maasumi et al. 2015).

In a study observing the long-term course of chronic migraine after several injection cycles with botulinum toxin in 27 patients, depressive symptoms measured by the BDI dropped stably and significantly throughout the yearlong study (Kollewe et al. 2016).

In a cohort of 60 patients with chronic migraine treated with botulinum toxin according to the PREEMPT protocol, BDI scores were significantly reduced 3 months after the treatment (Demiryurek et al. 2016).

A modified version of the PREEMPT injection scheme for the treatment of chronic migraine led to improvement in symptoms of depression and anxiety as measured by the HAM-D or Hamilton Anxiety Rating Scale (HAM-A) after 1 month. This prospective open-label study included 30 (in most cases) female patients with chronic daily headache (Zhang et al. 2017).

It is unclear if the effects on depression and headache in these studies were independent or if one occurred as function of the other.

However, other studies did not confirm improvement in psychiatric symptoms in migraine patients by botulinum toxin treatment.

In an open-label study, 190 patients were monitored for migraine symptoms and for negative emotional states measured by the Depression, Anxiety and Stress Scale (DASS-21) for almost a year. While migraine improved, there was no significant reduction in the DASS-21 scores. However, the interpretation of these findings is limited because a considerable proportion of patients was lost to follow-up (Aydinlar et al. 2017).

Psychiatric symptoms measured by the Zung self-rating anxiety and depression scale did not change significantly, while all headache-related parameters increasingly improved over 13 treatment cycles, according to the PREEMPT protocol in a study with 90 patients (Guerzoni et al. 2015, 2017).

4 Mechanisms of Action

While there is growing evidence for the efficacy of botulinum toxin injection as a treatment for depression, the mechanism of action by which it accomplishes the improvement in mood is still unknown. There are several possibilities how botulinum toxin may exert its mood-lifting effect:

Because of the obvious cosmetic change or the lack of, it is impossible to effectively blind patients for their treatment allocation in RCTs. Moreover, a facial injection has a high suggestive power and the targeted phenotype is subjective. Therefore, it is difficult to judge to what extent placebo and nocebo effects may inflate the differences between the botulinum toxin and placebo groups, respectively. However, some findings strongly argue against a predominant role of placebo effects: Patient's expectations and credibility regarding botulinum toxin therapy did not predict the outcome in one of the RCTs (Wollmer et al. 2012). In another RCT, patients' hit rate when guessing group allocation was only barely above chance level, and correct or incorrect guessing was not associated with outcome (Finzi and Rosenthal 2014). In a third RCT, the improvement in the symptoms of depression outlasted the cosmetic effect of the treatment (Magid et al. 2014). An Inverse-Frequency Analysis comprising millions of reports of the FDA Adverse Event Reporting System (FAERS) database showed that use of botulinum toxin as a cosmeceutical was associated with marked underreporting of depression and depressive symptoms as a side effect of the treatment with Log ORs of around 2.5 below the benchmark (Cohen et al. 2017). This is an indirect proof of the antidepressant effect of botulinum toxin treatment. Given the naturalistic application of botulinum toxin for cosmetic indications, it is unlikely that specific expectations of an antidepressant effect could induce placebo effects. Very recently, a study showed antidepressant-like effects of facial botulinum toxin injection in a mouse model of depression (Li et al. 2019). After stress induction by space restriction, mice showed prolonged immobility time in behavioral despair tests like the forced swim test and the tail suspension test. This may be looked upon as a correlate of learned helplessness associated with depression. A single facial injection of botulinum toxin improved this depression-like behavior and was associated with hippocampal increase in serotonin levels as well as activation of BDNF/ERK/CREB pathways. These findings argue against a predominant role of placebo effects but also raise questions regarding the possible mechanisms of action discussed below.

Cosmetic changes associated with glabellar muscle relaxation may improve body image, enhance self-esteem, and in the end elevate mood (Molina et al. 2015). Several findings speak against cosmetic improvement as the main mechanism of action: Recruiting for the first RCT tried to avoid attracting participants looking to receive a botulinum toxin treatment with its known cosmetic effects. The respective advertisements did not disclose that the study was about botulinum toxin injections. This information was given only at screening. In the same RCT, the appreciation of the cosmetic change did not correlate with treatment outcome (Wollmer et al. 2012). Also individual experiences of participants argued against the possibility that cosmetic changes mainly drove mood improvement: One patient with a structurally fixed severe frown line did not experience the expected cosmetic improvement after botulinum toxin injection, was convinced to having received placebo, but still attained remission of her depression. Another subject reported to dislike the "Mephisto sign" (lateral elevation of the eyebrows because of medial weakening of the frontalis muscle), that occurred after injection of botulinum toxin, but still her depression went into remission. The improvement in BDI scores did not correlate with the changes in self-esteem scores on the Rosenberg scale in an open-label study (Hexsel et al. 2013). In the second RCT, half the subjects had no frown at rest and therefore obtained no cosmetic benefit. Patients who had previously used BTT were excluded from the studies. In addition, the presence of frown lines at the baseline was shown not to be required for response (Finzi and Rosenthal 2014). In the third RCT, the improvement in the symptoms of depression outlasted the cosmetic change (Magid et al. 2014).

It is possible that a change in facial appearance from emotionally negative toward emotionally positive is associated with a more positive social feedback when interacting with a social partner or even the own image in the mirror. However, living on their own or living with a partner or families does not seem to predict the antidepressant effect of botulinum toxin treatment (Wollmer et al. 2012, 2014). Therefore, it is not very likely that altered social feedback is the main mechanism of action. An ongoing study investigates altered social interactions as a possible mediator of mood improvement by botulinum toxin (SNCTP000002474).

The most probable and most plausible mechanism that could explain how botulinum toxin may reduce the symptoms of depression is that it interrupts a feedback loop of "emotional proprioception" from the face to the brain that reinforces and maintains the negative emotions that are prevalent in depression (Finzi and Rosenthal 2016). Glabellar injection of botulinum toxin abolishes the contraction of the corrugator muscles, which is a key element in the expression of negative emotions. Patients suffering from depression show a relative overactivity of the corrugator muscles that would be corrected by botulinum toxin injection (Schwartz et al. 1976).

Back in the nineteenth century, William James and Charles Darwin formulated the facial feedback hypothesis (FFH). According to the FFH, the facial expression of an emotion generates a proprioceptive feedback signal from the face to the brain that enhances the respective emotion. This feedback turns it from an initially faint and cool semicognitive experience into a powerful and warm sensation (Fig. 3; Adelmann and Zajonc 1989; Al Abdulmohsen and Kruger 2011). The FFH has been validated in classical experiments. They show that the arbitrary contraction of facial muscles can modulate emotional appraisal of or emotional reactions to presented visual stimuli (Ekman et al. 1983; Coles et al. 2019). Facial feedback effects tend to be rather small and inconsistent under experimental conditions, but among them experiments using botulinum toxin injections with the resulting long-lasting and complete muscle paralysis seem to have the strongest effects. It is possible that a facial expression can reinforce a preexisting, matching emotion or produce a corresponding emotion. Conversely, lack of or nonmatching facial expression may prevent, weaken, or abolish an emotion: "Refuse to express a passion, and it dies," as William James put it, reflects exactly the rationale of BTT in the treatment of depression.

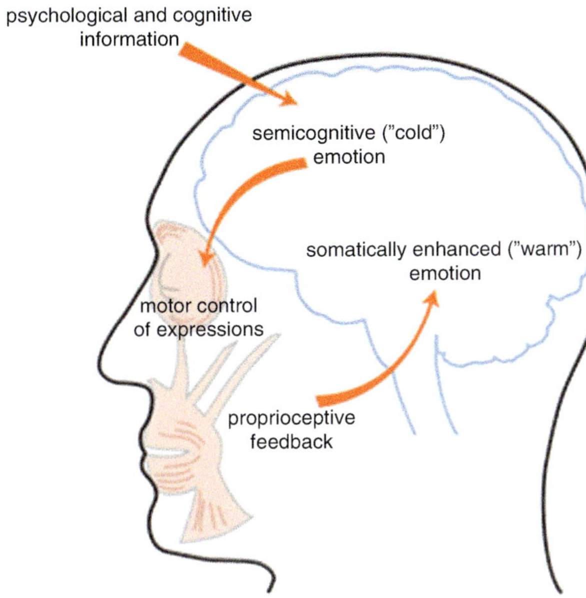

Fig. 3 The facial feedback hypothesis. Facial musculature activation expressing of emotional states generates proprioceptive feedback signals from the face to the emotional brain. This "emotional proprioception" reinforces the initially semicognitive, "cold" emotional state and turns it into a somatically enhanced, "warm" emotional experience. The antidepressant effect of glabellar botulinum toxin injections may be mediated by switching off proprioceptive feedback signals from this region that maintain negative emotions (Al Abdulmohsen and Kruger 2011)

The facial musculature is not equipped with typical proprioceptors like muscle spindles, and it is unknown how proprioceptive signals are picked up. Mechanical receptors in the skin or connective tissue may play a role (Cattaneo and Pavesi 2014). Along proprioceptive fibers that run with the facial and trigeminal nerve, signals are conducted to the mesencephalic trigeminal nucleus and locus coeruleus (Cobo et al. 2017). Via projections from there, they may modulate the activity of the prefrontal cortex and the amygdala (Matsuo et al. 2015; Finzi and Rosenthal 2016). In rats, facial injection of botulinum toxin can alter the metabolism of monoaminergic neurotransmitters in limbic brain regions (Ibragić et al. 2016).

There is some experimental evidence of axonal and even transsynaptic transport of locally injected botulinum toxin into the CNS, when injected in high doses (Caleo and Schiavo 2009). Central effects may be clinically relevant in humans, and it can't formally be excluded that they may be involved in the mood-lifting effects of botulinum toxin treatment (Marchand-Pauvert et al. 2013). Besides SNARE complex proteins, there are other central nervous substrates of botulinum toxin like the RAS-related C3 botulinum toxin substrate 1 (Rac1) that play a role in depression and other mental disorders (Golden et al. 2013). Theoretically, they might be involved in the antidepressant action of botulinum toxin.

5 BTT in the Clinical Management of Depression

BTT has several favorable aspects in the management of depression: A single treatment has a long-lasting effect averaging 3 months, which is practical for both physician and patient and may improve therapy adherence. With the long treatment intervals, BTT is even an economic therapeutic option, if the costs are calculated per treatment day (Beer 2010). Finally, botulinum toxin injections to the glabellar region have an excellent safety and tolerability record (Brin et al. 2009). All these positive aspects render BTT an attractive treatment option for patients, physicians, and the public health system. This may particularly apply for regions with limited resources in mental health-care supply (Chugh et al. 2018). Glabellar injection of botulinum toxin is not yet registered as a treatment for depression or any other mental disorder. However, it is registered for the treatment of frown lines. Thus, it is possible to treat psychiatric patients featuring such lines on label for this indication, but with the aim to induce improvement in the affective symptoms as a side effect.

The goal of BTT for psychiatric indications is not to obtain an optimum esthetic result. It rather aims to prevent the expression of negative emotions and the resulting proprioceptive facial feedback that may reinforce and maintain them. The glabellar muscles, i.e., the procerus and the corrugator muscles, are key muscles for the expression of negative emotions, which universally comprises a contraction of the eyebrows (facial action unit 4 in the Facial Action Coding System, FACS; Ekman and Friesen 1978). The paralysis of these muscles is the most parsimonious way to prevent the expression and, at the same time, the experience of negative emotions via interruption of the corresponding proprioceptive feedback loop. This paralysis should be complete, as a residual activity may be enough to keep up the feedback loop. Therefore, the doses applied for psychiatric indications may be above those used to obtain the "natural look" desired in cosmetic treatments (Carruthers et al. 2007). The injection scheme used in the first studies on BTT for depression provides 29 units of onabotulinumtoxinA at a concentration of 40 or 100 U/ml 0.9% saline. They are distributed to five injection points (7 U *m. procerus*; 6 U *m. corrugator supercilii* medial, bilaterally; 5 U *m. corrugator supercilii* lateral, bilaterally; Fig. 4) for women. To account for their usually higher muscle mass, men received two more units at each injection point. These doses are sufficient to achieve a complete

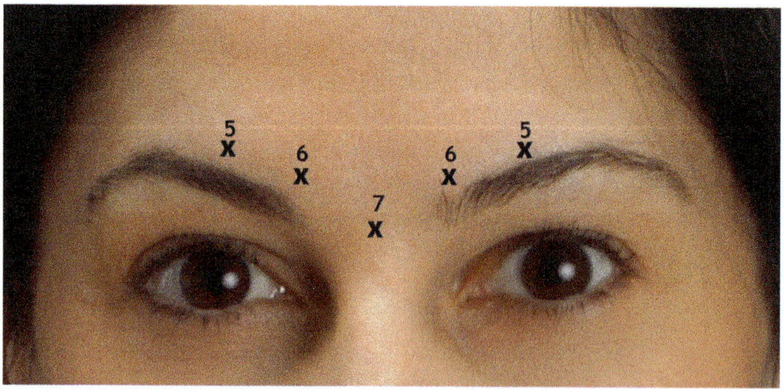

Fig. 4 The injection scheme used in the first studies on BTT for depression. For women, 29 units of onabotulinumtoxinA were distributed to five injection points (7 U *m. procerus*; 6 U *m. corrugator supercilii* medial, bilaterally; 5 U *m. corrugator supercilii* lateral, bilaterally). Men received two more units at each injection point, to account for their usually higher muscle mass

paralysis of the glabellar musculature in most cases. In clinical practice, the dose and its distribution may be adopted to the individual anatomic conditions. The facial action unit 1 that corresponds to the medial proportion of the *m. frontalis* is frequently activated together with the facial action unit 4. This activation pattern produces the "*omega melancholicum*," a wrinkle relief of the Greek letter omega. Patients with agitated depression frequently display this clinical sign (Greden et al. 1985; Fig. 1). The *m. frontalis* may be involved in the expression of negative emotions like sadness (only the medial proportion) or fear; yet, it is also involved in the expression of surprise, which can also be positive. Thus, extending the treatment to the *m. frontalis* to enhance the effect on specific negative emotions may be an option but may have a downside, too. A sad facial expression may involve chin dimpling ("popply chin") and depression of the corners of the mouth. They can occur habitually in depression. In this case, the injection of small doses of botulinum toxin into the *mm. depressores angulorum oris* (e.g., 2–3 U onabotulinumtoxinA, bilaterally) and the *m. mentalis* (e.g., 4–6 U onabotulinumtoxinA distributed to 1–3 injection points) may reinforce the mood-lifting effect of a glabellar treatment. Conversely, the injection of botulinum toxin into the *mm. orbiculares oculorum* to treat crowfeet may have detrimental effects on mood. These muscles are essential to the Duchenne smile, and their paralysis may not only weaken the expression of happiness but also interrupt the proprioceptive feedback that is involved in the experience of the happiness expressed by smiling.

To date botulinum toxin is not registered for any psychiatric indication. Thus, it is currently used on a compassionate basis for depressed patients who either did not improve sufficiently with established therapies or did not tolerate them. BTT may help many patients to attain substantial improvement or even remission of depression that was previously treatment resistant. The majority of patients need regular repetition of BTT to maintain an antidepressant effect, but some may stay well after a single treatment. So far, the majority of patients have been treated with onabotulinumtoxinA. However, first impressions from BTT with other botulinum toxins, especially incobotulinumtoxinA, are equally good.

6 Outlook

A phase III trial announced by Allergan will probably be pivotal for registration of botulinum toxin as an antidepressant drug. A great methodical challenge for future trials will be to control placebo effects. Comparator studies testing BTT against an established antidepressant may facilitate the estimation of the true effect sizes and help to overcome the obvious problems associated with blinding, expectations, and placebo/nocebo effects.

Very few men took part in the hitherto conducted clinical trials. In the first RCT, there seemed to be a gender effect in favor of women (Wollmer et al. 2012). However, in a recently published case series, BTT was equally effective for men (Chugh et al. 2018). Future RCTs will need to show if BTT is effective for depression in men.

RCTs on BTT in the treatment of severe, psychotic, or bipolar depression are still missing.

Future RCTs could also show if including other facial muscles involved in the expression of negative emotions, specifically sadness, like the *m. mentalis* or the *mm. depressores angulorum oris*, may enhance the effect of glabellar injections.

If BTT works by inhibiting negative emotions in general, then one would predict that BTT could be successfully used for a wide variety of disorders with an excess of negative emotions. Preliminary studies suggest this to be the case. A case series demonstrated that BTT may reduce the symptoms of borderline personality disorder (Kruger et al. 2016). Recently, BTT has been reported to induce and maintain remission of moderate to severe social anxiety disorder (Finzi and Rosenthal 2019).

Identifying predictors of response to BTT may allow for its stratified or even personalized application. A high level of agitation may be such a predictor (Wollmer et al. 2014). Moreover, BTT may be customized to patients' individual facial muscle activity patterns.

A great challenge for future research will be to uncover the neurobiological correlates and mechanisms of the mood-lifting effect of BTT.

Conceptually, BTT is fundamentally different from most established psychiatric treatment approaches: BTT tackles emotional processes in the CNS by their expression in the face, probably via interruption of a reinforcing proprioceptive feedback loop, reversing the therapeutic process from inside out (top down) to outside in (bottom up). BTT can be looked upon as a drug-mediated relaxation exercise that interrupts a behavior, i.e., negative facial expression, which maintains a negative emotional state. Unlike other psychiatric treatments, BTT is not syndrome-oriented or disorder-specific. It is rather a comprehensive approach as it targets a basic condition of mental suffering: the excess of negative emotionality.

References

Adelmann PK, Zajonc RB (1989) Facial efference and the experience of emotion. Annu Rev Psychol 40:249–280

Al Abdulmohsen T, Kruger TH (2011) The contribution of muscular and auditory pathologies to the symptomatology of autism. Med Hypotheses 77:1038–1047

Aydinlar EI, Dikmen PY, Kosak S et al (2017) OnabotulinumtoxinA effectiveness on chronic migraine, negative emotional states and sleep quality: a single-center prospective cohort study. J Headache Pain 18:23

Baumeister JC, Papa G, Foroni F (2016) Deeper than skin deep – the effect of botulinum toxin-A on emotion processing. Toxicon 118:86–90

Beer K (2010) Cost effectiveness of botulinum toxins for the treatment of depression: preliminary observations. J Drugs Dermatol 9:27–30

Boudreau GP, Grosberg BM, McAllister PJ et al (2015) Prophylactic onabotulinumtoxinA in patients with chronic migraine and comorbid depression: an open-label, multicenter, pilot study of efficacy, safety and effect on headache-related disability, depression, and anxiety. Int J Gen Med 9:79–86

Brin MF, Boodhoo TI, Pogoda JM et al (2009) Safety and tolerability of onabotulinumtoxinA in the treatment of facial lines: a meta-analysis of individual patient data from global clinical registration studies in 1678 participants. J Am Acad Dermatol 61:961–970

Caleo M, Schiavo G (2009) Central effects of tetanus and botulinum neurotoxins. Toxicon 54:593–599

Carruthers A, Cohen JL, Cox SE et al (2007) Facial aesthetics: achieving the natural, relaxed look. J Cosmet Laser Ther 9(Suppl 1):6–10

Cattaneo L, Pavesi G (2014) The facial motor system. Neurosci Biobehav Rev 38:135–159

Chugh S, Chhabria A, Jung S et al (2018) Botulinum toxin as a treatment for depression in a real-world setting. J Psychiatr Pract 24:15–20

Cobo JL, Sole-Magdalena A, Menendez I et al (2017) Connections between the facial and trigeminal nerves: anatomical basis for facial muscle proprioception. JPRAS Open 12:9–18

Cohen IV, Makunts T, Atayee R et al (2017) Population scale data reveals the antidepressant effects of ketamine and other therapeutics approved for non-psychiatric indications. Sci Rep 7:1450

Coles NA, Larsen JT, Lench HC (2019) A meta-analysis of the facial feedback literature: effects of facial feedback on emotional experience are small and variable. Psychol Bull 145:610–651

Davis JI, Senghas A, Brandt F et al (2010) The effects of BOTOX injections on emotional experience. Emotion 10:433–440

Demiryurek BE, Ertem DH, Tekin A et al (2016) Effects of onabotulinumtoxinA treatment on efficacy, depression, anxiety, and disability in Turkish patients with chronic migraine. Neurol Sci 37:1779–1784

Dodick DW, Turkel CC, RE DG, PREEMPT Chronic Migraine Study Group et al (2010) OnabotulinumtoxinA for treatment of chronic migraine: pooled results from the double-blind, randomized, placebo-controlled phases of the PREEMPT clinical program. Headache 50:921–936

Ekman P, Friesen WV (1978) Facial action coding system: a technique for the measurement of facial movement. Consulting Psychologists Press, Palo Alto

Ekman P, Levenson PW, Friesen WV (1983) Autonomic nervous system activity distinguishes among emotions. Science 212:1208–1210

Finzi E, Rosenthal NE (2014) Treatment of depression with onabotulinumtoxinA: a randomized, double-blind, placebo controlled trial. J Psychiatr Res 52:1–6

Finzi E, Rosenthal NE (2016) Emotional proprioception: treatment of depression with afferent facial feedback. J Psychiatr Res 80:93–96

Finzi E, Rosenthal NE (2019) Botulinum toxin therapy of social anxiety disorder: a case series. J Clin Psychopharmacol 39:410–412

Finzi E, Wasserman E (2006) Treatment of depression with botulinum toxin A: a case series. Dermatol Surg 32:645–649

Finzi E, Kels L, Axelowitz J et al (2018) Botulinum toxin therapy of bipolar depression: a case series. J Psychiatr Res 104:55–57

Golden SA, Christoffel DJ, Heshmati M et al (2013) Epigenetic regulation of RAC1 induces synaptic remodeling in stress disorders and depression. Nat Med 19:337–344

Greden JF, Genero N, Price HL (1985) Agitation-increased electromyogram activity in the corrugator muscle region: a possible explanation of the "Omega sign"? Am J Psychiatry 142:348–351

Guerzoni S, Pellesi L, Baraldi C et al (2015) Increased efficacy of regularly repeated cycles with OnabotulinumtoxinA in MOH patients beyond the first year of treatment. J Headache Pain 17:48

Guerzoni S, Pellesi L, Baraldi C et al (2017) Long-term treatment benefits and prolonged efficacy of OnabotulinumtoxinA in patients affected by chronic migraine and medication overuse headache over 3 years of therapy. Front Neurol 8:586

Havas DA, Glenberg AM, Gutowski KA et al (2010) Cosmetic use of botulinum toxin-a affects processing of emotional language. Psychol Sci 21:895–900

Heckmann M, Teichmann B, Schroder U et al (2003) Pharmacologic denervation of frown muscles enhances baseline expression of happiness and decreases baseline expression of anger, sadness, and fear. J Am Acad Dermatol 49:213–216

Hennenlotter A, Dresel C, Castrop F et al (2009) The link between facial feedback and neural activity within central circuitries of emotion-new insights from botulinum toxin-induced denervation of frown muscles. Cereb Cortex 19:537–542

Hexsel D, Brum C, Siega C et al (2013) Evaluation of self-esteem and depression symptoms in depressed and nondepressed subjects treated with onabotulinumtoxinA for glabellar lines. Dermatol Surg 39:1088–1096

Ibragić S, Matak I, Dračić A et al (2016) Effects of botulinum toxin type A facial injection on monoamines and their metabolites in sensory, limbic and motor brain regions in rats. Neurosci Lett 617:213–217

Kim MJ, Neta M, Davis FC et al (2014) Botulinum toxin-induced facial muscle paralysis affects amygdala responses to the perception of emotional expressions: preliminary findings from an A-B-A design. Biol Mood Anxiety Disord 4:11

Kollewe K, Escher CM, Wulff DU et al (2016) Long-term treatment of chronic migraine with OnabotulinumtoxinA: efficacy, quality of life and tolerability in a real-life setting. J Neural Transm 123:533–540

Kruger TH, Wollmer MA (2015) Depression--an emerging indication for botulinum toxin treatment. Toxicon 107:154–157

Kruger TH, Magid M, Wollmer MA (2016) Can botulinum toxin help patients with borderline personality disorder? Am J Psychiatry 173:940–941

Lewis MB, Bowler PJ (2009) Botulinum toxin cosmetic therapy correlates with a more positive mood. J Cosmet Dermatol 8:24–26

Li Y, Liu J, Liu X (2019) Antidepressant-like action of single facial injection of botulinum neurotoxin A is associated with augmented 5-HT levels and BDNF/ERK/CREB pathways in mouse brain. Neurosci Bull 35(4):661–672

Maasumi K, Thompson NR, Kriegler JS (2015) Effect of OnabotulinumtoxinA injection on depression in chronic migraine. Headache 55:1218–1224

Magid M, Reichenberg JS, Poth PE et al (2014) Treatment of major depressive disorder using botulinum toxin A: a 24-week randomized, double-blind, placebo-controlled study. J Clin Psychiatry 75:837–844

Magid M, Finzi E, Kruger TH et al (2015) Treating depression with botulinum toxin: a pooled analysis of randomized controlled trials. Pharmacopsychiatry 48:205–210

Marchand-Pauvert V, Aymard C, Giboin LS et al (2013) Beyond muscular effects: depression of spinal recurrent inhibition after botulinum neurotoxin A. J Physiol 591:1017–1029

Matsuo K, Ban R, Hama Y et al (2015) Eyelid opening with trigeminal proprioceptive activation regulates a brainstem arousal mechanism. PLoS One 10:e0134659

Molina B, Grangier Y, Mole B (2015) Patient satisfaction after the treatment of glabellar lines with Botulinum toxin type A (Speywood Unit): a multi-centre European observational study. J Eur Acad Dermatol Venereol 29:1382–1388

Parsaik AK, Mascarenhas SS, Hashmi A et al (2016) Role of botulinum toxin in depression. J Psychiatr Pract 22:99–110

Schwartz GE, Fair PL, Salt P et al (1976) Facial muscle patterning to affective imagery in depressed and nondepressed subjects. Science 192:489–491

Sommer B, Zschocke I, Bergfeld D et al (2003) Satisfaction of patients after treatment with botulinum toxin for dynamic facial lines. Dermatol Surg 29:456–460

Wollmer MA, de Boer C, Kalak N et al (2012) Facing depression with botulinum toxin: a randomized controlled trial. J Psychiatr Res 46:574–581

Wollmer MA, Kalak N, Jung S et al (2014) Agitation predicts response of depression to botulinum toxin treatment in a randomized controlled trial. Front Psych 5:36

Wollmer MA, Neumann I, Magid M et al (2018) Shrink that frown! Botulinum toxin therapy is lifting the face of psychiatry. G Ital Dermatol Venereol 153:540–548

Zamanian A, Ghanbari Jolfaei A, Mehran G et al (2017) Efficacy of botox versus placebo for treatment of patients with major depression. Iran J Public Health 46:982–984

Zhang H, Zhang H, Wei Y et al (2017) Treatment of chronic daily headache with comorbid anxiety and depression using botulinum toxin A: a prospective pilot study. Int J Neurosci 127:285–290

Correction to: The Use of Botulinum Toxin for Treatment of Depression

M. Axel Wollmer, Michelle Magid, Tillmann H. C. Kruger, and Eric Finzi

Correction to: Chapter "The Use of Botulinum Toxin for Treatment of Depression" in: M. Axel Wollmer et al., Handbook of Experimental Pharmacology, https://doi.org/10.1007/164_2019_272

The chapter was inadvertently published without a more specific title according to SEO guidelines. A chapter title needs to be understandable when seen as a stand-alone item, e.g. on PubMed. The chapter title has now been corrected as 'The Use of Botulinum Toxin for Treatment of Depression'.

The original article was corrected.

The updated online version of this chapter can be found at https://doi.org/10.1007/164_2019_272

S. M. Whitcup, M. Hallett (eds.), *Botulinum Toxin Therapy*,
Handbook of Experimental Pharmacology 263, https://doi.org/10.1007/164_2020_411

Correction to: The Use of Botulinum Toxin for Treatment of Spasticity

Sheng Li and Gerard E. Francisco

Correction to:
Chapter "The Use of Botulinum Toxin for Treatment of Spasticity" in: S. Li and G. E. Francisco, Handbook of Experimental Pharmacology, https://doi.org/10.1007/164_2019_315

The chapter was inadvertently published without a more specific title according to SEO guidelines. A chapter title needs to be understandable when seen as a stand-alone item, e.g. on PubMed. The chapter title has now been corrected as 'The Use of Botulinum Toxin for Treatment of Spasticity'.

The original article was corrected.

The updated online version of this chapter can be found at https://doi.org/10.1007/164_2019_315

S. M. Whitcup, M. Hallett (eds.), *Botulinum Toxin Therapy*,
Handbook of Experimental Pharmacology 263, https://doi.org/10.1007/164_2020_412

Correction to: Use of Botulinum Toxin in Ophthalmology

Michael J. Wan, Sara AlShaker, and David G. Hunter

Correction to:
Chapter "Use of Botulinum Toxin in Ophthalmology"
in: M. J. Wan et al., Handbook of Experimental Pharmacology,
https://doi.org/10.1007/164_2019_325

The chapter was inadvertently published without a more specific title according to SEO guidelines. A chapter title needs to be understandable when seen as a stand-alone item, e.g. on PubMed. The chapter title has now been corrected as 'Use of Botulinum Toxin in Ophthalmology'.

The original article was corrected.

The updated online version of this chapter can be found at
https://doi.org/10.1007/164_2019_325

S. M. Whitcup, M. Hallett (eds.), *Botulinum Toxin Therapy*,
Handbook of Experimental Pharmacology 263, https://doi.org/10.1007/164_2020_413

Correction to: The Use of Botulinum Toxin for Treatment of the Dystonias

Alfredo Berardelli and Antonella Conte

Correction to:
Chapter "The Use of Botulinum Toxin for Treatment of the Dystonias" in: A. Berardelli and A. Conte, Handbook of Experimental Pharmacology, https://doi.org/10.1007/164_2019_339

The chapter was inadvertently published without a more specific title according to SEO guidelines. A chapter title needs to be understandable when seen as a stand-alone item, e.g. on PubMed. The chapter title has now been corrected as 'The Use of Botulinum Toxin for Treatment of the Dystonias'.

The original article was corrected.

The updated online version of this chapter can be found at
https://doi.org/10.1007/164_2019_339

S. M. Whitcup, M. Hallett (eds.), *Botulinum Toxin Therapy*,
Handbook of Experimental Pharmacology 263, https://doi.org/10.1007/164_2020_414

Correction to: Botulinum Toxin and Pain

Zdravko Lacković

Correction to:
Chapter "Botulinum Toxin and Pain" in: Z. Lacković, Handbook of Experimental Pharmacology, https://doi.org/10.1007/164_2019_348

The chapter was inadvertently published without a more specific title according to SEO guidelines. A chapter title needs to be understandable when seen as a stand-alone item, e.g. on PubMed. The chapter title has now been corrected as 'Botulinum Toxin and Pain'.

The original article was corrected.

The updated online version of this chapter can be found at
https://doi.org/10.1007/164_2019_348

S. M. Whitcup, M. Hallett (eds.), *Botulinum Toxin Therapy*,
Handbook of Experimental Pharmacology 263, https://doi.org/10.1007/164_2020_415

Correction to: The Use of Botulinum Toxin in the Management of Headache Disorders

Hsiangkuo Yuan and Stephen D. Silberstein

Correction to:
Chapter "The Use of Botulinum Toxin in the Management of Headache Disorders" in: H. Yuan and S. D. Silberstein, Handbook of Experimental Pharmacology, https://doi.org/10.1007/164_2020_365

The authors noticed that unfortunately in the chapter, in Section 9 on the second paragraph lines 1 and 3, 2 numbers were replaced by 2 citations. The injection protocols commonly used are (GBD 2016 DALYs Hale Collaborators 2017) the fixed-site approach, which uses fixed, symmetrical injection sites and a range of predetermined doses (GBD 2016 Disease and Injury Incidence and Prevalence Collaborators 2017); the follow-the-pain approach, which often employs . . . .. The sentence should read as 'The injection protocols commonly used are: (1) the fixed-site approach, which uses fixed, symmetrical injection sites and a range of predetermined doses; (2) the follow-the-pain approach, which often employs. . . .

The original article was corrected.

The updated online version of this chapter can be found at
https://doi.org/10.1007/164_2020_365

S. M. Whitcup, M. Hallett (eds.), *Botulinum Toxin Therapy*,
Handbook of Experimental Pharmacology 263, https://doi.org/10.1007/164_2020_416